Entrepreneurship in Power Semiconductor Devices, Power Electronics, and Electric Machines and Drive Systems

Entrepreneurship in Power Semiconductor Devices, Power Electronics, and Electric Machines and Drive Systems

R. Krishnan

First edition published 2020
by CRC Press
6000 Broken Sound Parkway NW, Suite 300, Boca Raton, FL 33487-2742

and by CRC Press
2 Park Square, Milton Park, Abingdon, Oxon, OX14 4RN

CRC Press is an imprint of Taylor & Francis Group, LLC

ISBN: 978-0-367-55502-3 (hbk)
ISBN: 978-1-003-09379-4 (ebk)

Typeset in Times
by codeMantra

Contents

Preface xi
Acknowledgments xv
Author xvii

Chapter 1 Introduction 1

Chapter 2 Fundamentals of Entrepreneurship and Company Startup 7

2.1 Introduction 7
2.2 Entrepreneurship and Its Parts and Functions 7
2.3 Business Idea 8
2.4 Technology 11
2.5 Products 16
2.6 Manufacturing of Products 17
2.7 Market and Scope for the Products 20
2.8 Product Testing 23
2.8.1 Functionalities 23
2.8.2 Adherence to Standards 23
2.8.3 Testing 24
2.8.4 Certification 25
2.9 Capital (Funding) Sources for Startup 25
2.9.1 Institutional Investors 26
2.9.1.1 Venture Capital Funds 27
2.9.1.2 Finance Companies 28
2.9.1.3 Venture Arms of Established Companies 29
2.9.1.4 University Venture Funds 29
2.9.1.5 Philanthropic Charity Foundation Entrepreneurship Funds 30
2.9.1.6 Government Grants 31
2.9.2 Non-institutional Funds 32
2.9.2.1 Self (own) 32
2.9.2.2 Family 33
2.9.2.3 Friends 33
2.9.2.4 Crowd Source 33
2.9.2.5 Angel Investors 33
2.9.2.6 High Networth Individuals (HNIs) 34
2.10 Business Organization 34
2.10.1 Company Incorporation 34
2.10.2 Selection of Company Location 34

2.10.3 Selection of Attorney Services 35
2.10.4 Recruitment of Leadership Team 35
2.10.5 Recruitment of Engineering and Other Team Members 36
2.10.6 Selection of Chartered Public Accountant 36
2.11 Business Plan 37
2.12 Perspective for the Founders 38
2.13 Discussion Questions 38
2.14 Exercise Problems 39
References 40

Chapter 3 Introduction to Power Semiconductor Devices and Power Electronics 45

3.1 Introduction 45
3.2 Power Semiconductor Devices 46
3.2.1 Diodes 46
3.2.2 Bipolar Junction Transistor (BJT) 47
3.2.3 Thyristor Family 48
3.2.4 MOSFET 49
3.2.5 Insulated Gate Bipolar Transistor (IGBT) 50
3.2.6 Silicon Carbide Power Devices 51
3.2.7 Gallium Nitride (GaN) Power Devices 52
3.2.8 Future 52
3.3 Major Power Converters 53
3.3.1 Dc to dc Power Conversion 53
3.3.1.1 Step-Down (Buck) Converter 54
3.3.1.2 Step-Up (Boost) Converter 60
3.3.1.3 Step-Down/-UP (Buck Boost) Converter 60
3.3.2 Dc to ac Converters 62
3.3.2.1 Single-Phase Half-Wave Inverter 62
3.3.2.2 Single-Phase Full Wave Inverter 66
3.3.2.3 Inverter Control 68
3.3.2.4 Three-Phase Inverter (2 Level) 84
3.3.2.5 Multilevel Inverters 94
3.3.2.6 Resonant Converter Circuits 102
3.3.3 Ac to Dc Power Converters 102
3.3.3.1 Uncontrolled Rectification 102
3.3.3.2 Controlled Rectification 106
3.3.4 Direct ac to ac Conversion 108
3.3.4.1 Cycloconverter 108
3.3.4.2 Matrix Converter 110
3.4 Applications 112
3.4.1 Active Power Filter 113
3.4.2 Solar and Grid Interface 116

3.4.3 Uninterruptible Power Supply (UPS) ... 116
3.4.4 DC Power Supplies ... 118
3.4.5 Motor Drives ... 118
3.4.6 Future in Applications ... 118
3.5 Exercise Problems ... 119
3.6 Class Projects ... 120
References ... 121

Chapter 4 Entrepreneurship in Power Semiconductor Devices ... 129

4.1 Introduction ... 129
4.2 Objectives of the Study ... 129
4.3 Semiconductor Power Devices ... 130
4.3.1 GaN Devices ... 131
4.3.1.1 Wide-Bandgap Materials ... 131
4.3.1.2 Depletion Mode GaN ... 132
4.3.1.3 Enhancement Mode GaN FET Structure and Its Operational Characteristics ... 132
4.3.1.4 Cascode (with the D-Mode GaN) Structure and Its Operational Characteristics ... 134
4.3.1.5 Targeted Applications ... 139
4.3.2 Silicon Carbide Devices ... 140
4.3.2.1 SiC Material and Its Advantages ... 140
4.3.2.2 Normally-On JFET ... 141
4.3.2.3 Normally-Off JFET ... 142
4.3.2.4 SiC MOSFET ... 144
4.3.2.5 Targeted Applications ... 147
4.3.3 Market ... 148
4.4 Gallium Nitride Device Entrepreneurship ... 149
4.4.1 Avogy, Inc. ... 149
4.4.2 Cambridge Electronics, Inc. ... 150
4.4.3 Efficient Power Conversion Corporation ... 152
4.4.4 Exagan ... 155
4.4.5 Flosfia, Inc. ... 157
4.4.6 GaN Power International, Inc. ... 159
4.4.7 GaN Systems ... 163
4.4.8 MicroGaN GmbH ... 164
4.4.9 Navitas Semiconductor ... 166
4.4.10 NexGen Power Systems, Inc. ... 169
4.4.11 Nitronex Corporation ... 171
4.4.12 Transphorm, Inc. ... 173
4.4.13 VisIC Technologies ... 177
4.5 SiC Devices and Related Startup Companies ... 180
4.5.1 Anvil Semiconductors ... 180

4.5.2 Arkansas Power Electronics International (APEI) 181
4.5.3 Ascatron 183
4.5.4 GeneSiC Semiconductor, Inc. 184
4.5.5 SemiSouth Laboratories, Inc. 186
4.5.6 United Silicon Carbide, Inc. 188
4.6 Device Control Oriented Entrepreneurship 190
4.6.1 Amantys 190
4.6.2 General Observations on Startups in Power Devices 191
4.7 Exercise Problems 195
4.8 Class Projects 196
References 199

Chapter 5 Entrepreneurship in Power Electronics 209

5.1 Introduction 209
5.2 Control Circuits 210
5.2.1 Powervation Ltd. 210
5.3 Solid-State Circuit Breaker 212
5.3.1 Atom Power 212
5.4 Microgrids, Home Energy Management Systems (HEMS) 215
5.4.1 Technology 215
5.4.2 Market for Home Solar Power System 218
5.4.3 Startups 219
5.4.3.1 Mera Gao Power 219
5.4.3.2 Me SOLshare Ltd. 221
5.4.3.3 Cygni Energy 225
5.4.3.4 Zola Electric 228
5.4.3.5 AlphaESS 230
5.4.3.6 Electriq Power 234
5.4.3.7 Geli 236
5.4.3.8 Sunverge Energy 238
5.4.3.9 Bboxx 241
5.5 Mobile Power Platforms 244
5.5.1 Multicon Solar AG 244
5.5.2 FreeWire 245
5.6 Battery Energy Storage Systems 249
5.6.1 Advanced Microgrid Solutions 250
5.6.2 Greensmith Energy Management System 254
5.6.3 Stem 256
5.7 Grid Edge Power Electronic Systems 259
5.7.1 Introduction 259

5.7.2 Technology of Grid Edge Control with Power Converters (Mainly Inverters) 260
5.7.3 Market Scope 261
5.7.4 Startups 261
5.7.4.1 GridBridge 261
5.7.4.2 Gridco Systems 264
5.7.4.3 Varentec 267
5.7.4.4 Smart Wires 271
5.7.4.5 Envelio 275
5.7.4.6 Faraday Grid 277
5.8 Charging of EVs 280
5.8.1 Introduction to Power Transfer Schemes for Charging 280
5.8.1.1 Conductive Power Transfer for Charging 281
5.8.1.2 Wireless Power Transfer for Charging 284
5.8.1.3 Market Potential for EV Chargers 289
5.8.2 Startups 290
5.8.2.1 ChargePoint 290
5.8.2.2 Wiferion 294
5.8.2.3 Momentum Dynamics 296
5.8.2.4 WiTricity 300
5.9 Low-Power Plug-In and Wireless Chargers 303
5.9.1 FINSix 304
5.9.2 PowerSphyr 307
5.10 Conclusion 310
5.11 Discussion Questions 311
5.12 Exercise Problems 316
5.13 Class Projects 318
References 319

Chapter 6 Introduction to Electric Machines and Drive Systems 337

6.1 Introduction 337
6.1.1 Electrical Machines and Their Classification 337
6.2 PM Synchronous and Brushless DC Machines 342
6.2.1 Radial Flux Machines 342
6.2.2 Circumferential Flux Path Machines 343
6.2.3 Axial Flux Machines 344
6.2.4 Model of the PM Synchronous Machine 345
6.2.5 PM Brushless DC Machine and Its Control 353
6.3 Switched Reluctance Machine (SRM) 356
6.3.1 Description of Machine 356
6.3.2 Principle of Operation 357

6.3.3 Converters ... 360
6.3.4 Control of SRM ... 362
6.4 Discussion Questions ... 363
6.5 Exercise Problems ... 364
References ... 364

Chapter 7 Entrepreneurship in Electric Machines and Drive Systems ... 367

7.1 Introduction ... 367
7.2 Present and Future Market Size ... 367
7.2.1 Market for PM Motors ... 367
7.2.2 Market for SR Motors ... 367
7.2.3 Market for Power Electronic Converters ... 368
7.2.4 Controllers ... 368
7.3 PM Motor Drive Startups ... 368
7.3.1 Axiflux ... 369
7.3.2 Linear Labs, Inc. ... 370
7.3.3 Magnax ... 372
7.3.4 YASA ... 374
7.3.5 Guina ePropulsion ... 376
7.3.6 QM Power ... 377
7.4 Thermal Cooling Technology Based Electric Motor Startups ... 380
7.4.1 Zero E Technologies ... 381
7.4.2 LC Drives, Inc. ... 382
7.5 Printed Circuit Stator-Based Motor Startups ... 385
7.5.1 ECM Software and Technology ... 385
7.5.2 Infinitum Electric ... 388
7.6 Electric Aircraft Based Motor Drive Startup ... 391
7.6.1 Magnix ... 391
7.7 Switched Reluctance Motor Drive Startups ... 393
7.7.1 Panaphase Technologies ... 394
7.7.2 ePower Motors ApS ... 397
7.7.3 Ramu, Inc. ... 399
7.7.4 Software Motor Corporation ... 402
7.8 Electrostatic Machines ... 404
7.8.1 C-Motive Technologies ... 405
7.9 Conclusion ... 408
7.10 Discussion Questions (Beware: The Discussion May Lead to Innovations and Possible Startup Ideas!!) ... 409
7.11 Projects ... 410
References ... 412

Chapter 8 Conclusions ... 421

Index ... 427

Preface

The author directed an undergraduate course in 1997 at Virginia Tech on entrepreneurship named Virtual Corporation (founder: Dr. Leonard Ferrari, the department head) with a lot of enthusiasm from students all over the university. The course tried to educate students from across the university on entrepreneurship and in the course of it achieved a number of socially oriented major projects in electric transportation using linear motor drives and magnetic levitation, among others. The course closed in 2004 for lack of support from the university administration. Since 2015, the same university is making a big investment on entrepreneurship education as they must have realized that it is important for the students and their future! The author has been working for more than 23 years to find effective ways to educate electrical engineers and engineering students in entrepreneurship related to startups resulting in this first book of its kind.

Entrepreneurship as a subject and its practice are very much promoted within university campuses around the globe in recent times. Rapidly changing technologies demand such a skill set in their engineers. Engineering students, in general, and electrical engineering students, in particular, as well as professional engineers hardly take a course in entrepreneurship due to their heavy course load and workload, respectively, and it may also be due to the fact that the offerings in the business colleges are not tied directly to their specialization and hence of no immediate interest to them. This book aims to address that concern for electrical engineering students and professional engineers with the following approach:

i. The course is structured with an introduction to engineering entrepreneurship (Chapter 2) in less than four lectures covering the basics, but some topics related to services that usually are outsourced such as accounting, both company and patent legal advice and services, advertisement, public relations, etc. only briefly touched here.
ii. Three subdomains (also referred to as fields or subjects hereafter) within electrical engineering are considered for illustration, learning and study of entrepreneurship, and they are power semiconductor devices, power electronics and electric motor drives and in all their emerging applications.
iii. The three fields are reviewed (Chapters 3 and 6) including the latest research results to pave the way for understanding the startups and their technologies and how they are different from the currently available knowledge base that the readers are familiar with.
iv. The study of entrepreneurship is initiated then for the chosen fields (Power semiconductor devices in Chapter 4, Power electronics and applications in Chapter 5 and Electric motor drives in Chapter 7) with the following steps that are outlined briefly below:

- Study of the current market size and its predicted scope in the future for each field
- Assembling data for the existing and more importantly emerging applications in them
- Assessing technology requirements to meet the emerging applications
- Recent startups within the past two decades in each field are considered for study based on a number of facts discussed in the entrepreneurship basics and newer methodology (explained in Chapter 1)
- A total of 64 startups are studied in all the three fields with general observations for learning
- The final chapter contains the conclusions learned from the study of these fields (Chapter 8)

The book is in textbook format to enable undergraduate and graduate students learn as part of a self-study (in senior year for undergraduates), if necessary, until a regular course on this topic is offered in electrical engineering departments. It contains discussion questions, exercise problems, and team and class projects to broaden their understanding of entrepreneurship and also for an in-depth probing of various issues that confront startups in the chosen fields. Practical aspects of startups are guided through various team and class projects and that is aimed toward startups and attempts to startup as part of the course work. Engineers in the field will benefit also from these exercises as they would open their imagination that may lead to startups also. Even though currently there is no structured course in electrical engineering departments at this time on the subject matter of this book, it can be easily introduced as part of the major in power systems, electronics, power electronics and electric motor drives and can be handled by faculty members as some of them have enough industrial experience to lead the students toward one of the important and fulfilling goals of their electrical engineering education.

It is hoped that electrical engineers and students from other fields will learn entrepreneurship from this book and apply the same or similar methodology to study startups in their area of expertise and benefit from them.

The book can also be used by engineering managers to keep abreast of the technology development, newer movements in the market sphere which has the potential to disrupt their products and their market share, if not immediately but in the future. The senior management in companies, small and big, will gain by looking at the startups in their arena of interest:

i. to learn of their convergence with their future plans, to explore a collaborative arrangement with the startups in some products exclusively or in a limited way,
ii. to form joint ventures with the startups in case of significant interest from inside their organizations, and

iii. to acquire (i.e., buy and merge) them eventually in case of high interest to their companies and to existing and potential new/future clients and to broaden their clientele base.

The author is very grateful for all the illustrated startups and their managements for sharing the information about their companies over their websites in many cases, very generously with the public. The startups consist of great electrical engineering professionals and management teams, and we wish them great success in their efforts and personal satisfaction. Hopefully, all of them talk and interact with students and professionals through various forums they are invited and honored with to share their immense experience and knowledge.

The author will always be grateful to readers for suggestions to improve the book that will benefit readers in successive editions.

This book would not have been possible without the constant encouragement of my dear wife, Mrs. Vijaya Krishnan and words are inadequate to express my eternal gratitude to her for all the responsibilities that she takes upon and bringing happiness and well-being into my life.

R. Krishnan
June 25, 2020

Acknowledgments

1. The author thanks the executives and their marketing executives and media consultants of the startups, and also two established companies that responded to my requests (almost 90%) with permission to use their figures for illustration of their technologies/products. They are given in the following:

Chapter 4:
Electric Power Conversion Corp. (EPC)
GAN Systems
GPI
Navitas Semiconductor
Transphorm Inc.
United SiC
VisIC Technologies
Infineon Technologies AG

Chapter 5:
Atom Power
Mera Gao Power
MeSOLshare
Cygni Energy
AlphaESS
Bboxx
FreeWire
Advanced Microgrid Solutions
Wartsila Energy Business (for GreenSmith)
Stem Inc.
GridBridge
Varentec
Smart Wires, Inc.
ChargePoint
Wiferion
Momentum Dynamics Corp.
WiTricity
PowerSphyr, Inc.

Chapter 6:
Taylor & Francis Group

Chapter 7:
Magnax, Belgium
QM Power
LC Drives
ECM Software & Technology
Infinitum Electric
Magnix
ePower Technology ApS, Denmark
C-Motive Technologies

2. Founders, executives and their teams make possible innovation of the technologies in the market place with immense work, and their contributions through the 64 startups are acknowledged. Their work provides the window through which their ecosystems of entrepreneurship (in power semiconductor devices, power electronics and electric motor drives) can be viewed in this book. It provides entrepreneurship education to electrical engineering students and practicing engineers, and it is hoped that will further advance both the technical and business innovations in respective fields. The author wishes success to all the startup founders and their teams.
3. The author gratefully acknowledges Mathcad for their generosity in giving access to their software (mainly MATLAB) for his personal simulations that are included in the book.
4. The author is very grateful to Ms. Nora Konopka, Global Editorial Director, Engineering & Environmental Sciences, CRC Press (Taylor & Francis Group) and to Ms. Prachi Mishra, Editorial Assistant, for their guidance, encouragement, immense support and fast response. Nora has taken big risks, over two decades, with my former books, and I hope that this work exceeds the previous ones in sales and number of editions including translations to justify her trust in my work. It has been a pleasure to work with them.

Author

Dr. Krishnan Ramu (writes under the name of **R. Krishnan**) is a Professor Emeritus of electrical and computer engineering at Virginia Tech, Blacksburg, VA, USA. His research interests are in electric motor drives, electric machines, and power electronics.

Books

a. *Single Authored* in the field of electric machines, drives and control:
   i. *Electric Motor Drives* (also translated into Chinese, Greek), Prentice Hall, Inc., 2001.
   ii. *Solutions Manual to Electric Motor Drives*, Prentice Hall, Inc., 2001.
   iii. *Switched Reluctance Motor Drives*, CRC Press, 2001.
   iv. *Permanent Magnet Synchronous and Brushless DC Motor Drives* (translated into Chinese also), CRC Press, 2009.
   v. *Entrepreneurship in Power Semiconductor Devices, Power Electronics and Electric Machines and Drive Systems*, CRC Press, 2020.

b. *Co-authored One Book*:
   vi. *Control in Power Electronics*, Academic Press, 2002.

c. *Chapter Contributions*: Three other books.

Publications: 170+ technical papers

Patents: 31 patents (28 US and 3 Japanese) granted, and 5+ pending.

Awards:

i. IEEE Industrial Electronics Society's Dr. Eugene-Mittelmann Achievement Award for outstanding technical contributions to the field of industrial electronics,
ii. Anthony Hornfeck award for outstanding service to IEEE IE Society,
iii. IEEE Fellowship for contributions to ac and switched reluctance motor drives, and currently IEEE Life Fellow,
iv. Archer award for IP contributions from Regal Beloit Corp.,
v. IEEE Industrial Electronics Distinguished Lecturer,
vi. Life membership of AdCom, IEEE IE Society,
vii. Best book award from Ministry of Education and Sport, Poland, in 2003 for my coauthored book, Control in Power Electronics,
viii. Six Best Paper Prize Awards from the IEEE Industry Applications Society, 1980–2009,

ix. First Prize Paper Award from the IEEE Transactions on Industry Applications, 1983, and
x. Best Paper Award from the IEEE Industrial Electronics Magazine in 2007,

Industrial Experience: Served in two US companies full time: (i) Principal engineer in Gould, Inc., Rolling Meadows, Ill (1982–1985), (ii) Founder and CEO, Ramu, Inc., (2008–2010), and CTO (2010–2011) {mostly on leave and permission from Virginia Tech}, and (iii) Business Leader and CTO of Ramu Inc., a division of Regal Beloit America, Blacksburg, VA (2011–2014), (iii) Founder and CEO, Panaphase Technologies, Inc., 2001, and Chairman of the Board of Directors, 2002–2007.

Founder of Two Motor Drives Companies: Two companies in *switched reluctance motor drives*: (i) *Panaphase Technologies*, which was acquired by Delta-Gee in 2007, and (ii) *Ramu, Inc.*, which was acquired by Regal Beloit Corporation in 2011. Served as founding Chief Executive Officer and Chief Technology Officer for them. (He is the only faculty in his area of research in the world to have started, operated and sold two successful companies in electric motor drives.)

Teaching Experience: (i) College of Engineering, Guindy, Anna University, 1972–1979, (ii) Concordia University, Canada 1979–1982 and (iii) Virginia Tech 1985–present with intermittent service in Industry with permission/leave, and currently Professor Emeritus with no teaching responsibility.

Directed the first entrepreneurship course for mainly engineering majors (but inclusive of business and other students) *in 1997 at Virginia Tech* under the course name of Virtual Corporation for 7 years.

Short Courses: (i) Delivered GIAN lectures on PM Synchronous and Brushless Motor Drives, at NIT, Hamirpur, December 2017. (ii) Delivered and organized a large number (>20) of 3-day short courses to the industry in electric motor drives, power electronics, permanent magnet synchronous motor drives, induction motor drives and switched reluctance motor drives (all in the USA, France, Italy and Denmark).

Consulting: Served as a consultant to more than 18 companies in the USA and Europe.

Media Coverage: (i) *Wall Street Journal*, Profiling of my motor invention, July 22, 2006; (ii) Full-length interview in *MIT Technology Review*, December 12, 2001; (iii) MSNBC Lead in article on my interview with *MIT Technology Review*, December 13, 2001; (iv) Peter Barnes and Alina Mesenbourg, "Transportation technology expose with my interview and CRTS laboratory prototypes and research" TECH TV, Washington DC, aired multiple times on February 12 and 13, 2002; (v) Mike Gangloff, "Reaping from research", Front-page article, *The Roanoke Times*, May 11, 2002 on virtual corporation and my research in the Center for Rapid Transit Systems (CRTS) {2.25 full-page article}; (vi) Cassius A. Harris, "New Transportation System Being Designed to Ease Traffic Congestion", *American Society of Civil Engineers Magazine*, p. 31, March 2002 {based on an interview on my research in personal electric rapid transit systems}; (vii) Taped interview, KNX News Radio, LA, aired on December 19, 2001; and (viii) Doug Koelemay, "Take Your Car Skip the Traffic", The Voice of Technology, *Northern Virginia Technology Council Journal*, Cover-page article (interview on Personal Electric Rapid Transit System), pp. 8–9, March 2001.

Current Interests: (i) Inventions in permanent magnet (PM) and switched reluctance (SR) machines, power converters and their controls, and power electronics for diverse applications, (ii) Linear SR motor drives for transportation and elevator applications, (iii) Startups, and (iv) Collaboration with ECE faculty members around the globe.

1 Introduction

The book and its subject matter are introduced in this chapter. The subject matter of entrepreneurship as applied to startups is developed with a special application focus in three fields (or branches) of electrical engineering: power semiconductor devices, power electronics and electric motors and their drive systems. It is usual to study entrepreneurship as an independent subject, if at all, included in many curricula and they most usually are confined to business schools. Many engineering students hardly venture in that direction of learning from where it is traditionally taught during their study. It is to an extent understandable as the engineering students have little common ground with the business school and its offered courses of study. Also their specialization in engineering is grounded in physics and mathematics giving way to a high level of mathematical modeling, analysis, design, and most importantly laboratory testing and verification. The result is that outcomes are predictable with a high amount of accuracy, whereas in business, that may not be the case and in entrepreneurship that is the case very much too. The unpredictability in outcomes in workplace is neither desirable for the engineering employers nor to the careers of engineers, and that may keep them off from learning about entrepreneurship which is unfortunate. This book is the first of its kind in attempting to bring entrepreneurship as it applies to startups for electrical students and professionals alike in their own atmosphere and as a natural concomitant of their own studies and research, and illustrated with the three fields chosen to be described. Further, it can also be learned in the engineering school itself with more authenticity and in a known environment making use of the knowledge base that is acquired in their previous courses and research and also that is equally applicable to the practicing engineers as well. Note that well-known electrical engineering departments such as Stanford University's has an average of three startups per faculty, whereas that may not be the case in other engineering schools, but there are some faculty members with that experience to teach students in the regular curriculum and the professionals through short courses. It is believed that electrical engineering professors with knowledge in the chosen fields will use this book to introduce entrepreneurship to their students. This book's topic can be a three-credit subject offering at senior and graduate levels to those majoring in electrical engineering but specializing in the three fields. Accordingly, the book is organized on the following lines.

The basics of entrepreneurship and startups are presented in Chapter 2 containing basic definition of the terminology, and fundamental ingredients of the startup such as business idea, technology, manufacturing, market, product testing, funding sources – both institutional and non-institutional, fundraising, business organization ending with a business plan. The emphasis is placed on general knowledge of these subject matters but with sufficient practical details to assist any new entrepreneur, when ready, to begin the process of launching a company without having to go through any business courses in this regard. It is a self-study chapter for practicing engineers, but engineering students taking the course may have it augmented with general comments from an engineering faculty member or an external experienced professional and/or startup entrepreneur entertaining a question-and-answer session and to enthuse the students/audience with their experience.

Power devices, power electronics and electric motor drives are offered in the curriculum in major universities in the USA and across the globe. These are the subject matters considered for illustration of startup dynamics. Students take these courses in their junior and senior years of study, and professionals may also have taken them if they had graduated within the last 40+ years or so. They are taught more rigorously in both undergraduate and graduate levels, the latter with a high emphasis on current research in these fields. They are (both the undergraduate and graduate parts) covered here in review/recap form with an emphasis on certain important aspects that may be applicable to current startup activities. Mathematical modeling, extensive analysis and design are left to readers so that they can go back to their textbooks (and research articles), if necessary, that they have studied from so that most of their energy is focused on the startups and their ecosystems. Power devices and power electronics are presented in Chapter 3, and the advanced state of development with regard to machines and their drive systems is presented in Chapter 6.

Applications of power electronic converters in active power filter, solar and grid interface, uninterruptible power supplies, dc power supplies, motor drives and future applications are also introduced in Chapter 3 apart from various power semiconductor devices and power converter topologies.

A **general methodology** for the study of startups and their evolution in the twenty-first century, with regard to all three fields, is summarized here before entrepreneurship in the three chosen fields is taken up. The methodology discussed here is applied to startups of the three fields, and it is believed that it can be used to study other fields in electrical engineering and probably other fields of engineering. The basic theme of the book is to learn entrepreneurship from looking at the ecosystem in which it is prospering in the fields of interest considered in the book. There is no alternative to practical learning as in all engineering in general and electrical engineering, in particular, and that is regardless of the amount of exposure to theory. Noting that entrepreneurship is not a science makes it absolutely essential that practice is a better environment for learning. From that point of view, a large number of startups in each domain of the fields are taken up for illustration and study. The advanced state and innovations in the fields are presented so that the innovation of each startup is differentiated from the current state of affairs in the technology arena and becomes evident to the readers. Then, the startups having similar technologies or belonging to the same category are separated in each subsection. The information about the startups hardly appear in technical journals and only available in their own websites, IEEE conference presentations, trade journals, business websites, patent sources and discussions sometimes in the internet infosphere. They are carefully assembled and organized in the form of the startup location, its founder(s) and their background and experience, its technology and products with which they make efforts to address certain markets, intellectual property, funding and sources of them, collaborations/partnerships with big and small entities, whether exit has happened and also their current state of development. Finally based on these information, observations are made that may be helpful to understand the startup in many other aspects such as the strength of founders and its management team and their accomplishments and experience in the field of current startup, the strength of its intellectual property in terms of patents and state of that in comparison to its competitors, market potential for their products, and some fundamental questions and issues for which answers are not provided by the startups in their websites and/or other means so that reader can critically evaluate the startup's strengths and, if any, weaknesses also. Such a set of observations provide further learning to readers and aspiring entrepreneurs, and with their team members, if they take the opportunity to go through a rigorous mental exercise and discussion to justify or refute some of them and then complement them with their own set of observations based on their further research. Their conclusions will address and answer whether there are areas that can be covered in a new startup with regard to the startup(s) under study, thus leading them, depending on their interests, toward entrepreneurship and opening their minds to technical opportunities in new directions. At the end of each chapter, general underlying features of the startups and their various application objectives are summarized for all the three fields. The study is further continued by discussion questions which are intended to provoke the readers to ask the most difficult questions driving them to find answers so that gaps in the knowledge and technology can be evaluated furthering their knowledge base and that may hopefully result in their own pursuit of a startup. Each of the topics in the chapters contains exercise problems and team and class projects to further an in-depth probing of the topics and in relation to that of the startups in their fields of interest. Noting that the startups usually never discuss their strategies and tactics regarding building from the ground up to a significant presence in their respective markets, the critical aspect of that learning to some extent comes from raising issues in observations made and also through exercises and discussions. The startups may have answers to all the questions and issues raised in these aspects but they are not available to the public has to be recognized.

Entrepreneurship in power semiconductor devices is covered in Chapter 4. It mainly deals with rising gallium nitride (GaN) and silicon carbide (SiC) power devices, their unique features, operation, and advantages with emerging and existing applications and their market scope in the future. The impact of these new devices in power electronics and motor drives will be enormous as their prices start coming down, if not now but in course of time, with respect to existing devices. With that in view, emerging applications of these devices are tracked in this chapter from the startups. In each category of the devices, startups are presented and analyzed with the methodology described above. A total of 20 startups are considered in this chapter. Finally, general observations relating to the startups in this field are given to get a picture of the activities in them and also intended to further the investigation by the readers.

Power electronics startups are presented in Chapter 5. The applications and control of the power converters for specific applications dominate in power electronics startups. Innovations in power electronics converter topologies emerge all the time but most of the power electronics related innovations arise from applications with existing converter topologies is to be noted. Many application areas such as control circuits, solid state circuit breaker, micro grids, home energy management systems, mobile power platforms, battery energy storage systems, power grid control and/or enhancement, wireless charging for EV and other applications (this section could also be in motor drives section as a significant part of them is magnetics design), and low-power plug in power devices are identified to study the startup activities. A total of 28 startups are taken for study in these applications. These are not necessarily the full spectrum of the application areas and that has to be noted. Even though the methodology is the same for studying the startups and their environment, what is different is that they are mostly application-oriented. General observations on the power electronics startups are summarized at the end of this chapter.

Permanent magnet synchronous and brushless dc motors, and switched reluctance motors and their drive systems are considered in Chapter 6 as most of the innovations coming through them in this arena result in startups. The basic motor topologies based on flux paths such as radial, axial and circumferential flux paths are introduced to understand some of the innovations in machine structures, their advantages and unique applications with the resulting impact on emerging startups in this field.

Electric motor and motor drive startups are considered in Chapter 7. The market for the motors and drive systems in the present and its future forecast are given. Motor drive startups are broadly divided into permanent magnet (PM) synchronous motors and switched reluctance motors, and their drive systems are considered for the study of startups. One of the areas that is emerging in the motors is the demand-driven and/or foreseen applications with the high-power-density and high-efficiency requirements such as in electric aircraft's main motor turbine, and electric vehicles from 4-wheel cars to 22-wheel trucks. A significant number of the startups fall under PM synchronous variety in this regard and one with partial superconductor type. One category that is surging forward is the printed circuit board stators for the motors with resulting compactness and low weight, and this category of startups are presented. Also, innovation to increase power density is through novel cooling techniques, and one startup is chosen for the study. For high-volume, variable-speed applications requiring low cost but high performance in terms of efficiency, switched reluctance motor (SRM) startups have made their entry. The competition for these motors and motor drives is fierce from well-established PM synchronous motor (PMSM) and brushless dc motors and drive systems startups as well as from established large companies. General observations on startups and emerging environment are discussed and presented to enhance the knowledge base and to provide insight so that interested entrepreneurs can find ways to find the lacunae in these specific startup domains and have the desire to address them through their own entrepreneurial efforts. A total of 16 startups are studied in this chapter.

Overall conclusions of the startup scenario in the three chosen fields of study, markets, and funding sources are made in Chapter 8. The pursued approach and methodology for the study of entrepreneurship related to startups are equally applicable to various fields of electrical engineering and possibly other engineering areas as well. Readers are encouraged to apply these techniques to pursue their interest in other fields as well, as startup activities are not limited to one field or a few fields.

Mostly startups in the USA are selected for presentation as the data for them is available over open media. It is not easy to find such information needed for learning from startups directly or about them through other sources in countries such as India, China, Japan, Korea, Russia, Australia and much of eastern European countries. It is hoped that readers in these countries may have access to such information that may further their database and knowledge base, and hopefully they make it available either to the author or through the author to the audience or through publications.

Note about references used in the book: A small subset of technical publications from technical societies such as IEEE and others contribute to startup technologies, even though thousands of research papers are published in their journals and conference proceedings every year. Note that many startup founders do not come from this research background. Technology basis for some startups can be found in patent applications and granted patents in the USA and other countries. Most of the time, there is a lag between the startup's birth and ensuing patents from it, which makes it difficult to exactly pin down the uniqueness of its technology vis-à-vis others'. Therefore, careful wading of these mountains of pieces of information has to be done to get to the technology basis of the startups. Likewise, the same techniques as pursued by the startups have to be imitated to hide when one starts a venture as well. This process hinders studies, but the startups' aim is not to educate the audience but to get to the market and capture a slice of it! Therefore, the reference sources are limited to learn of the strategy and tactics pursued by the startups and known only to their inner circle consisting of CEO, President, CTO, VPs and Board of Directors. Their bios help to gauge something about the startups as well and their capability to raise funds. Wherever possible, they are identified for the presented startups, to the extent that is relevant.

Sources of funding and quantum of funding are given by public sources including the websites of the startups, trade journals, and commercial sites that allow such information to be accessed such as Crunchbase, Pitchbook and others cited in the reference sections. Patents can fully be accessed through national patent organizations at no expense at all, and the other sources that give only a brief description about the startups' patents such as in Justia's website are also given in our references.

2 Fundamentals of Entrepreneurship and Company Startup

2.1 INTRODUCTION

The basic definition of entrepreneurship and various kinds of entrepreneurship are the starting points to get to know the aspects of the company startup. There are many aspects, but only topics essential to understanding business startups for engineering students and professionals are presented in this chapter. They include: conceiving a business idea and where does that idea emanate from, business setup, technology behind it, how to get the technology if not invented by the entrepreneur and his/her cofounders, realization of business with the concomitant requirements of the product, its specification, market and its scope, manufacturing, testing and agency approval. In addition, the most important aspects including raising capital with its well-known multiple avenues and the pros and cons of each of the fundraising routes, and forming a leadership team and an organization to realize the business idea are all discussed in detail. It is important for engineers to realize that in engineering the outcome is highly controlled and predictable because of their basis in science and mathematics. The same is not applicable to startups due to various factors such as market and its control which are outside of their control parameters. It results in their outcomes being not controllable and hence in their success rate. The chance of success on an average for a startup enterprise in the USA is included in this chapter.

2.2 ENTREPRENEURSHIP AND ITS PARTS AND FUNCTIONS

Entrepreneurship means many things to many people, and in order to avoid the confusion, one definition of that will be considered that captures in some ways its modern usage. The Oxford dictionary gives its meaning as the activity of setting up a business or businesses, taking on financial risks in the hope of profit. One manifestation of it then is creating a business. The business dictionary defines it as the capacity and willingness to develop, organize and manage a business venture along with any of its risks in order to make a profit. This goes beyond not only creating a business that can be done but may also be handed over to others to run, and therefore, this precludes such a standoff relationship between the founder/ entrepreneur and the enterprise. It further implies that the founder/entrepreneur is personally involved from founding to seeing that it is developed, run and makes a profit, and this definition is more realistic in many of the entrepreneurs of this day and this will be considered as and when the term entrepreneur is used from now on in this text. The founders are usually entrepreneurs and the two terms are intermingled and used extensively hereafter. It is generally true in the case of engineering graduates starting their enterprises, and these are the graduates to be served in this book and, therefore, it is adopted as the working definition.

The components of entrepreneurship then include, at least from engineers' point of view, the key functions given below:

- Development: Creating or buying into an idea for the product, not necessarily the science and engineering behind it or innovation underlying the idea, and this part is only concerned with identifying the product. In the case of engineering founders, business ideas and innovation usually come with them.
- Organization: Raise capital, bring human resources, find a place of work, laboratory, etc.
- Management: Coordinating product development, marketing, manufacture, day-to-day finance and sales, to name a few.

2.3 BUSINESS IDEA

The sources of the business ideas can be broadly classified into two tracks:

i. Inside Track: Meaning people very knowledgeable about a company's product lines or have access one time or another with companies in those lines and their thinking leads to new product lines and hence business ideas. Consider the following cases:
   a. A company's quest for new products in their lines for continued revenue growth and pursuit of that will result in new products and possibly a business idea. Note that only insiders to this industry sector know such possibilities or push for such efforts and hence it belongs to the inside track. Anyone involved in this kind of effort is at a risk of litigation to take this idea or knowledge outside that industry for violation of nondisclosure agreements (NDAs) and infringement of intellectual property (IP). All ideas that come from inside the company may not get traction with the upper management as it may be of peripheral interest or outside of its interest. In that case, some large corporations may encourage the inventor or idea generator to pursue that as a separate entity with the majority share being held by the parent company.
   b. Competition and market forces in a capitalist society, say, for example, the drive among many companies to create an electric vehicle with the lowest cost.
   c. Cost reduction goals in standardized products with high volume market sales force the corresponding industries to look for ideas and

inventions to achieve that. This route requires a sound knowledge of the current products, the ferocity of competition and what would be a significant cost reduction for a company or individual to pursue this goal.

   d. Looking for features that will endow old products a new lease of life and an expanded market resulting in business ideas. Again this requires very high familiarity with an inside working knowledge of industry sectors and its products.

ii. Outside Track: Not necessarily connected to knowledge of product lines in their background but outside forces such as developing and evolving technologies impacting their imagination resulting in new business ideas. Consider the various sources for the business idea generation in the outside track:

   a. Keen observers of a market place in a geographical area.
   b. Researchers and students in universities.
   c. General drive of technologies and spillover in the form of products, for example, artificial intelligence applications to autonomous driving in the cars.
   d. Newer technologies opening up newer vistas, for example, use of cell phones in various applications in the social domain, and in the medical domain for sensing body temperature, blood pressure, and all statistics related to physical movement such as minutes walked, distance covered, etc. Classic examples include connecting the internet and the cellular services resulting in WhatsApp, Hangouts, etc. An endless stream of business startups is rolling out in many applications due to cell phone technologies.
   e. Keen desire to make new products by engineers, marketing leaders, keen observers of industries and societies, for example, connecting solar power to the grid but in the process increasing power delivery to customers without additional conventional power plants and in the process increasing operational reliability of the grid itself.
   f. The drive of human beings to identify something that does not exist, to go after that to get a sense of fulfillment, with plenty of examples in the field of electrical engineering, say, from computers to cell phones.
   g. To satisfy the needs of under-privileged societies and communities such as providing power for charging and lighting the dwellings in rural areas using solar and/or wind power.
   h. Usually from the arising needs of a company and segments of the population, for example, the aging population and their need for robotic assistants in western countries where there is a shortage of human workforce.
   i. Users' preferences and what was acceptable to them once becoming unacceptable now leads the way in many startups. A case in point is the new internet browsers eschewing privacy invasions, such as

DuckDuckGo browser and many others in the internet domain and cell phone apps.

j. Replace old technologies with the ones that deliver high efficiency and/or more controllable features, for example, move from traditional rail transportation with maglev and/or bullet trains, and ropeless elevators in the place of roped ones.

The business idea when it is available requires a technology base and some IP with concrete details. An entrepreneur (hereafter we use founder to mean the same in this text) need not be an inventor or have even technology background and experience (and note that usually is not the case with engineers). Therefore, various ways to obtain technology for the startup are taken up in the following section.

Viability of Business Idea: Many ideas are easy to float about but it is important to test their viability at a level that is satisfactory for it to move forward via various steps in realizing it through a startup. The initial vetting of the business idea usually involves, but not necessarily for every startup, the following steps:

i. Testing all the way by realizing a product, talking and reaching some agreement with potential customers, and testing the product in their premises is an ideal way to go. But many (almost all) startups cannot take this path because the startup itself may not be in place with a team and other resources to achieve it at this stage.

ii. Therefore, at what level of viability test is sufficient to move forward is to be determined that will make sense to potential investors and customers. The easiest step is to meet with an inner circle of knowledgeable friends, discuss the idea and see whether there is positive or negative traction based on seasoned reasoning, verifiable facts and figures. The chosen audience must be highly trustworthy and of demonstrated confidentiality in their dealings so that the idea does not get leaked outside of this circle. In addition to this, a mentor, preferably retired from active roles in industry, of proven entrepreneurial experience can be approached for discussion, guidance and assessing the viability of the business idea. Many such volunteer mentors are available in cities and some in towns and their identification is a challenge to the beginning entrepreneur and universities and/or other entrepreneurs may be able to connect them. A number of such mentors and advisors will give their initial assessment of the idea, most of them free of cost. The key is that reasoning-based evaluation with as many facts and figures as can be gathered has to be enforced at this stage of viability testing to save time and resources. The entrepreneur has to do independent research with limited resources available on hand at this time and compose a report for discussion with the circle of friends, mentors or their own professors (in case of people with ties to their university faculty).

After all is said and done, the entrepreneur has to take the risk and push the business idea to a startup launch. There is and will always be a lot of unknowns in the technology behind the business idea, such as its realizable products with capabilities that could supersede existing products in the market place in their performance but with low cost, among other things. The decision to move forward with the business idea is the first test for entrepreneurs whether they have the ability to take risks and handle them fairly well. Note that not everyone is interested and inclined to take on the entrepreneurial path.

Caution has to be exercised in sharing the contents of the IP behind the business idea and the details of its implementations, if available, and they need to be kept off the limits in discussions with friends, mentors and advisors. It will endanger any possibility of filing the IP for patents. In case they need to be revealed to get a meaningful test for business idea viability, protections in the form of NDAs have to be executed with anyone to whom the information is shared. In general, it is a good idea to have NDA executed before discussing it with anyone. This requires the services of an attorney to draft the NDA that is appropriate for specific uses including this one, and it will involve some expenditure on the part of the entrepreneur. In order to avoid the expense involved, some may resort to some freely available NDAs over the internet but it is not advisable to take that route as it is not tailored to fit your circumstances. Having finalized the business idea, the next step is to make sure that technology is available to proceed further. That is considered in the following section.

2.4 TECHNOLOGY

All companies are based on technologies relevant to their product lines. Startups in engineering and in many other domains are no exception to this rule. The technology and related IP is obtained through many routes and they are taken up in the following:

i. Entrepreneurs can by research pursue inventions once the business idea is concretized. Such an attempt may or may not end up with inventions as they may not have sufficient background in the chosen technology and/or training. There are other alternatives then of acquiring the inventions and IP and they are briefly treated below.
ii. The technology and related inventions can come through the engineering founders due to their experiences and specialization in the field and laboratories. This is the easiest to manage and provides a clear path to technology with the least expense and trouble for the startup. If the engineer makes inventions and wants to start a company immediately leaving the employer, there is a potential conflict of interest with the previous employer in case the technology is an offshoot or related to the employer's products, IPs, technologies, manufacturing methods and/or requires their customers for sale of the startup's products. The NDA which engineers have to sign with the employer lays out the terms and

conditions of the potential conflicts. They must be carefully observed so that the founder engineer does not have to face the legal consequences of conflict with the previous employer. The terms and conditions in the NDAs vary with the employers but usually a wait time of 2 years (and sometimes even more for senior employees) is required for the engineers before they can come up with their inventions and use them for a startup. This is the case in the USA but as to the entrepreneurs from other countries planning their startups may refer to their laws and practices before they venture.

iii. Researchers and students (post-doctoral fellows, graduating doctorates, master's degree holders and sometimes bachelor students as well) from universities generate IP and the technologies through their thesis work mostly as a byproduct of it. This set of people on an entrepreneurial path based on their own research has a big advantage compared to the engineers from industries in regard to conflict of interest with the previous employers. Sources of research funding for their work and quantum of university facilities used in the course of research will determine who owns what in the resulting inventions and IP. They are briefly given below:

Case (i): If their research is funded by private companies, the results of the research belong to the funding companies and they have a right to license the inventions arising out of them. In case their research is not funded by companies, the IP out of the research entirely belongs to the researchers and university, and they can license them from the university. This also provides a clear path for the startup but many times require extra steps to acquire this technology.

Case (ii): Note that most of the research funding comes from federal and state agencies in the USA and they almost account for 90% of the total research funding in universities and there is a high likelihood that the researchers may be funded this way. In such a case, the university owns the IP from the research and then the founders can get a license from the university for use in the startup.

Case (iii): If the researchers and students do not use any university property such as laboratory, materials, computing facilities, etc. not to the tune of greater than a minimum fixed amount such as $1,000 (note, this upper limit may vary with universities), then the property usually belongs to them and in such a case they do not need the permission of the university to use their own inventions to form the basis for a startup or for outright sale or for licensing to others. If the university facilities are used beyond the specified minimum limit, then the property belongs to the university together with the inventor. Then, in that case, the startup has to acquire this technology through a license, exclusive or otherwise.

License from Universities: Universities in general have a separate office for IP licensing. The proceeds from the sale or license go partly to inventors (as much as

50%), and the remaining go to the university directly or through a foundation; the later route is common in state universities. License is granted for a fixed number of years, say 10–20 years for certain fees or for a share in a startup or a percentage of the product sales made possible through the licensed IP, and many such options may be available. The licensing process usually is an easy process with a university but becomes complicated if an outside company is also interested in the IP and bids on it even though the inventor owns part of that property. As the inventor entrepreneur may be starting and may not have much funds or financial backing at this instant, and to attract investment funds will require the license without which no one will finance the startup. This puts the inventor-founder under a difficult situation as the universities usually go for quick money and the competing company may buy the license. Sometimes, the consent of the inventors is required for licensing it off to a buyer and that can be used as a leverage to get the university to license it to the inventor. If the offer on table from the competing company is sizeable, then it may be in the financial interest of the inventor to go along with the university and license the IP off.

Given such scenarios, it is advisable to go through a lawyer to negotiate a license even if one knows very well the IP administrators in the university because the beginning entrepreneur may not, in all probability, know the rules, regulations and legal aspects of a binding agreement in a license. The lawyer will assist to negotiate favorable terms on the entrepreneur's behalf. Note that the university side is well lawyered and well versed and knowledgeable in all these matters than the beginning entrepreneur which makes an additional compelling reason to have the lawyer on the side of the entrepreneur. The author had such a positive personal experience in this regard and among other legal matters in his startups thanks to outstanding representation and guidance by a great attorney (Mr. Hugh Wellons, Spillman Thomas & Battle, PLLC, Roanoke, Virginia) starting from licensing, incorporation, leadership recruitment, patents and successful exit among other things.

Case (iv): Acquire an existing business and absorb its IPs and the technology. Such an approach is not further explored in the text as that is not the route taken by most of the beginning engineering founders.

Case (v): Another route to acquire is by finding relevant IP in the domain of the business idea which can be searched in the patent database. If the inventor and assignee are individuals and not connected to a company, then it is worth contacting them for a license or outright purchase of the rights to it. If the assignee is an existing operational company, and that too a large one, the effort may not go anywhere. It is because the beginning entrepreneur will not have enough resources to offer an amount that may attract their interest. Also, most of the large companies usually do not license their IPs to individuals or startups but only cross-license to larger and well-established companies. Some companies, for example, Tesla, offer a limited number of their patents for public use. Then, the problem arises in that the same IP can be used by others as well thus killing the IP being the exclusive property of the intended startup. Dependence and use of such a patent license may not attract venture funds and investors as long as the IP is not the proprietary property of the planned startup.

Case (vi): There is possibly another route to acquiring IPs from foreign companies, individual inventors and universities. Relationship with foreign entities in the form of functioning companies poses some challenges as follows:

a. Their reluctance and even unwillingness to license off their IPs will be as high as the native companies, if not higher.
b. Understanding their legal issues connected to a license governed by foreign country's laws, their interpretations in their judicial system in case a dispute arising in the future is very critical. A startup has neither the resources, i.e., financial and legal, nor the time to deal with it.

But here is a slight chance to navigate a license from foreign companies provided:

a. When no product or plan is dependent on the IP, then the IP is not in their interest to keep without earning some revenue.
b. In such a case, the entity may entertain a joint venture (JV) with a major stake in it as well as having the first right of refusal when the JV becomes successful and is positioned to exit or being bid by other companies for purchase.
c. Such an opportunity for a startup comes with many major advantages, such as IP to start with, and its protection, possible capital investment from it, and possibly a potential buyer when the venture shows signs of success.
d. JV lifetime is usually limited to a term of 3 to 4 years. This is a great advantage to the entrepreneur because by that time the success or failure of the company will be known and hence freeing the entrepreneur from this commitment as well as the relationship with the entity to pursue other opportunities.
e. In order for the entity to be interested in the JV idea, the entrepreneurs have to demonstrate that they have what it takes to make a success of it and the advantages that they bring uniquely to the JV. The items may include a market place and customers in a location that the entity has no presence but wants to have a foothold in future, possible connections to political establishment (in case of public sector-based applications such as maglev and high-speed transportations), venture capitalists and organizations, etc.

Case (vii): Consider the case of foreign individual inventors having IP related to the startup. This is relatively the simpler case to deal with compared to an entity discussed earlier. Steps and caution must be taken to ensure that the IP is free of encumbrances from erstwhile colleagues and former employer of the foreign individual inventor, it is not pledged or licensed to any other parties in that country or anywhere, and free from pledged debts or from spouses going through a divorce and the property is not settled, etc. and therefore requires an extensive due diligence. Then, an agreement can be reached observing the laws of the countries involved of all parties. If the IP is not filed for the patent, then adequate steps have to be taken to file in countries of interest to the business for which it will form a base. Buying IP from east Asian inventors will be less expensive than buying from western inventors. The majority of the inventors may be from the universities, and it is easier and safer to buy from them than dealing with the individual inventors as the universities have an established process and procedure in regard to IPs and their license, sale and transfer. Legally this is a safe route to acquire IP from abroad.

Key Technology Personnel: Getting IP is only one part of technology for the startup. IP by itself is not productive unless people who have inside knowledge of how it can be embedded in a product to make it attractive for customers in a specific market space are with the startup. Therefore, the next most important phase in the technology acquisition is to bring on the people of such capabilities. They can be broadly classified into two categories as,

i. Inventors and researchers
ii. Product developers

The first category realizes the working model from the IP, refine the IP and advance it further by additional inventions surrounding the IP so that others cannot circumvent the IP by any other means. The inventors of the IP are the ideal people to have onboard for this category. The second best personnel of choice are the graduates with their research theses on this and/or related topics. Many times, most startups in USA get them on board. The role of universities in producing such inventors and researchers cannot be underestimated in USA. Many with such backgrounds may be available also from various other countries, for example, from European countries, China and India.

The product developers are a different category. They are not easy to spot as many do not publish in journals and conferences and hence are hidden from public view and deeply embedded in established companies possessing the industrial experience related to the domain startup. They are the key to turn and embed the IP in newer products to give an edge and lead in the market place for the startup and produce a viable product that will satisfy a customer's needs while meeting all the regulations and standards in that product domain. Many of the product developers are not well known to inventors and researchers. The most likely way to reach them is through the personnel search firms (commonly known by the slang, headhunters, in USA). Persuading the right product developers to join a startup is a Herculean task and they are approached with incentives in the form of a higher salary, stock options in the startup, and vacation policy, to name a few.

For many product developers and researchers, the startup's location is a major factor in their consideration even to talk to start with. A case in point is the Silicon Valley (California, USA) where the venture capital companies have a dominant presence in USA enabling startups and where the most skilled product developer and researcher pool in plenty are available.

At this stage, or even an earlier stage, there needs to be an extended discussion on finance and methods and sources to raise it to move forward the startup. It is deferred to later subsections in order to keep the continuity from technology and IP to products, market, its scope and marketing the products, etc.

2.5 PRODUCTS

Product plans principally determine the revenue basis for the startup and hence will occupy a central role in it. The business idea clearly outlines the business domain's technology with a sketchy view of the individual products coming out of it. This section brings out the various possible products with their principal features, and the company's plans as to the order in which the products will be realized, differentiated from the competitors' offerings if they exist at this time and introduced in the market. The rationale for picking a certain order in realizing the products has to be clearly thought out and put in the plans in regard to revenue, ease of penetration in the market space and also based on some potential customers' reactions and feedbacks if and when they are available.

Founders during the process of business planning stage through various sources may establish contact with some potential customers in the startup's product space and get the drawbacks of the competitors' products in the market and what kind of improvements or radically a different product with features of their choice will attract their interest in the startup's products. While this process is fairly easy for experienced engineers/entrepreneurs because of their vast network of contacts, newer entrants should not hesitate and be bold enough to knock on the doors of potential customers. Say a disinterested (or so so) response from potential clients is expected when there is no display of functioning products since they may not yet be ready from the startup. But if in the course of the meetings, if their views on futuristic products can be obtained, then it is a big gain for the startup. This goes to prove that the team must have at least one experienced person with contacts in the technology domain of the startup to get it through this stage. The investors value the potential customers' opinions and possible engagement with the startup even without any firm commitment but having the interest to watch the progress of the startup. This is discussed more later in relevant sections.

It is emphasized here that the basic product and its functionalities will be there for quite a time with the company but its features to suit various customers' specifications and particularities will change from one version to the next version of the product. Accordingly, incorporating many features beyond the current expectation of the market strengthens the marketing of the product to a wide range of customers.

Differentiation of a company's products from that of its competitors is very central to assessing the value of the startup and is based on the following aspects:

i. Functionalities,
ii. Fundamental differences in performance, such as in an electric motor drive, speed of response to desired command references, rejection capability of disturbances on the fundamental function of the system to which the product is applied, etc.,
iii. Competitors' IP portfolio which may be disrupted or superseded by the startup's IPs in capabilities and claims,
iv. Cost comparisons between the products of the startup and that of the competitors, and
v. Based on (iv), assessing the price differences and their significance to the potential customers which will play a big role in the high-volume products in the market.

Assuming that the startup has finalized the products, the next item for consideration is manufacturing of the products. It is considered in the following section.

2.6 MANUFACTURING OF PRODUCTS

The manufacturing plant for engineering companies particularly is highly desired. A number of factors have to be considered, and the principal ones are as follows:

i. Location of the manufacturing site,
ii. Required space for office, laboratory and manufacturing plant,
iii. Plant machinery,
iv. Test beds and test facilities,
v. Assembly plant requirements,
vi. Power system infrastructure, and
vii. Security.

Briefly, they are touched upon in the business plan but addressed in detail in the appendix of a business plan to the extent that enables funding requirement calculations for this setup at various phases/stages of business evolution. This requires greater attention only if the business plan includes an aggressive path to place and distribute the products in the market within 18–24 months of the launch of a startup.

A company's effort in manufacturing becomes major if the products are hardware-based than software-oriented. Many engineering companies fall under the category of having hardware as well as software components in their products and therefore, there is no way out for them except plan for manufacturing and related plant and operation. Software-only products require no broad-based manufacturing in terms of materials, physical assembly, packaging and shipping compared to hardware products. Also, all of them can be handled by developers and worked with the website, email and online automated sales and support with hardly any human interaction with the customers. An example is software sales of all large companies such as Microsoft and MATLAB®. Such is not the case for hardware product-based companies and that translates into higher cost via a higher number of personnel, plant and workers, and other infrastructure and recurring expenses. The trend of the entrepreneurs slanted toward software products is increasing with all the lower amount of money to be raised to launch the startup compared to an engineering startup having both software and hardware components in its products.

Volume of Products: Depending on the volume of products for manufacture, various options come into play. If the volume is small and can be produced within the startup unit, then the manufacture is not an issue. For example, consider the startup's focus is on megawatt variable frequency inverter with a custom motor controller, the market for which in a country, say, is about 20 units/year. In such a case, the workforce required is small and the founding engineering team itself will be able to deliver them without much capital and workforce infusion. In the case of high-volume product manufacture, the startup has to draft a plan and create an organization to handle it.

Paths to Manufacture: There are many routes to have products manufactured and they are:

i. Contract Manufacturing: The advantage of this approach is that the startup does not have to invest, burden itself with major management responsibility and product quality control. If the contract manufacturer is of high reputation and well known, the startup's customers will likely approve the product with much ease. Thus, the startup can save itself from the iterative process of addressing the quality control issues from the customers in its new plant. Hence, this particular option saves valuable time and resources for the startup at the early stages of the startup when both of these items are least affordable.

   There are some disadvantages to this contract manufacturing route:

   a. Part of the profit from products goes to the contract manufacturers, thus reducing the startup's profits and not getting a high valuation from the investors.
   b. Having a legally binding contract to utilize the contract manufacturer's services for a certain number of years constrains the startup when an interested buyer turns up and may see this contract as a handicap for them and much more if the buyer has own manufacturing facilities.

In recent times, foreign contract manufacturers have become a major option. They give significant price advantage that allows the startup to lower the product price to make it very attractive for big customers. On the flip side, a concern comes to the fore as to illegal copying around their environment and not by them which may threaten the business of the startup. If not with the present customers, the startup's chances of selling the same products from the contract manufacturer may become remote as it could have been copied by another local unit. The concern could become a strong dissuading factor for big customers. But major multinationals such as Apple and others have managed to put to sleep this concern and they are able to do because their contract manufacturers are behemoths like themselves and ward off the potential copiers and infringers with their immense financial and legal powers. A startup may not have the luxury of having a big contract manufacturer for its products. Ways to protect the IP in their products have to be found and implemented in the foreign land for such a foreign manufacturing relationship to work. Otherwise, the only other alternative is to have its own manufacturing facility.

ii. Manufacture by the Startup: Main advantages of this route are it is much easier to incorporate changes required by the customers in the shortest possible time, higher revenues to book and hence higher profits, close to or around the startup's location and the IP and manufacturing techniques are easier to protect compared to the contract manufacturing arrangement located in the native as well as foreign lands. The positive among other factors is that the customers may feel assured that there is only one link to production from the startup instead of two or more in contract manufacturing alternative. The major disadvantages of manufacture by the startup itself as mentioned elsewhere are investment capital, recruiting the most qualified personnel, developing infrastructure for the plant, and management of the facility, all of which may take the time and efforts of the limited management team of the startup.

iii. Licensing: In this approach, the product, its underlying IP and any related information can be licensed to major customers who may have their manufacturing facilities or even competitors on different and advantageous terms. The terms can be onetime price or royalty based on a percentage of annual sales of the licensed product for a certain number of years or life of the product. It can be licensed to a joint venture with interested parties, say big customers, and this entity manufactures and sells the products. The only disadvantage under licensing is that the sales revenue will not get into account book as revenues for the startup. Usually, higher revenue in the book gives a higher book value in case of a startup's exit (its acquisition by another entity) and thus possibly a higher return to the investors.

One of the manufacturing routes considered will become the only option depending on the following:

i. Investment capital and its availability influence whether the path is in-house or contract manufacturing or licensing.
ii. Highly experienced management and skilled staff availability in a location of choice for manufacturing.

iii. Time constraint to deliver to the customers, shorter it is, may lead to contract manufacturing as sufficient time may not be there to organize in-house manufacturing unit.
iv. Customers' requirements that the production plant be near to have no delivery delays.
v. Tax incentives to the startup from town/city, county and state agencies which may make raising the capital to build an in-house production unit for the startup.
vi. Investors reaction to options as some may be more interested to create jobs in a particular locality and similarly other factors influencing the decision.
vii. Bandwidth of the upper management to take on this huge responsibility.
viii. The objectives of the founders and what their motivations are that may tip the decision to one or other choices.

2.7 MARKET AND SCOPE FOR THE PRODUCTS

Product finalization for the long term requires the search of the market space, and from it, the relevant and related products for the startup are identified. The information on the market and its scope very rarely come from one individual or source. Then how to go about them including competitors' position inside the market has to be examined. The following pathways may be traversed to get them:

i. Public Information Sources: Under this category fall social sites and internet search engines to locate the information. The social sites both on and off lines include trade magazines, papers published or presented in conferences in the area of startup's technologies, university studies/theses mainly based on their research, research reports from federal agency-funded studies and patents granted and pending. Almost all of them are open to the public. Access to these sources mostly is free over the web, and for some reasonable fee, many technical societies allow their members to access their publications. Market reports of the companies that are public have a lot of revenue-related and sometimes products-/divisions-/technology-based information. They can be harnessed from available discussion forums on the internet. Altogether, public sources provide a bulk of the required market information.

   Another source of very reliable information is many national governments and their respective agencies. They publish annually a complete product sales information on each and every industrial product with the number sold, revenue, etc., in each market domain. It does not break down the data for each company but gives for the whole nation. This source is a valuable one as the data is not biased by any private entity and the reports are freely available and can be accessed over the web.

ii. Market Research Firms: Many over the web firms for market research can be accessed for a fee and sometimes the cost of their reports runs into a few thousand dollars. The ones with the high fees may be difficult to meet for the beginning entrepreneurs. But parts and parcels of the report are made available by these firms for advertisement reason and sometimes collating the publicly available information from many such firms provide almost a full picture of the market and its scope for the startup.

Also from search and study, the competitors and their revenues from the related product sales, the unique features of each and every competitors' products are extensively collected and analyzed. The significant differences between the startup's products and that of the competitors' products and how they can be accentuated with the customers and from them to end users charts the marketing strategies and sales tactics for the startup. These functions are all related to the study of the market for the intended products. Further, the distribution channels for the products in the market and possibly the cost associated with the product sale are all identified. This information is useful to get to the cost that will befall the startup when it introduces the products. The final picture with all the accumulated data moves forward the startup if all point to a definite space in the market place for its products.

The most important aspect of this study is the market scope in dollars for the products. The investors' interest in the startup will be directly proportional to this data. If the market potential, for example, is in the range of $10M/year for the next five years, the established venture capital companies will not touch this startup under any circumstance. Given the market scope and if it is worth the effort from the founders and investors, then the next step is to evaluate whether the market can be accessed by the startup's products.

"You (the startup) build, they (customers) come", is a feel-good thing only and mostly applies to vested interests in the highway construction industry in the USA. In reality, this does not fall on one's laps and if so, only rarely. The entrepreneur cannot (or must not?) bank on luck but must ensure and exhaust all avenues to reach/penetrate the market through various customers. The general way to go after this objective consists of one or many of the following approaches and associated actions:

i. Develop channels of communication to potential customers first by identifying the competent marketing leader and/or a team for the startup with extensive connections and name recognition in that sector. This may be possible but not affordable initially by the startup. Suitable incentives may net such a marketing leader/team.
ii. Investors (individuals and organizations) may have connections to customer companies and their boards. They need to be tapped to get an audience with them. Note that the high-level contacts will open meetings with the customer company's technical teams to assess the startup's

products, their capabilities, and the value that they may bring to the customers.

iii. A website launch is almost essential to advertise the products and its essential features (without comparing them to competitors' products by name for legal reasons and not for offending the users at this stage or any time for that matter). The advertisement by this means is rarely sufficient, and other means have to be resorted to as well.

iv. Trade conferences/meetings in the market sector have to be identified, and there are plenty of them all around the world, particularly in western countries throughout the year. A booth can be set up, and if the paper presentation is part of the conference, then advantage must be taken of it to bring the products to the attention of attendees. A quasi technical marketing paper is the way to go in such meetings. Attractive features and technical specifications of the products and capabilities have to be made available to the participants passing through the booth. Customers passing through the booth must be engaged as much as possible and their contact information and their technical interests have to be collected for analysis by the marketing team. Hosting special by invitation only receptions at the conference sites for potential customers also is very helpful to get the message across. The contacts developed through all these means can be pursued for face to face meetings with their teams preferably at their company premises. When that happens, the startup is taking initial steps to broach the customers. This is a very positive event for it.

v. Publications in trade magazines about company's products help to introduce them to customer's engineering and marketing professionals. They will also reach across customers' various divisions as many of them will not have opportunities to travel to trade conferences and even very large companies put a travel ban for all their employees during downturns. This helps them to give strong feedback to their management in case of interest in the startup's products.

vi. All of the contact methods through various channels require material preparations and packaging to suit a fairly diverse media and for different levels of the customer base of their top management, research and development engineering staff and wide cross sections of marketing and manufacturing engineers. Therefore, separate presentations and brochures to address each of these audiences have to be put in place before marketing is embarked upon.

vii. At this final stage, on contacts with customers, fully functional prototypes at least and preferably final products in industry-specific packaging must be ready both for demonstration in the startup's premises and eventually for testing at customers' laboratories with a possibility for testing at their respective clients' sites as well.

These steps apply to both hardware and software engineering products.

2.8 PRODUCT TESTING

Once the product is developed, the next stage is to have a comprehensive testing of it and some aspects of product testing pertain to:

i. Functionalities
ii. Adherence to standards
iii. Testing under controlled conditions
iv. Testing at customer sites
v. Testing at customer's client sites
vi. Certification

A brief description of these items is given below. They tend to be more focused on hardware products with their software control, but such an approach to an extent is applicable to software products.

2.8.1 Functionalities

Product tests for specified inputs and their corresponding responses (or outputs) with allowable variations in inputs are measured. The allowable variations in input voltage and frequency, say a fixed percentage from its nominal values, from the power grid are specified, and they are not universal but vary from nation to nation. These allowable variations from nominal values are strictly enforced in USA and other western nations. The same is not true in developing countries where the swing in power grid voltage at consumer terminal outposts can be as much as 25%–50% of the nominal voltage of 240V. This deviation is in the range of more than 5 times allowed values in western countries. Many times end users under that circumstance install voltage stabilizers before connecting, say, a refrigerator thus increasing the cost to the consumers. If these voltage swings are factored into a product development such that a voltage stabilizer is made superfluous, then the startup has an upper edge in the market place. The product also has to be tested for the maximum listed ambient temperature and for higher elevations. They are all part of a routine product testing. Federal communication commission (FCC) requires that the products conform to radiation emission standards for user safety, such as in personal computers. Vibration and physical drop tests are some of the critical tests for electrical equipment used in certain sectors.

2.8.2 Adherence to Standards

Two organizations in USA, National Electrical Manufacturers Association (NEMA) and IEEE standards association, issue standards and guidelines on electrical products and related topics. NEMA has standards and guidelines, whereas those of the Institute of Electrical and Electronics Engineers (IEEE) belong to the category of guidelines. For a good product to be trustworthy, in the eyes of the end user, particularly that of large companies, it is important and usual to consider the guidelines as important as standards for incorporation into products and in their development. Almost every country has its own standards and guidelines. International Electrotechnical Commission (IEC), an organization in Europe, has standards that are widely adopted in Europe and many other countries around the world.

As the business is global, it becomes important and profitable to incorporate as many standards in a product so that design, development and testing can be achieved at one time for a product instead of developing products for each country. Many times such an approach is possible if not for the entire global market but at least for many countries at a stroke. Therefore, there is a necessity for such a business vision, planning and execution of the products and their testing.

Note that NEMA makes available its standards for free, while other organizations such as IEC and IEEE offer them for a price that may run to a few hundred dollars for a standard. Until the startup gets up to a level, it may be wise to use the freely available standards for product development and testing.

2.8.3 Testing

Different environment testing will be encountered before a product hits the market. They may be:

i. Testing under controlled conditions with specified inputs and disturbances but within the laboratory. This is the first phase of product testing.
ii. Testing in customer's laboratories happens when the customers are big companies and they have their own specific requirements over and beyond the standards. Even though their expectation would have been communicated to the startup, there may be newer minor requirements that would crop up in the course of the relationship. This is anticipated whenever humans interact and the customer's requirements are also dynamic and impacted by their competitors as well as the startup's competitors, among other factors. The environment for this testing is friendly as it is under controlled conditions.
iii. Testing in many cases of customers with high-volume end users may be required by large customers at their end-user (consumer) sites and data gathered to ensure that mean time between failure (MTBF) satisfies the standards and/or specifications. This testing is not under controlled conditions but in real-life conditions for the product and hence its success counts a lot in the eyes of the customers and invariably for the startup. Examples of such on-site testing of the products are motor drives in compressors for air conditioners.

All these tests are very time consuming for hardware (even with software controllers) products, and therefore, accordingly, the plan must take into account the time and resources required to go through them successfully. Relatively, software-only products have shorter testing times with much less capital requirement.

Testing for the first time or more times may reveal shortcomings in the product's functioning. The feedback from the customers and their end users have to be incorporated in the next iteration of the product and its testing. Many such iterations will downsize the confidence in the startup's products and hence it is critical that the iterations are kept to the bare minimum. This in turn necessitates that the startup has the most professional and experienced design, development, testing and manufacturing engineers in their teams.

2.8.4 Certification

As and when the product passes internal and customer laboratory tests, and prior to end-user installation and testing, certification for the product has to be phased in. In most of the end-user home appliance products and other electrical products, Underwriters Laboratory (UL) certification is expected for product sales in USA and likewise similar certifications in other countries. UL certification means that a specific set of samples of the product, the facility in which it is tested, and its manufacture and process to industry-wide safety requirements meet UL standards. This certification is not legally required for the sale of the product but almost all of the big customers invariably expect and insist on UL's certification and therefore, it is an almost universal requirement and there is no way to skip this for a startup. UL is a global organization and functions in 40 countries. Similar to the UL, Europe has a CE certification requirement to ensure adherence to European health, safety and environmental protection legislation/laws and directives.

UL certification is a long drawn-out process. The testing is comprehensive and experienced engineers having taken products through UL are the most ideal candidates for the startups. Interaction of the design, development, manufacture and test teams from day 1 of the product development will enable smooth sailing with the UL in minimum time and with the lowest expenditure.

2.9 CAPITAL (FUNDING) SOURCES FOR STARTUP

Assume that a business idea has been formulated that is a necessary condition, and the founder/entrepreneur has sufficient high-level management team members or committed personnel, a desirable but not necessary condition, to move ahead with a business plan for the startup. Crafting of the business plan is given elsewhere. Then, the next step is to launch the startup that requires resources the most important of which is capital at this juncture. Capital determines the number of personnel to execute the business plan of building the company involving sizeable expenditure among other items such as rent for the office, laboratory, test and workshop facilities, travel, advertisement, etc. The capital in most cases has to be raised except when the entrepreneur is endowed with it through family inheritance or other means. There are many sources to raise capital, and each of them comes with advantages and some disadvantages. The various avenues for the founder to raise the capital and what it requires and involves are considered here.

Investors rarely come knocking on the founder's doorsteps to get involved with the startup. Almost always, it is the founder to go and seek investors. Invariably, the investors are given an oral presentation by the founder with or without a very few key personnel and a copy of the business plan. The presentation time varies from one set of investors to another, shorter in the range of less than an hour in the case of very well-known venture capital firms to as many hours as it takes to win individual investors.

Various sources of investment capital are broadly classified into either institution-based or private non-institution-based. The expectation of investors varies significantly from one another and it will be pointed out when these investors are described.

2.9.1 Institutional Investors

In general, funding from this kind of investment firms is highly desirable for the following reasons:

i. These firms are run by professionals well versed in high risks associated with investments in startups and it is their business for managing the risks. Because of this fact, there is no spilling of blood or bad blood when and if the startup goes bust or underperforms the expectations.
ii. Their expectations from the startup will be very clear and their communications with the startup's management will be timely and professional.
iii. These organizations are headed by well-known people in the business and that enables the opening of the doors and connections to other influential investors, and potential customer companies' top management. They are very hands-on in guiding and helping in the business strategies and tactics at each and every development phase of the startup.
iv. For founder-based startups, they can bring the best management team with their choice of CEO.
v. Being well versed in legal matters relating to startups, their advice and help from incorporating the startup to finally selling it to a buyer will be crucial to the startups.
vi. They do not usually sign NDAs with the startups and therefore caution must be exercised as to disclosing IPs that are not protected yet.

The above-stated benefits of working with investment firms apply only to some of them and not necessarily to all and they will be indicated in related cases appropriately.

There are many institutional investors and they are:

i. Venture capital companies (commonly known as VCs)
ii. Finance companies
iii. Venture arms of well-established companies
iv. University venture funds

v. Philanthropic charity funds
vi. Government funds through federal government agencies and state and city government agencies

The broad characteristics of these various organizations in this set will be discussed. There may be some peculiarities among these institutions in one category, say VCs, and the founders will have to become familiar before they meet with them and mostly they will be communicated of these requirements by the concerned institution itself.

2.9.1.1 Venture Capital Funds

Funding from them is highly preferred because they bring with them all the advantageous features given in the above. In addition to that:

i. Their main business does not stop with the investment only but to make the success of the startup and they work for it every instant.
ii. VCs have partners, one of whom takes responsibility for the startup's portfolio and helps to run and grow it almost on a daily basis.
iii. The partners from VCs are professionals armed with cutting edge knowledge of business, finance and in securing vital connections from their customers to further investment capital as and when required to scale up the operation.
iv. Their knowledge and navigation during the sale of the startup (called exit in these circles) are essential and of immense value to the startup and its founders.
v. Great recognition is bestowed on the startup when well-known VCs fund it. It puts the startup immediately in front of its would-be large competitors and gives a kind of notice that their bet on the startup is likely to create tremors in the near future.
vi. From the day of their investment, the startup becomes a potential candidate for watch and observation and potentially leading to a takeover by large companies, provided the startup's IP is of immense value to them.

All of these benefits have a price. They are:

i. Significant ownership of the startup has to be given to the VCs. If it is the first investment for the startup, it may even be that the majority of the shares will go to them. For many a founder, giving a majority or significant ownership to VCs is a heartache. It is better to realize that VCs give instant stature, value and a greater potential and high probability for success of the startup unsurpassed by any other entity's help to the startup. Moreover, it is better for the founders to have some significant share of the real pie (with VC funding) than having the full imaginary pie (startup by itself with no funding and recognition). Therefore, the founders have nothing to lose with this and everything to gain only!

ii. A majority appointment of directors on the board and it is not a major concern for the startup.
iii. Their right to control and implement decisions. Not much to worry for the founders as their interests and VCs are one and the same, i.e., to make a success of the startup and make money.

With multiple rounds of funding by VCs for the startup, the founders and original shareholders will see a progressive dilution of their share in the stock holding. Note that the value of the company is increasing with additional rounds of funding, and even with a lower stock share, the worth of it in real terms has increased. The founders only gain with this development.

Location preference of VCs: VCs come in different sizes from millions to billions of investment capital in their chests. They all prefer that the startups locate very near to them, say, within a radius of 100 miles of their offices. It is to enable their partners to interact with the startup CEO and founders at a moment's notice for face to face meetings to fine-tune and influence the direction of the startup and to respond faster to sudden developments such as that of a customer, market space, potential buyer, threats from competitors, further funding needs and how to go about them, to name a few. Wherever there is a cluster of VCs, the startups crowd around them. A prime example is Silicon Valley in California, USA. The physical nearness to their location is insisted upon or expected due to the fact that they have limited partners and it is imperative for them to use their prime human resources to maximize the return of their investments and it is understandable.

Some Features of VCs: VCs specialize in certain fields, and in that way, they accumulate expertise over time. Seed funding of startups is avoided by many VCs but not by all. They usually are not interested in basic sciences and research but are more interested in the product's realizable outcomes of them. Many hold the portfolios for <5 years, and within that time, they expect a successful exit or initial public offering of the company stock or multiple rounds of financing with other VCs to expand the startup's product lines and hence their market and sales for achieving a higher value for their investments. A general rule of thumb is that VCs expect a reward of ten times their investments. Such a high expectation on return on investment (ROI) is placed because not all of their portfolio companies succeed as some will fail and in spite of that, the VCs strive to be profitable and be responsible and answerable to their own investors.

2.9.1.2 Finance Companies

Some Wall Street finance companies such as Goldman Sachs are into venture funding of startups. Their funding basis is comparable to traditional VCs. Their current interest is for newer software that will help their business in the finance and banking sectors of the economy and hence they look for startups in that line. Uber was funded initially by them and many such ones. Startups headed by well-known former CEOs (note that the business idea comes from founders) inspire the finance venture arms' confidence and tend to get their funding. In many ways, they do not hold the hands of the startup's top management team and do not contribute to strategies and tactics on a day to day basis like traditional VCs. Because of their deep pockets, they are capable of bringing multiple rounds of funding for the growth of successful startups. A startup with a market scope of billion dollars or more and that too in the near term of 5–10 years becomes a potential candidate for adoption by the finance companies venture arms. To get to that point, the startups have to grow with a few rounds of investment funding from traditional sources of VCs and others. This is not the rule but a key to success with this sector of venture funding.

2.9.1.3 Venture Arms of Established Companies

Many large corporations have venture arms that are constantly searching for startups in their market domains of specialization or domains that they want to expand to and with their interests aligned for funding them. The funding is applied in a focused way only in their areas of activity as a way of complementing their in-house research and development efforts and they do not want to be blindsided by new developments should it happen outside of their organizations and without their knowledge. This is a low-cost means to avoid any unpleasant and threatening surprises to their product lines and business and a potential loss of some customers. Examples of such venture arms of the established companies are Siemens, General Electric, ABB, Intel, Schneider Electric, Yaskawa Electric, Mitsubishi, Volkswagen, Toyota, Panasonic, Sony, Hitachi, General Motors, Honda, Johnson & Johnson and many others.

These venture institutions invest at the seed stage of the startups and very rarely go beyond that stage to higher levels of funding in further rounds of investment. The startup founders going through the venture institutions' due diligence, even if not funded, will learn and still gain a lot by getting to know the expectations of the investors which may put them better equipped to fine-tune their business plan and rethinking their technical plans all of which may pave the way for success with other investors.

2.9.1.4 University Venture Funds

Many big and well-known universities in the western world have accumulated immense wealth, from the contributions of the public, alumnae and founders of very successful companies. That wealth is sitting in their foundations to the tune of tens of billions of dollars. The foundations then invest these funds in the stock market and other options. In recent times, they are setting aside a small proportion of their foundation money (as much as $100 M in some universities) to fund the startups around them. Examples of such venture funds from universities are Cambridge University, Oxford University, Imperial College London, all in UK, University of Chicago, Duke University, MIT, Stanford University and many others in USA. Universities with much smaller foundation money (such as Virginia Tech, VA, USA) start these venture funds by themselves and sometime aligned with other organizations and call these funds as innovation funds. The aims of these university venture funds are manifold and they are to present an ecosystem of startups engendered by their alumnae, attract through them large corporations to place their units around them to get funding for their faculty in many of their disciplines, to harvest a rich payout when their funded startups make it big in the market, attract brilliant students and researchers from around the globe through these activities and to attract top-notch students from their own nations to further their base for future gifts to their foundations. All of these outcomes enable them to grow in standing and stature, both nationally and internationally and to grow their foundation's wealth. Regardless of the motives, an important point for the founders of startups is that there is one more avenue for obtaining their funding. Many times, but not all times, these venture funds prefer to invest in their own graduates. It is also because their own graduates know about them and know the funds management as they are on the same campus and get to know what they are looking for in the startups. The university's students when they have significant thesis research outcomes may find a home with that university venture funds. In this regard, it is advisable that innovation- and entrepreneurship-driven students to seek universities with venture funds for their graduate studies and research. Then they have a better chance compared to one with no association to the university and/or faculty to launch a startup and get its seed funding also. It goes without saying that going to top universities has many advantages including this one.

2.9.1.5 Philanthropic Charity Foundation Entrepreneurship Funds

A large number of philanthropic foundations (mostly in western countries) are in place to support, seed fund socially impacting startups. They are guided by their principal donors, their board of directors and the stated focus of the foundation. A few examples are cited below.

The Draper Richards Kaplan Foundation is a global venture supporting high impact social enterprises for their early-stage investment needs. Evolve Foundation targets startups currently directed to relieve human suffering spread through technology in the form of loneliness, fear and anger. It is being recognized as a growing problem affecting children as well as adults. Cystic Fibrosis Foundation (CFF) gave money to a biotech startup in the year 2000 to encourage it to find a treatment for cystic fibrosis. With its success, CFF received $3.3 billion as part of its share when the firm was acquired by Vertex Pharmaceutical. Therefore, charitable investments also make money for the foundation which goes a long way to fund future startups and their charitable work. Bill Gates (founder of Microsoft) Foundation invested in a sizeable amount in CureVac, a biotech in Germany. Thiel Foundation (founder Peter Thiel) has funded 22 startups since 2012.

Social entrepreneurship is supported by Schwab Foundation helping to develop means and methods for sustainable social innovation. An example is developing and promoting technologies for use by entrepreneurs in Africa to run profitable enterprises. Skill Foundation and Ashoka Foundation for developing countries to aid and support social entrepreneurs to solve world's problems. Many such foundations can be found around the world. A number of organizations are there to help minority (race-related and gender-related) startup founders so as to increase diversity in entrepreneurship. Some organizations give grants to startups which emerge winners in competitions held by them. One such organization is WeWork. Direct application-based grants are also given by some charities, and one of them is Kaufmann Foundation, but its activities are confined to a specific geographical area.

Many of these foundations take a slice of the shares in the startup or enterprise at the time of investment and that is a small price to pay for and nothing to be concerned about by the founders as the future proceeds will go to seed more startups.

2.9.1.6 Government Grants

These grants come through two channels as follows:

i. Federal agencies
ii. State agencies and City/Town administrations and agencies

Federal Agencies: They (almost all of them) give grants to companies and universities for research and development activities on topics of national interest. These grants are based primarily on the competitive strength of the proposals containing the technology being proposed and demonstrating their applicability in some of the agency's concerns. There is a federal agency to cater to small businesses and its proposals and here the grants are not open to competition from larger enterprises. The startups are small and have little or no revenue at all and they fit the bill of small business definition for all rhyme and reason. Grants are usually from $100 K to $2 to $3 M depending on the agency's funding for that year and the number of grants being assigned. The lower amounts are for the first round of the project and if the results of the first phase are acceptable to the agency then the company can go to the next higher stages of the project. Up to three levels of funding are through the small business administration. Many electrical engineering companies that are an offshoot of university research have been funded through these competitive grants and they benefited in their early stage of development. They are useful for these startups to build prototypes and demonstrate their performance. With this, the startup is ready to take the business plan to various investment organizations as it can answer many questions as to the practical proof of the technology to their satisfaction and by surviving the first year or two with these grants they have demonstrated that they passed the seed stage and ready for major investment round and growth. VCs will have more to see and feel into what they are investing at this stage.

The major advantage of government agencies funding is that startup does not have to give a share of its company to them in return for their grants. Further, the results of the research and any resulting IP belong to the startup company. There is a clause in federal grants that they can take over the IP in case of national interest and emergency. They have not done so far in history, and if and when that happens in the future, which has the lowest probability, it is quite likely that the company will be compensated.

State and Local Government (Town/City) Agencies: They hardly ever give research grants, but they give incentives to startups if they get located in an economically backward part of the state or if the startup makes a commitment to locate its manufacturing unit and its office in there. These grants expect compliance of such an agreement by the startup and no ownership of the company needs to be shared in return for the grants. Primarily these grants are intended for economic development in their regions.

In general, the disadvantages of all government agency grants are:

i. It is time-consuming to chase these grants.
ii. The time between submission of company's proposal and the arrival of grants, if at all they are granted, may be 6 months to even a year plus.
iii. The grant requirements may not coincide exactly with the company's objectives and goals and many times they will partially fit only.
iv. Because of all these factors, the startup's goal of getting into the market space with its products may be delayed, but the consolation is that it is on the path, at least, even though with delays.

2.9.2 Non-institutional Funds

Some noted sources of funding under this category are:

i. Self (own)
ii. Family
iii. Friends
iv. Crowdsource
v. Angel investors
vi. High net worth individuals

Each of the sources for startup's capital is described in the following.

2.9.2.1 Self (own)

A significant number of startups are initially funded by the founders. Usually the target is to have the company incorporated, a prototype to show to potential investors and a website for advertising. Founders coming from well-to-do background may have the resources to invest to move the startup all the way to its final goals. It is advisable even then to bring in external investors for the following reasons:

i. Very difficult to have a perspective with founders alone trying to steer the company in the unchartered territory of business with little or no previous experience in startups.
ii. Many support sources and connections to finance, marketing, potential customers, legal and accounting experts are absolutely required for success and they may not be with the startups' founders.

All the above requirements for success will be made easy and available through external sources such as VCs, for example.

2.9.2.2 Family

This constituency is the easiest to pitch for funding by the founders. Family is the least to be critical and most supportive to one of their own. It has the disadvantage in that unbiased outlook will not be imparted to the founders and they need it to retune their business ideas and to take the startup forward. If the startup succeeds, all is good and well with the relationship between the family and founder. If it does not succeed, the founder loses the goodwill of the family forever creating emotional problems for the rest of the life. Important for the founders to realize that family members in almost all cases are not aware of the risks associated with investments in startup and hence taking their investments is not right. But if the family is wealthy enough and its investment will not dent their holdings, then it may be considered by the founder.

2.9.2.3 Friends

Similar to the family category and therefore, more or less the same discussion applies.

2.9.2.4 Crowd Source

This is a recent phenomenon with small contributions from a large number of people around the globe over the internet-enabled websites. The contributions are in the categories of a donation to a cause or a project, equity purchase in a new enterprise seeking funding and reward-based return. The startups unless they are for nonprofit cannot rely on donation-based crowdfunding and they have a better chance of attracting sizeable investment in return for equity/shares in the startup. This is the category relevant for the engineering-based startups should they choose to pursue crowdsource funding.

The engineering founders have a disadvantage with crowdfunding source as they will not be able to reveal details about their inventions over the internet and underlying IPs as they may not have been protected at this early stage as that requires funds to do so which they may not have. Because of this handicap, the engineering founders may not widely use the crowdfunding sources. Further information is available in the references under this heading.

2.9.2.5 Angel Investors

A few well to do in a town or locality join together to pool their contributions with the sole intention of investing in startups in return for partial ownership in them. Such entities are known as angel investors or networks in USA. These organizations are in almost all cities and even a large number of towns. The amount of funding is at the seed level (<$100 K) and very rarely exceeds that. It is easy to approach them and make presentations to garner their interest. Some of their members have an entrepreneurial background and that usually is a helpful factor for the startup as they will assist in getting its technology better discussed among themselves. The founders have to be cautious in towns particularly as the angel investors are known to demand a large share of the company for a small investment.

2.9.2.6 High Networth Individuals (HNIs)

They are of a different category from angel investors in that their wealth is in the range of hundreds of millions to billions. Many times they act independently and sometimes they join together. They are usually located in major cities and have their business managers to vet the startups (unlike the angel investors where they are directly involved). Their due diligence is somewhat similar to that of the VCs. The big advantage of getting HNI support and investment is further rounds of investments become quite possible when the startups' development and progress are on the target. Further, they have connections to further the interests of the startups. The HNIs are all across the globe. It is a challenge for the first time entrepreneur to find the communication channels to HNIs and get the business references to get to know them. Many HNIs also invest through traditional VCs, a case in point is Bill Gates working with Khosla Ventures, CA, USA.

2.10 BUSINESS ORGANIZATION

The practical aspects of the business organization require the following activities to be addressed by the startups:

i. Company incorporation
ii. Location selection
iii. Selection of legal firm or individual attorney
iv. Recruitment of leadership team
v. Recruitment of engineering and other team members
vi. Selection of chartered public accountant (CPA)

A brief description of them is considered here.

2.10.1 Company Incorporation

The startup has to be registered with appropriate government departments/agencies in the country of its launch. Tax identification is also obtained with the registration which is essential for the company to comply with internal revenue service (IRS) regulations regarding tax return in USA and likewise for corresponding agencies in other countries. In USA, there are three choices for incorporation of the startup. They are Limited liability corporation (LLC), S corporation and C corporation. The most advantageous choice emerges in discussion with the company's attorney.

2.10.2 Selection of Company Location

Factors that influence the choice of the company location are as follows:

i. Investors have a big say in this matter. For example, VCs prefer that their funded startups be very close to them geographically. The same may be the case with HNIs and angel investors.

ii. Closeness to customers and workforce availability are other factors.
iii. Founders' preferences.

2.10.3 Selection of Attorney Services

Two types of legal services are required for the startup and they are corporate and patent attorney services.

Corporate Attorney Services: The services of a corporate attorney are essential in the following categories:

i. Incorporation of the company.
ii. Advising founders as to the implications of the terms and conditions given by the investors.
iii. Developing NDAs between the startup and potential customers when confidential information is being shared between them.
iv. Developing employment contracts and NDAs between the employees and the startup.
v. Developing license agreements in case the startup's IP has to be licensed off to another entity or getting some IP from an outside source.
vi. Advising startup's management on legal terms and conditions for working relationships with startup's customers, suppliers, contractors and consultants.
vii. Advising the management when the startup is given an offer of purchase on terms and conditions and to help the legal aspects of a deal.

Initially, very limited service is required of the corporate attorney but as the company grows the need for legal services also expands in the areas outlined above. Many law firms or individual attorney practices offer these services on an hourly rate basis with the rates differing for the attorneys within a practice. Corporate attorney services can be obtained in most of the places, small and big, in the USA.

Patent Attorney Services: In engineering- and science-related startups, the major asset is in the IP and that needs to be protected through patents and other means. This invariably requires the services of a patent attorney. They are not commonly located in smaller towns but only in major cities and prominent towns. A significant amount of expenses is involved to get the patent attorney's services and subsequently for fees associated with filing, processing and maintenance (only when the patents are granted and issued) with the government patent office.

There are all kinds of web-based legal help available and their effectiveness for startups is not known.

2.10.4 Recruitment of Leadership Team

The founder starts alone initially but with the funding, the leadership team has to be recruited. The team at a minimum has to have the following members:

i. Chief Executive Officer (CEO) and President: Usually the founder assumes for a time the functions at the start and can be changed, if necessary, when funding flows as investors have control to change in the best interests of the startup. The engineering founders usually go and look for a CEO after a year or two of functioning to devote their time to technology and product development.
ii. Chief Technology Officer (CTO): Many engineering founders automatically fit into this position. In case the founder is not the inventor of technologies and is not a technologist at all, then a suitable person has to be recruited with the background and achievements in the startup business' specialization.
iii. Vice Presidents (VP) in Engineering and Marketing: These are critical positions for the company and its growth. CTOs can fill in as VP engineering for a while until the startup starts to grow but eventually a full-timer has to be searched and recruited. A founder with no technology background, say occupying initially the role of CEO, can fill in the role of VP marketing for some time and then has to resort to finding a full timer for this position with the growth of the company. Note that all these VP positions need not be filled at the beginning stage of the company and can be at the manager's level. As the company grows, the managers' contributions become visible and it is possible that they may grow into the VP positions.
iv. Chief Financial Officer (CFO): This position is not very important for engineering startups in the beginning or until the company produces significant revenue. If external accounting services are procured, they will be able to do the required work.

In the case of startup having been funded, these positions (CEO, CTO and VPs) can be filled and their recruitment is achieved through search firms. Their service comes at a steep price. Therefore, it is in the interest of the founders and investors to use their contacts and networks to find suitable professionals for these positions and in the process save a significant amount of funds.

2.10.5 Recruitment of Engineering and Other Team Members

They are planned by the leadership team members and will be recruited using their own personal contacts, advertisements and search firms.

2.10.6 Selection of Chartered Public Accountant

Services of a CPA are very crucial for handling personnel payroll and for filing the tax return with internal revenue service. There is no need to have a full-timer at the beginning of the startup when funds are not plenty and therefore, accounting services can be outsourced to a local CPA firm for a frugal operation. CPAs are located from small towns to large cities in USA.

2.11 BUSINESS PLAN

One of the important and formidable tasks in the startup is to raise capital for operation and expansion. That requires a presentation of the various aspects of the startup in a crisp manner. That presentation in the report format is known as the business plan. Many references as to what a business plan constitutes can be found. But the key is to have the overall plan rather than compiling a book for the intended audience, i.e., investors. The investors have very limited time as they are bombarded with a large number of potential startups. This fact has to be taken into account in having a short presentation not more than 30 minutes, in general.

The presentation cannot have the details of the IP as most of the VCs and investor organizations will not sign NDA. If details are presented without the NDA, it will pose a danger to the filing of patents as the materials have been made public.

Important ingredients and practical aspects of startup business have been described in earlier sections and again topic wise given below for easy reference:

i. A brief summary of the plan.
ii. Business Idea: Startup and what it intends to bring out, products and/or services.
iii. Business Organization: Current status regarding incorporation, level of product development, customer contact and marketing efforts, etc.
iv. Technology and IP behind the startup, their principal features, strengths and weaknesses.
v. Market analysis, competitors and their share of the market, nationally and globally.
vi. Comparison of startup's products and their features vs. the competitors'.
vii. Marketing strategy or approach.
viii. Products, their manufacture and distribution with their timelines.
ix. Financial Plan: Budget and funding requirements (including people, place, facilities, materials, testing, certification if needed) for a length of time such as year 1 to year 3, at the minimum.
x. Leadership team and their background with a short bio of the founder and possible candidates for CEO, CTO, VP engineering and VP marketing, if not in place already.

For the first time meeting, the business plan of the startup to the investors is a lean one both in the report and presentation form. In the case of investors' interest, further meetings are bound to take place. In them, more data to back up the claims being made in the presentation and significant concerns that came out in the form of comments and questions have to be addressed satisfactorily to gain the support of the investors. There the effort required will be much more but not in terms of adding topics but going to the depth of the topics being covered. Therefore, the business plan is not static but will be dynamic in its details, not only up to the funding stage but even after funding when the market imposes some changes to the original plan.

2.12 PERSPECTIVE FOR THE FOUNDERS

It is important that the founders have a perspective not only with their startups and their chances of success and about life in general. The startups are created and function in a society which is governed by the basic laws of capitalism. The society bestows only unfettered rights to individuals to do anything within the society's norms and laws in pursuit of profit and wealth. As there is no coordination of these individuals, as each one can pursue regardless of what others pursue, with the end result that there will always be duplications. That is termed as market chaos in society. Engineers' education and practice are science-based, but business is not at all. Therefore, it is impossible to guarantee the success of a venture in the business domain with any certainty. This is important for engineers to recognize so that they can have the perspective to look up at their own startup and be realistic that they alone cannot control the destiny of it but forces outside of their control will prevail in its success or failure.

That leads to the personally painful question of what are the chances of success based on some available statistics of the startups. The general belief is that the success rate is from 10% to 25% of all the startups that are launched but having 60%–65% success rate for startups with the VCs funding. The more visible and well-known VCs backed startups have considerable success rate much closer to the upper limit of 65%. The general rate of all startups' success is not encouraging viewed by itself but it is better than that of the Olympic athletes, say in running 100m where only three in the world get to win a medal and where the success rate is very negligible! That success rate statistic should not discourage potential entrepreneurs from pursuing their ideas.

A good idea for the entrepreneurs is to have the moral courage and mental strength in dissecting their own idea to see whether it has a chance of success under all possible scenarios and conditions that they can think of, surround themselves with intelligent and dedicated people with experience to get their counsel and most importantly having a good perspective of life that neither it begins nor ends with their startup but life is to live and bring happiness to one and all around and be helpful and be productive members of the society.

2.13 DISCUSSION QUESTIONS

1. What are the pros and cons of various fundraising avenues?
2. When to start and launch the company? Is there an optimal time and if so, explain?
3. Advantages of single founder vs. multiple founders based startup?
4. Disadvantages of single founder vs. multiple founders based startup?
5. Is there a special country to locate a business? Give examples for a couple of hardware- and software-based companies.
6. Is there a special city/town to locate a business given the best country to start a business? Give examples for a couple of hardware- and software-based companies.

7. Is there a special city/town to locate a business in India to start a business? Give examples for a couple of hardware- and software-based companies.
8. Is there a special city/town to locate a business in China to start a business? Give examples for a couple of hardware- and software-based companies.
9. Is there a special city/town to locate a business in the USA to start a business? Give examples for a couple of hardware- and software-based companies.
10. What are the primary factors that will decide in finding the location of the company?
11. How will work force availability going to impact the location of the startup?
12. How will cost of living impact the location of the startup?
13. How will city/town administrations influence the location of the startup?
14. Discuss the share of the transportation and air connection to influence the startup.
15. Discuss the influence of the customer base on the location of the startup.
16. Where to locate and why a particular town/city and site for success, what factors they are dependent on for this decision, and many more such topics.
17. What is the success rate amongst motor drives startups in your country?
18. What is the success rate amongst power electronics startups in your country?

2.14 EXERCISE PROBLEMS

1. Identify: (i) leading VCs in your country, find their specialization, what levels of funding they go to, their success rate, national ranking (if available), and (ii) in your area of interest which will be the most ideal VC to go for fundraising.
2. Identify the HNIs in your country involved in startup funding. Discuss their portfolios and success rates.
3. The problems are project-oriented so that teams can work and then interact throughout the semester to gain startup experience. The objective of these projects is to give students the skill to learn about how the following big-name companies did come about and compile a report on them covering all the topics covered in this chapter such as business idea genesis, initial launch, funding from beginning, scaling up, marketing, revenue base, management structures, business growth and how well managed they are, among other things:
   i. Facebook
   ii. Google
   iii. WhatsApp
   iv. Hangout
   v. Twitter

vi. Tesla
vii. Uber
viii. Other companies of the project team's and/or instructor's choice
4. Each team has to identify a business idea in the electrical engineering domain and develop a business plan for it.
5. The teams can exchange their business plans and come up with a critique of them and then how to improve on them for a higher success rate.
6. The final examination may consist of the evolved business plans' presentations in class and grading by all the students on each and every project with the evaluation criteria to be formulated by the faculty member.

REFERENCES

ENTREPRENEURSHIP

1. P. Drucker, *Innovation and Entrepreneurship*, Harper Row, Publishers, Inc., 1985.
2. R.D. Hisrich and M.P. Peters, *Entrepreneurship*, 5th Edition, Tata McGraw-Hill, New York, 2002.
3. P. Swamidoss, *Engineering Entrepreneurship from Idea to Business Plan*, Cambridge University Press, Cambridge, 2016.
4. K. Uchino, *Entrepreneurship for Engineers*, CRC Press, Baca Raton, FL, 2010.
5. M.B. Timmons, R.L. Weiss, D.P. Loucks, J.R. Callister, J.E. Timmons, *The Entrepreneurial Engineer*, Cambridge University Press, Cambridge, 2014.
6. Lecture Notes, *Entrepreneurship Development*, EIILM University, Sikkim, India.
7. S.S. Khanka, *Entrepreneurial Development*, Chand & Company Ltd., New Delhi, India, 2012.
8. Harvard University, Office of Technology Development, Startup guide, https://otd.harvard.edu/upload/files/OTD_Startup_Guide.pdf.
9. T. Oppong, Failed startup lessons, 50 founders reveal why their startups failed, https://ats-alltopstartups.netdna-ssl.com/wp-content/uploads/2015/05/Failedstartups_ebook.pdf.
10. Entrepreneurship Development and Enterprise Management, BBA course 301, directorate of distance education, Guru Jambheshwar University, Hisar, 2008, http://www.ddegjust.ac.in/studymaterial/bba/bba-301.pdf.
11. P. Natarajan, Entrepreneurship management, http://www.pondiuni.edu.in/sites/default/files/Entrepreneurship%20Managementt200813.pdf.
12. W. Aulet, H. Anderson and M. Mar, New Enterprises, MIT course no. 15.390, Spring 2013, All materials available in https://ocw.mit.edu/courses/sloan-school-of-management/15-390-new-enterprises-spring-2013/index.htm?utm_source=OCWCourseList&utm_medium=CarouselSm&utm_campaign=FeaturedCourse.

FINANCE COMPANIES INVESTMENT ARMS

1. Business Insider, Companies backed by Goldman Sachs, https://www.businessinsider.com/companies-backed-by-goldman-sachs-2015-10.
2. Quartz, Goldman Sachs (GS) is the biggest tech-startup investor of any fortune 500 company, https://qz.com/1059294/goldman-sachs-gs-is-the-biggest-tech-startup-investor-of-any-fortune-500-company/.

3. Principal Strategic Investments, Goldman Sachs | Investing and lending, https://www.goldmansachs.com/what-we-do/investing-and-lending/principal-strategic-investments/psi.html.
4. Wikipedia, List of unicorn startup companies, https://en.wikipedia.org/wiki/List_of_unicorn_startup_companies.
5. Bloomberg, How Goldman Sachs became a tech-investing powerhouse, https://www.bloomberg.com/news/features/2015-07-28/how-goldman-sachs-became-a-tech-investing-powerhouse.
6. Wall Street Prep, What is investment banking? How the investment banking industry works, http://www.wallstreetprep.com/knowledge/about-investment-banking/.
7. Wall Street Prep, Investment banking vs. private equity, https://www.wallstreetprep.com/knowledge/investment-banking-vs-private-equity/.
8. Morgan Stanley, Morgan Stanley selects 10 startup companies for 2nd cohort of innovation lab targeting multicultural and women founders, https://www.morganstanley.com/press-releases/morgan-stanley-selects-10-startup-companies-for-2nd-cohort-of-in.
9. The Fintech Finance 40 | Institutional investor, https://www.institutionalinvestor.com/article/b16fj05mr7lzvs/the-fintech-finance-40.
10. WSJ. Startups that never grow up, https://www.wsj.com/articles/startups-that-never-grow-up-1509442201.

Venture Arms of Large Companies

1. MarketWatch, Cisco eyes early-stage startups with new venture fund, https://www.marketwatch.com/story/cisco-eyes-early-stage-startups-with-new-venture-fund-2018-05-15.
2. GLM Co., Ltd. Supercar acquired 2017, http://glm.jp/.
3. ABB Investments, https://new.abb.com/about/technology/ventures/investments.
4. ABB Technology Ventures (ATV), https://new.abb.com/about/technology/ventures.
5. Gigaom | ABB Buys Controlling Interest in Data Center Power Company Validus, https://gigaom.com/2011/05/12/abb-buys-controlling-interest-in-data-center-power-company-validus/.
6. Press Releases - Siemens Global Website, https://www.siemens.com/press/en/pressrelease/?press=/en/pressrelease/2014/financial-services/sfs201402002.htm.
7. Why Work with us | GE Ventures, https://www.ge.com/ventures/why-work-us.
8. Japanese venture capital investment hits record levels | Financial Times, Panasonic joins Toyota and Sony in bid to fund innovative tech start-UPS, https://www.ft.com/content/927a9d14-1d21-11e8-aaca-4574d7dabfb6.

University Venture Funds

1. How many universities have venture capital funds and which ones are most prominent? https://www.quora.com/How-many-universities-have-venture-capital-funds-and-which-ones-are-most-prominent.
2. University venture funds must reach beyond the Golden Triangle, http://iettn.ieee-ies.org/university-venture-funds-must-reach-beyond-golden-triangle/, May 11, 2018.
3. Universities are venturing into new territory: Funding start-up businesses, https://www.washingtonpost.com/business/capitalbusiness/many-universities-weigh-risk-and-reward-as-they-create-venture-funds/2015/02/14/81d865aa-b2c0-11e4-827f-93f454140e2b_story.html?noredirect=on.

Crowd Sourcing/Funding

1. Raising money and awareness online, https://blog.fundly.com/crowdfunding/.
2. What is crowdfunding? Clear, simple answer here. https://www.fundable.com/learn/resources/guides/crowdfunding-guide/what-is-crowdfunding.
3. 10 Top crowdfunding websites for entrepreneurs, https://www.entrepreneur.com/article/228534.
4. The 40+ best crowdfunding websites, https://blog.fundly.com/crowdfunding-websites/.
5. Crowdfunding, The basics of crowdfunding, https://www.investopedia.com/terms/c/crowdfunding.asp.
6. Crowdfunding, https://en.wikipedia.org/wiki/Crowdfunding.
7. Equity crowdfunding | Invest in startups, https://equity.indiegogo.com/.
8. Crowdfunding investments for startups - Over $200MM raised | crowdfunder, https://www.crowdfunder.com/.
9. Top 10 equity crowdfunding sites for investors and entrepreneurs, https://www.moneycrashers.com/equity-crowdfunding-sites-investors-entrepreneurs/.
10. Local investing clubs and networks | Local investing resource center, https://www.local-investing.com/how-to/local-investing-clubs-and-networks.

Philanthropic Foundations for Capital

1. Sparkplug foundation | Funding start-up organizations and new projects in music, education and community organizing, https://www.sparkplugfoundation.org/.
2. Global Entrepreneurship, https://casefoundation.org/program/global-entrepreneurship/.
3. 5 Foundations supporting social good entrepreneurship, https://mashable.com/2011/10/27/online-social-entrepreneurship-foundations/#8BzsauH7yEqY.
4. Grants | Kauffman.org, https://www.kauffman.org/grants.
5. 5 Organizations helping minority startup founders succeed, https://www.entrepreneur.com/article/282529.
6. WeWork commits $20 million in grants to startups, nonprofits | News | PND, https://philanthropynewsdigest.org/news/wework-commits-20-million-in-grants-to-startups-nonprofits.
7. Charities are making big money by acting like venture capitalists | Fortune, http://fortune.com/2015/05/27/charities-acting-like-venture-capitalists/.
8. Evolve Foundation launches a $100 million fund to find startups working to relieve human suffering | TechCrunch, https://techcrunch.com/2017/11/03/evolve-foundation-launches-a-100-million-fund-to-find-startups-working-to-relieve-human-suffering/.
9. Portfolio | DRK foundation | Supporting passionate, high impact social enterprises, https://www.drkfoundation.org/portfolio/.

Company Corp Structures

1. Compare S Corporation vs. LLC, Differences and benefits | BizFilings, https://www.bizfilings.com/toolkit/research-topics/incorporating-your-business/s-corp-vs-llc.
2. What's the difference between a C Corp, S Corp, and LLC? https://www.startupdocuments.com/incorporation/what-s-the-difference-between-a-c-corp-s-corp-and-llc.
3. LLC vs S Corporation vs C Corporation: Best small business structure? https://fitsmallbusiness.com/llc-vs-s-corp-vs-c-corp/.

4. LLC vs S Corporation | CT Corporation, https://ct.wolterskluwer.com/resource-center/articles/llc-vs-s-corp-advantages-and-disadvantages.
5. Compare Business Structures, LLC vs. Corporation, S Corporation and C Corporation | LegalZoom, https://www.legalzoom.com/business/business-formation/compare.html.

Failures and Reasons

1. Conventional wisdom says 90% of startups fail. Data says otherwise, http://fortune.com/2017/06/27/startup-advice-data-failure/.
2. 134 of the biggest, costliest startup failures of all time, https://www.cbinsights.com/research/biggest-startup-failures/.
3. Business Insider, 10 startups that died in 2017: Despite $1.7 billion in funding, https://www.businessinsider.com/startups-that-raised-148-billion-have-shut-down-or-may-soon-2017-11#doppler-labs-2013-november-2017-10.
4. 12 failed startups in India 2017 and lessons learned, https://www.techinasia.com/insights-from-12-failed-startups-in-india-2017.
5. 5 lessons to learn from the 10 biggest startup failures so far in 2017, https://www.inc.com/jory-mackay/5-lessons-to-learn-from-10-biggest-startup-failures-of-2017.html.

Standards and Guidelines

1. All Standards by Product, NEMA, The association of electrical equipment and medical imaging manufacturers, https://www.nema.org/Standards/Pages/All-Standards-by-Product.aspx.
2. MotorandGeneratorStandards,NEMA,https://www.nema.org/Standards/Pages/All-Standards-by-Product.aspx?ProductId=f6107549-40c5-4110-9a4c-dd7215bf1e60.
3. Power Electronics, NEMA, https://www.nema.org/Products/Pages/Power-Electronics.aspx.
4. NFPA 79: Electrical standard for industrial machinery, https://www.nfpa.org/codes-and-standards/all-codes-and-standards/list-of-codes-and-standards/detail?code=79.

Certification

1. Certification Services | UL, https://services.ul.com/categories/certification/.
2. UL, https://www.ul.com/.
3. One-stop CE Marking (CE Mark) Information Source: Everything you need to know about CE marking - World's most comprehensive multi-lingual information guide on CE conformity marking. CE-Marking, CE Mark or EC Mark? http://www.ce-marking.org/.
4. The FCC Standard, http://www.goodhealthinfo.net/radiation/fcc_standard.htm.
5. Standards and Certifications, FCC certification, DoC, Declaration of Conformity, UL Mark, Underwriters Laboratories, CE certification, CE Mark, Safety Regulations MPR II, TCO, CSA certification, Canadian Standards Association, ISO standards, ISO 9000, ISO, https://idp.net/sysinfo/standardsand.asp.
6. What is the different between UL, CE, EMC, FCC and CSA certification, https://www.batteryspace.com/ul-ce-emc-fcc-and-csa.aspx.

Business Plan References

1. Top 10 business plan templates you can download free | Inc.com, https://www.inc.com/larry-kim/top-10-business-plan-templates-you-can-download-free.html.
2. Create a business plan | The U.S. small business administration | SBA.gov, https://www.sba.gov/tools/business-plan/1.
3. Simple business plan template for entrepreneurs, https://www.thebalancesmb.com/entrepreneur-simple-business-plan-template-4126711.
4. T. Candia, *Developing a Business Plan*, Book 1 in Starting Your Start-Up E-book series, IEEE, Piscataway, NJ, 2016.

3 Introduction to Power Semiconductor Devices and Power Electronics

3.1 INTRODUCTION

Power electronics is a branch of electrical engineering concerned with the study of combining power semiconductor devices in various configurations with passive elements such as resistances, inductors and capacitors to provide voltage/current and/or frequency conversion and their control in many industrial, home and other applications. This chapter is a recap and an introduction to widely used semiconductor power devices, power electronic converters, their operation and some applications. The primary focus of this chapter is to assist readers who have not taken a course on this in undergraduate or graduate studies or come from a different area of electrical engineering, and for those who may have studied it a while back. The power devices are discussed with reference to the cost and time in bringing out new devices and how that impacts applications possibly opening up new entrepreneurship opportunities. Rigorous mathematical analysis of the power converters, their modeling, simulation, analysis and design are contained in many textbooks and therefore will not be attempted here. The power converters are classified into categories of ac to dc, dc to dc, dc to ac, ac to ac, the number of phases and switches, and including resonant control. Applications of power converters in select high-volume applications that are in dc power supplies, electric motor drives, uninterruptible power supplies, solar and grid applications are introduced. Current interest in grid controlling power electronics applications and this technology implemented earlier not at grid edges but at central power systems in the 1960s as part of utilities with the current research shift toward the grid edge control is traced and discussed. Interested readers may access all other applications through books, internet websites and technical journals and conference proceedings. Recent developments are also discussed so that an appreciation of evolving opportunities with possible applications is gained and that may pave the path to future entrepreneurship. Almost all of the high-volume applications of power converters can be found in undergraduate textbooks and even the very few that are emerging from the research domain are quite the derivatives of the early practice in this field can be traced.

3.2 POWER SEMICONDUCTOR DEVICES

Power semiconductor devices provide the controlling elements for power conversion circuits. The physics of the devices is familiar to all electrical engineers and students and, therefore, it is not considered in detail in this book. Only the functionalities of these devices are recalled here to get to understand the power converters that can be designed and built with some of these devices as applications call for. For the practice of power electronics in the field, more in-depth knowledge and exposure to practical experience are required and the references will help toward that goal. This section deals mostly with the power electronic devices that one may come across in day to day practice of power electronics such as the following:

i. Diode
ii. Bipolar Junction Transistor (BJT)
iii. Thyristor (SCR, GTO, GCT) Family
iv. MOSFET
v. Insulated Gate Bipolar Transistor (IGBT)
vi. Silicon Carbide Power Devices
vii. Gallium Nitride Power Devices

3.2.1 Diodes

It is one of the essential components of almost all power electronic circuits. The semiconductor materials such as silicon and gallium arsenide have free electron density that falls in between metal conductor and an insulator. When the silicon is doped with an impurity such that it frees a hole in it, it is p doped, and likewise, when the impurity frees an electron, it is n doped. When such p n elements are joined, and a voltage is applied such that p (called anode of the diode) is positively and n (called cathode of the diode) is negatively connected, the device is forward biased and at that time current flows from anode to cathode part of the diode and the voltage drop across the diode is almost constant around 1 V or so, depending on the device. The diode, whose model is shown in Figure 3.1a, may be modeled as a constant voltage drop device with a small resistance which influences the voltage drop to a smaller measure and is shown in Figure 3.1b along with its symbol. In order to stop the current flow, a reverse voltage is applied (i.e., with the p connected to negative and n connected to positive of the voltage source) the

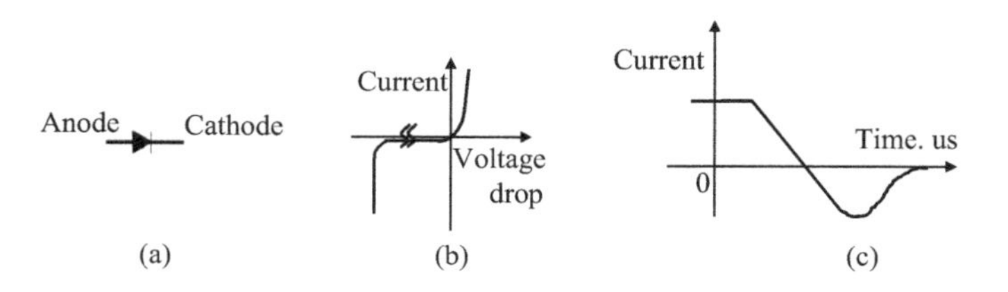

FIGURE 3.1 Diode: (a) symbol, (b) current vs voltage drop, and (c) turn-off.

current then goes from positive to zero value and then goes negative as well for a short time due to holes going toward the negative side and electrons going toward the positive side of the supply and this reversal of current lasts for a small time, known as reverse recovery time and the energy during the negative current duration is known as reverse recovery energy of the diode, shown in Figure 3.1c. The reversal of current in the device is undesirable as it limits its high-frequency applications, and it also creates higher energy loss during the reversal period resulting in lowering the thermal rating. The smaller the reverse recovery charge, the better is the device and ideal for all applications and that is always aimed in devices. If an increasing negative voltage is applied to the diode, then a small reverse current flows into it and at some point of increasing voltage, the current increases due to a phenomenon known as avalanche breakdown as shown in Figure 3.1b, and at that point the diode breaks down.

Many kinds of diodes are available intended for different applications. Regular diodes used in rectifier applications have ratings of 15 kA and few kV and their conduction drop is around 0.7 V or so. They are intended for use with standard utility ac voltages at 60 or 50 Hz. For fast switching applications intended for high-frequency converters, fast-acting diodes are used having the lowest reverse recovery time and energy and they come with higher voltage drops (>1.5 V) across them during conduction. Schottky diodes are used for low-voltage operation and for a low conduction voltage drop of about 0.3 V providing higher efficiencies compared to regular and fast-acting diodes in rectifier applications.

3.2.2 Bipolar Junction Transistor (BJT)

Power BJT is an outgrowth of the early transistor for power switching applications and it is an NPN device and its complement is the PNP device. The NPN device is much more in use than the PNP device. The symbol for the NPN BJT is shown in Figure 3.2a. The top N end is the collector, P end is the base and the bottom N end is the emitter. For operation, the collector is connected to the positive of the dc voltage and the emitter is at negative compared to the base and the collector. The base is applied a voltage about 10–12 V positive compared to the emitter for power devices which is much smaller voltage compared to the collector voltage. With such a voltage application, the transistor conducts current from collector to emitter and the physics of the operation is not discussed here for

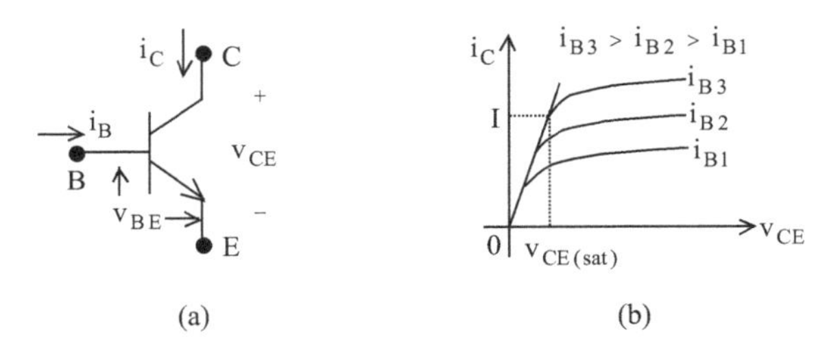

FIGURE 3.2 BJT: (a) symbol and (b) characteristics.

brevity. These devices cannot support a reverse voltage, for NPN device cannot sustain a negative voltage between collector and emitter, and note that voltage is with reference to the emitter. The characteristics of the NPN BJT is shown in Figure 3.2b in the form of collector current I_C vs. collector-emitter voltage, V_{CE} for various base currents, i_B. The device is operated in the quasi-saturation region that is around the knee point of the characteristics, as it gives the shortest turn-off time for the collector current thus enhancing the efficiency of the device and increasing the thermal capability. During turn-on of the device, the current increases from zero to steady value while the voltage across the device remains at full supply voltage and then after reaching its steady value, the voltage across the collector-emitter drops from supply voltage to conduction drop of 2 V or so. Likewise, during the turn-off time, V_{CE} rises from its conduction drop of 2 V to collector voltage while the current is at its steady-state value and then the current drops while the collector-emitter voltage remains at the collector voltage. During the turn-on and turn-off times, the power loss is very high even though it is for a very short time of a few microseconds in each case. The power BJT has a gain between the collector and base current in the order of 10-20. The transistor output is controlled by the base voltage and its frequency. In an ideal device ignoring the turn-on and turn-off times, the output is duplicated by the base control signal. The device is not in high use now because of its limitations in terms of the base signal current requirement and lower current gain and availability of better devices.

3.2.3 Thyristor Family

i. Thyristor (Silicon Controlled Rectifier): Known as SCR or thyristor in the literature is a device from 1957 (but General Electric announced it in 1956 in a conference in Bangalore, India). It is a PNPN device with the top P known as anode and the bottom P is the gate and the bottom N is the cathode. It can be considered as two PNP transistors in tandem with the emitter of the top transistor connected to the gate of the bottom transistor and the gate of the top transistor connected to the collector of the bottom transistor. Its symbol is shown in Figure 3.3a. It functions as a rectifier, but the onset of rectification can be controlled through its gate with a signal that is positive with respect to the cathode. It cannot be turned off like a BJT with the gate signal but only by reducing the current below that of a level so that it goes into blocking mode and applying a reverse voltage across the device and it is capable of sustaining that. During its conduction state, the voltage drop is around 2-3 V or so. The device will fail if a large positive or negative voltage is applied across it, given limits of V_{fbd} for positive voltage and likewise reverse breakdown voltage as the limit for negative voltage application, shown in Figure 3.3b. During conduction, it has a characteristic very similar to diodes in terms of its device voltage drop. SCR is available at a rating of 1.5 kA with a voltage rating of 12 kV [Mitsubishi] and in lower voltages at higher current ratings.

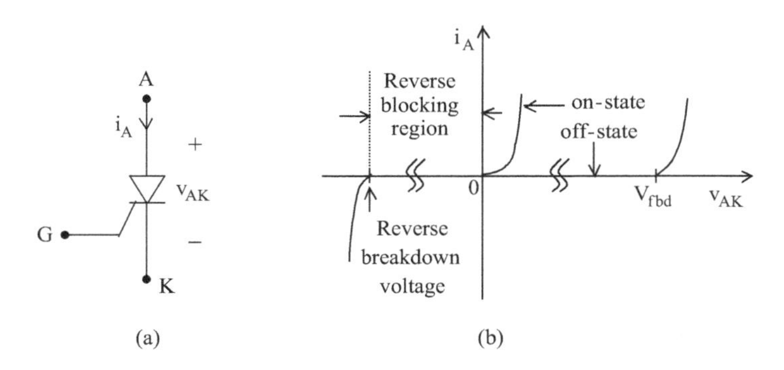

FIGURE 3.3 SCR: (a) symbol and (b) anode current vs voltage across anode and cathode.

ii. Gate Turn-Off (GTO) Thyristor: GTO thyristors are very much similar to SCRs except that these devices can be turned off by a negative gate signal. Therein lies its major advantage. Currently available ratings of these devices are at voltages of 2,500–4,500V with currents in the range of 1,000–4,000 A [Mitsubishi]. Its symbol is very similar to that of the SCR but with a vertical hatch line on the gate.
iii. Gate Commutated Thyristor (GCT): GCT is very similar to GTO except that it is capable of operating at a higher switching speed. It is available in the ratings of 6,000–6,500 V/400–6,000 A [Mitsubishi]. Later development of this device to reduce the on-state voltage and for higher switching frequency operation, lower gate requirements, optional turn-off protection of the device in the form of snubbers, but requiring protection at the turn-on to limit the current rise is known as insulated gate commutated thyristor (IGCT).

The family of thyristor devices is mostly used in applications with higher voltages such as in traction, static volt-ampere reactive system, inverters, etc., in the fractional to integral MW power range.

3.2.4 MOSFET

This is a member of the field-effect transistors (FETs), and it is known as metal oxide semiconductor FET with its acronym MOSFET. The symbol for a N-channel MOSFET and its characteristics, i.e., the drain current vs. drain-source voltage for various values of the gate-source voltage are shown in Figure 3.4a and b, respectively. The drain, gate and source of the device are denoted by D, G and S, respectively. Its unique features are:

i. Easy turn-on and turn-off of the device through its gate at the logic level and with hardly a noticeable power requirement.
ii. High switching frequencies in the range of 30–5 MHz.

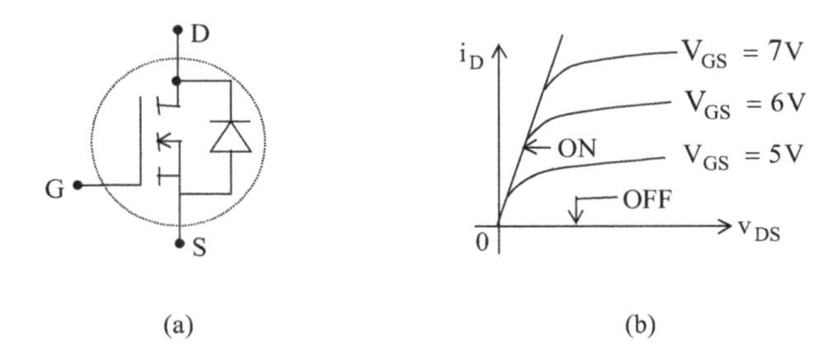

FIGURE 3.4 MOSFET: (a) symbol and (b) characteristics.

iii. When in conduction the voltage drop across its drain and source is equivalent to a resistive voltage drop and hence drain-source path can be modeled by a simple resistance.
iv. Having an antiparallel diode as an inherent part of the device structure and that helps very much with reverse current flow in a converter without needing an external diode as in the case of other switching devices.
v. It has a low drain to source resistance during conduction with the effect that the voltage drop is small and hence most of the input voltage is applied to the load such as in automotive accessory applications with 12–14 V batteries. Also, only a small fraction of the power is lost in this device thus facilitating a high current operation at these input voltages.

It is available with ratings of 100 A up to 200 V or so and 10 A at 1,000 V. Major applications are in appliances, automotive, uninterruptible power supplies of rating of <1 kVA, dc to dc power supplies, electric hand tools, small dc and ac motor drives, etc., to name a few.

3.2.5 Insulated Gate Bipolar Transistor (IGBT)

IGBT is a transistor with three elements but has a desirable mix of characteristics of a MOSFET as to its gating operation and of a BJT in terms of its current and voltage handling capacities. It has a high switching frequency compared to a BJT but not to the level of a MOSFET. IGBT has reverse voltage blocking capability like most of the devices but unlike MOSFET. Its symbol is shown in Figure 3.5. Its principal features are simpler gate drive requirements similar to MOSFET,

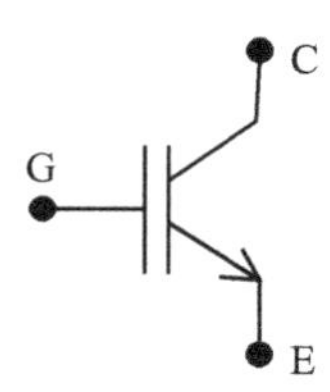

FIGURE 3.5 IGBT.

low on-state voltage drop, high efficiency of operation and reasonably high frequency as much as 30 kHz. It is available in ratings of 3,600 A, 1.7 kV range and 1,000 A, 6.5 kV and in the intermediate current range at 3.3 and 4.5 kV. Lower current ratings than 1 kA are common for less than the 1.7 kV range. IGBT is the most popular device for application in all the ranges except at low voltage ratings where MOSFET is predominant.

3.2.6 Silicon Carbide Power Devices

Silicon carbide (SiC) semiconductor-based power devices have been in research since the middle of the 1980s with a significant breakthrough in the commercial introduction in the late 1990s, though small compared to the silicon devices. The SiC semiconductor is a combination of silicon and carbon with the following desirable characteristics compared to silicon semiconductor:

i. Thrice the bandgap energy allowing higher temperature operation,
ii. Ten times the dielectric field strength, a higher breakdown voltage lending itself to high-voltage operation, and
iii. Three times the thermal conductivity making heat removal easier from the device.

They have, compared to silicon devices, the following performance advantages: lower on-state voltage drop contributing to high efficiency in circuit operation, faster-switching speeds resulting in compact packaging of the circuits, higher temperature operability at a junction temperature of 150°C or even more thus enabling higher pulsed currents, and lower leakage current.

SiC semiconductor is used in power electronic devices such as Schottky diodes, MOSFETs and junction transistors (BJT version) and thyristors. SiC Schottky diodes are available with a max current of 134 A at 1,700 V rating. SiC MOSFETs are in the market for some time at lower ratings compared to diodes, 100 A at 1,200 V for example but at higher voltages of 10 and 15 KV with a current rating of 10–15 A. SiC junction transistors are currently available in limited ratings from 160 A at 1,700 V. SiC thyristor with the rating of 85 A and 6,500 V is available. All these data are from GeneSiC product lines excepting the ones at 10/15 KV MOSFETS. Hybrid inverter power modules with SiC diodes and IGBT are available at 1,200 A and 1,200 V from Mitsubishi.

A number of applications for these devices in converters arise where smaller packaging sizes, smaller cooling requirements, enhanced efficiency, and high-temperature and high-frequency operability are predominant requirements at a comparable cost to silicon device based counterparts. Among other applications, these devices are touted for high-speed motor drives requiring a high fundamental frequency and also high switching frequency, electric traction, large uninterruptible power systems and electric vehicles. Because of the price differences between the silicon-based devices and SiC devices and due to limited availability, their volume of sale is yet to get to $1B level and it is being forecast that it will be a big player in a few years.

3.2.7 Gallium Nitride (GaN) Power Devices

GaN semiconductor has similar characteristics of SiC in the bandgap and breakdown field except it has lower thermal conductivity and higher electron mobility, and it is a wide bandgap material like SiC. It's on-state resistance is much lower and its frequency of operation is higher than SiC devices. GaN transistor product offerings are in low voltages of 30, 60, 100 and 200 V positioned mostly for high-volume end-user markets in computer power supplies and small automotive power supplies and accessory motor drives. One of the makers of GaN transistor (Efficient Power Conversion Corp.) is claiming that in this range of products, the prices are equal to silicon devices. If that be the case, a large number of applications are within reach in the coming few years. Devices rated for 120 A, 650 V and below that range in currents and voltages are available, say from the company, GaN Systems, among some others in the market currently. The market is defense-related presently apart from the power supplies and other markets and the market is projected to grow by $4.5B by 2025.

3.2.8 Future

Newer devices, both SiC- and GaN-based, encounter now in the market place the same difficulties that previously introduced but currently used popular devices did. It may be due to a number of factors and some of which are familiarity with the popular silicon devices, higher prices compared to them, not enough information as to the reliability and failure rates in applications specific to products, not many companies have introduced products made out of them in power electronics community and lack of competition in the market space to go for GaN-based systems, and still being in the research and/or development stage for use at a higher voltage and a higher power. Time and efforts from interested device manufacturers and power electronic and motor drives companies will turn the tide.

It is important to realize the time gap between research and device introduction is about 15 years and then product made out of these devices to enter a market place is around another 5 to 10 years. This is the case in previous devices as well as with the newer devices. The SiC and GaN devices research is from around the late 1980s and their product introduction is happening in and around the late 1990s at a smaller level but mostly in the power electronics industries. The interest in these devices is growing high since 2010. Yet they have a market share of $100M or so that is comparably <1% of the semiconductor devices market. It is important to appreciate the time lag between research to the production of devices and then to the introduction in products and realize the enormous amount of investment required to make things happen in this domain which may run to a few billion dollars. That is one of the principal reasons why newer devices do not crop up like circuit innovations in power electronics.

3.3 MAJOR POWER CONVERTERS

The devices that have been discussed are used in power converters to convert ac source to a dc source, dc from one level to dc another level of voltage, dc to ac as well as ac to ac for variable voltage, current and frequency output. The power converters and their many configurations are in prevalent use and they are briefly described in this section.

3.3.1 Dc to dc Power Conversion

Applications require many times to change a given fixed input dc voltage to another dc voltage of different magnitude. Let V_d and V_0 represent the input and output voltages, and the voltage conversion is made possible by a converter and let the ratio between the output and input voltages be

$$G = \frac{V_o}{V_d} \tag{3.1}$$

where G is the ratio or gain of the circuit. Different power converter circuits deliver for various values of G as follows:

i. For $G < 1$, the power converter circuit is known as buck or step-down converter.
ii. For $G > 1$, the power converter circuit is termed as a boost or step-up converter.
iii. For any value of G, i.e., covering <1 and >1, the power converter is called a buck-boost converter.

These three converters are described in the following. The switching devices in the converter including diodes are assumed ideal by neglecting the effects of voltage drops and switching effects of the devices so as to get to the fundamental concept of the converters without being clouded by them. Note for actual design and development of the converter circuits, these secondary effects should not be neglected. At that stage, such information may be accessed easily from any data sheets and application notes from device manufacturers.

Only basic types of dc to dc switchmode converters are discussed. A dc source is assumed for these circuits and they can be obtained, say, from battery packs, a rectified ac source, PV solar panels. A transistor is shown for the switching device in these circuits for illustration of the basic operational function of the circuits, but any other similar switching device can be substituted in the place of the transistor as in the practice. The outputs of the converter circuits are connected to a load more of a resistive nature, and there are inductors and capacitors in the circuits whose functions are to transfer energy from the dc source to the load and act as a filter to the current and voltage ripples. To derive the key relationship between the output and input voltages in these converter circuits, use is made of the fact that the integral of the inductor voltage over a full cycle of switching operation is zero. Only continuous current conduction in the inductor is covered in the following and for discontinuous current conduction in the inductor may be referred to in power electronic texts.

Within the three classes of dc to dc converters, i.e., step-down, step-up and step- down/up, various versions based on user requirements such as transformer isolation, number of switching devices and high efficiency have come into practice. They usually go by the name of switch-mode power converters, and a few of them are treated within this section of dc to dc converters as they are the natural evolutions of the basic types.

3.3.1.1 Step-Down (Buck) Converter

The converter circuit is shown in Figure 3.6a. When transistor Tr is on, the input voltage to the network is V_i which is equal to the dc source voltage V_d. There is a current flow through the inductor, L and that splits into the capacitor, C and load that usually is predominantly resistive in nature. The current in the inductor is considered to be continuous for the derivation between the output and input voltages. The inductor current rises during the on-time and falls during the off-time of the transistor. The voltage across the inductor is $(V_i - V_o)$ which is equal to $(V_d - V_o)$ as the switching device has been assumed to be ideal. When the transistor is turned off at time dT, the diode D takes over the current from the transistor. The voltage across the inductor during the off-time is equal to $-V_o$ with the result that the current in the inductor decreases until the end of the switching cycle given in time by T and it is related to the switching frequency, f_s as,

$$T = \frac{1}{f_s}, s \tag{3.2}$$

The switching signal, the voltage to the network V_i and output voltage V_o, and the voltage across the inductor are shown in Figure 3.6b. The volt-sec, i.e., the

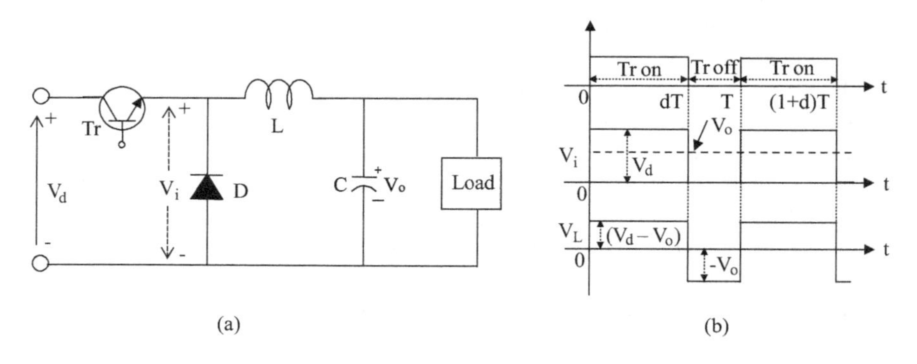

FIGURE 3.6 Buck converter circuit (a) and inductor voltage derivation (b).

integral of the inductor voltage is zero over a cycle and it is computed as from the figure as,

$$(V_d - V_o)dT + (-V_o)(T - dT) = 0 \quad (3.3)$$

From simplifying this, the ratio between the output to the input source voltage is obtained as,

$$\frac{V_o}{V_d} = d \quad (3.4)$$

where d is the duty cycle and its value will be less than or equal to one, and thus the circuit works to step down the input voltage for the output. Similarly, derivation for the case of discontinuous current in the inductor can be derived. Choices of L and C are dependent on many criteria such as current ripple, voltage ripple, continuous or discontinuous current, and load. All of the related information can be obtained from standard texts or derived by the readers.

Isolation between the input and output is required in many dc to dc converter applications and it is provided by a high-frequency transformer. Then, the circuit has to adapt to transformer characteristics such as to handle magnetizing current when the primary current is being turned off. High-frequency operation enables the smaller size of the transformer leading to compact packing and resulting high power density. An ideal transformer that ignores the leakage inductances of the primary and secondary windings is considered in the following discussion for conceptual understanding.

3.3.1.1.1 Step-Down Converters with Isolation

(i) Forward Converter: A buck converter configuration operating with a transformer is known as a forward converter. The forward converter and its operational waveforms are given in Figure 3.7a and b. The output side with the filter, L, C and freewheeling diode D_2, is the same as that from the conventional buck converter. The change is that the switching side is connected to the primaries of the transformer and the secondary is connected to the filter and freewheeling part with a diode D_1. The operation of the circuit is as follows: When T_r is switched on, the primary with N_1 turns is applied with the dc input voltage, V_d and the primary side voltage is V_1. V_1 is $<V_d$ due to switching device's conduction voltage drop. The turn-on of the transistor results in a secondary voltage of V_2 which is equal to $V_1*(N_2/N_1)$. When it is turned off, a path for the magnetizing current, i_m, has to be there and it is provided by D_3 and transformer primary winding with N_3 turns and note that the voltage applied is equal to the negative V_d. It also results in reversal of the voltage across the secondary winding and that turns off D_1 enabling D_2 to take over the current in the inductor, L. Note that the magnetizing current is very small compared to the rated current of the converter and accordingly shown in the Figure 3.7b. The output dc voltage is dV_2 for a duty cycle of d. As V_2 is equal to the turns ratio of the transformer times V_1 and approximating V_1 to V_d results

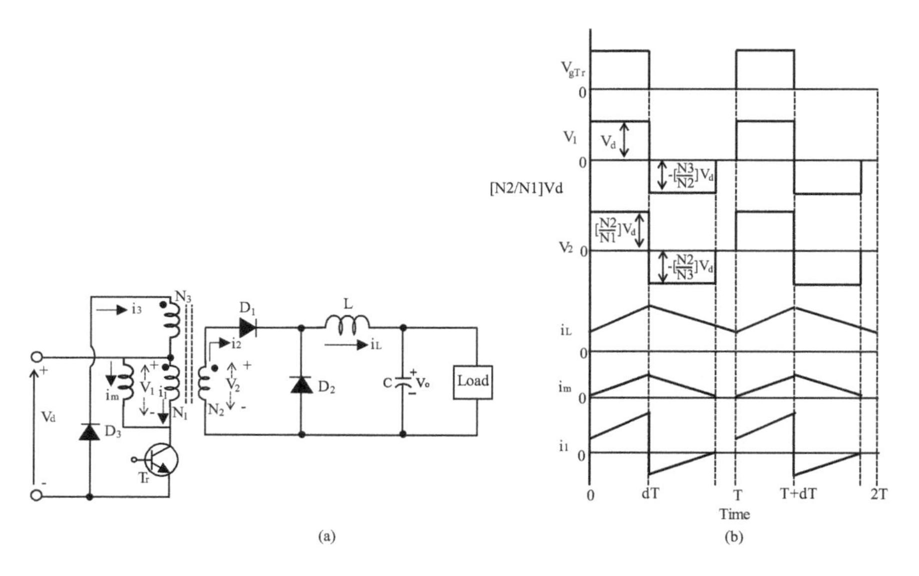

FIGURE 3.7 Forward converter (a) and its operational waveforms (b).

in the output voltage being equal to $d(N_2/N_1)V_d$. Note this is very similar to the relationship of the standard buck converter where the output voltage is given by the product of the duty cycle and input voltage and here it is further a function of the transformer turns ratio.

(ii) Push-Pull Converter: Various versions of step-down type converters with transformer isolations exist. They differ in the type of transformer used such as with four windings (two primary and two secondary windings), and the use of two or more than two switching devices (other than diodes). One such power converter is shown in Figure 3.8 with its functional waveforms assuming that the converter and transformer are ideal. It is known as a push-pull power converter. It has two switching devices with a transformer having two primary and two secondary

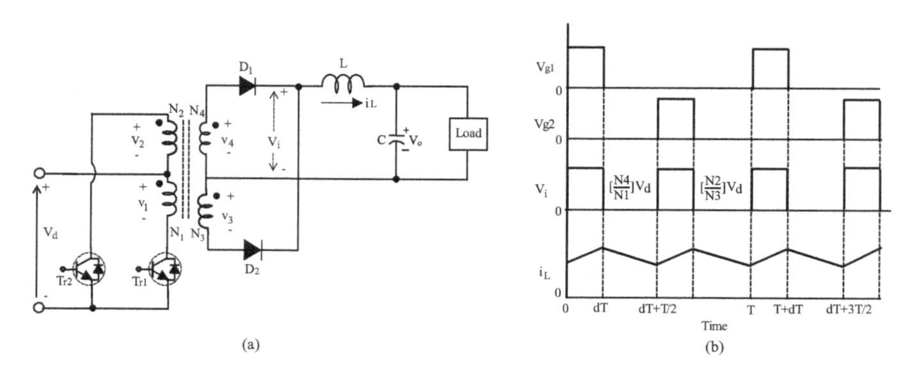

FIGURE 3.8 Push-pull converter (a) and its operational waveforms (b).

windings with different turns ratio (if desired) between each of the primary and its secondary. Each primary and its corresponding secondary with its switching device (T_{r1} or T_{r2}) works as a forward converter, but there is no separate demagnetizing winding. The antiparallel diodes in T_{r1} and T_{r2} are essential to carry the demagnetizing current. When T_{r1} is switched off, the antiparallel diode of T_{r2} conducts the current via the other primary with N_2 turns and likewise when T_{r2} is switched off, the antiparallel diode of T_{r1} carries the current with the other primary with N_1 turns. Consider T_{r1} is turned on by the gate V_{g1} leading to primary with N_1 turns energized in turn, giving secondary with N_4 turns with a positive voltage (note the winding direction given by the dot sign) V_4, thus enabling D_1 to conduct. At this time, the secondary with N_2 turns has a voltage V_2 that is positive and hence D_2 is reverse biased. For the other duty cycle, T_{r2} is energized by the gate signal V_{g2} that making V_2 negative and enabling D_2 to be forward-biased to conduct the current to the filter inductance, L. During off times of the devices T_{r1} and T_{r2} of the switching cycle, diodes D_1 and D_2 conduct and share equally the current in the inductance, L. Unbalance in the primary winding voltages during switching cause saturation in the transformer core and may lead to a converter failure. Measures have to be taken in the control system for both voltage and current controls.

Different versions of push-pull converter exist. They use a single primary winding. In addition, the primary side of the transformer has a half-bridge inverter with two switching devices and the resulting converter is known as half-bridge dc to dc converter or a full-bridge inverter with four switching devices known as full-bridge dc to dc converter. There is no change on the secondary side of the push-pull converter of the transformer and associated rectifier and dc filter with the load for these converters also. The half-bridge and full-bridge converters are introduced in the dc to ac converter section as these converters known as half-bridge inverter and full-bridge inverter are used on the primary side of the transformer in the step-down power supplies.

3.3.1.1.2 High-Efficiency Operational Basis

In the last two decades, many applications such as from phone chargers to data processing equipment requiring from a few hundreds of mA to few tens of A output at low dc voltages of 1.6–3.3 V for input dc voltages in the range of 150–400 V (obtained from rectification of utility ac supplies) have surfaced. These power supplies are required to have small footprints with a low profile for integration as well as high efficiency that are not obtainable with the standard dc to dc converter such as a standard buck converter. Obtaining higher efficiency requires lowering converter losses that occur in the transformer, switching devices, rectification part of the circuit with diodes, gating circuits of the devices and in the switching losses of the devices. Sizeable work resulted in tackling many of these challenges over the last 30 years via:

i. Schottky diodes in the rectifier part of the converter thus reducing the losses compared to the fast diodes.

ii. A better solution to (i) is possible by using a MOSFET that can conduct current in the reverse direction with proper switching, a property that had been rarely used in previous years. Then the MOSFET to be chosen has to have a very low drain to source resistance providing a voltage conduction drop for the load current considerably smaller than that of the Schottky diode. Such reverse conducting use of MOSFETs in the rectifier part resulted in a class of dc to dc converters known as synchronous rectifiers yielding high efficiency.
iii. Solution (ii) has to contend with gate drivers and control circuits for these synchronous rectifiers. They become more efficient when they are self-controlled and such self-control in synchronous rectifier-based converters is currently in vogue. This step also eliminates additional circuitry leading to compact packaging.
iv. In addition to steps (ii) and (iii), switching losses can be minimized by resorting to zero voltage switching (ZVS) and zero current switching (ZCS) using resonance in the circuit contributing to higher efficiency. Such techniques have been in vogue for a long time.
v. The transformer core losses are minimized by reducing the size of its core through high-frequency operation facilitated by MOSFETs and GaN devices to the range of a few to hundreds of MHz. Very high-frequency operation is made feasible by GaN devices.

A synchronous forward (buck) converter is presented in the following to appreciate some of the underlying steps outlined above.

3.3.1.1.3 Synchronous Buck Converter

Consider the buck converter in Figure 3.6a with its transistor and rectifier diode replaced with MOSFET switches, S_{w1} and S_{w2}, respectively and shown in Figure 3.9. The diode replacement switch S_{w2} is called a synchronous rectifier. When S_{w1} conducts for a time equal to dT, it charges the inductor as seen from Figure 3.6b. When S_{w1} is turned off, the diode across S_{w2} takes over and that could be bypassed when S_{w2} is gated on, it will be conducting the current from its source to drain, in the same direction as the diode used to conduct in the conventional buck converter. Bypassing the diode across the S_{w2} requires that the product of the current and the drain-source resistance is lower than the diode's conduction

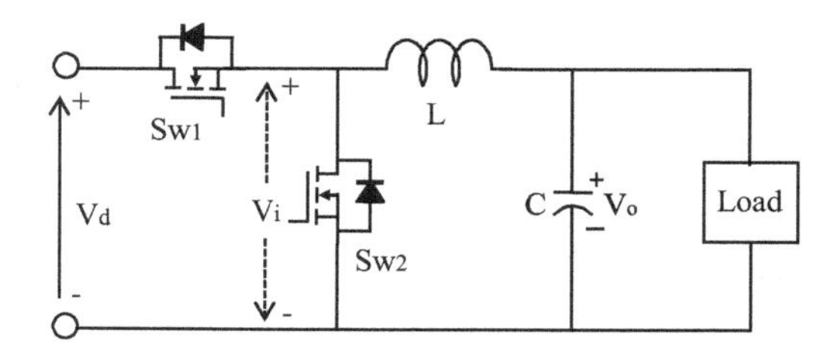

FIGURE 3.9 Synchronous buck converter.

voltage drop and therefore, the switch S_{w2} is accordingly chosen from a large pool of MOSFETs having a drain to source resistance of a few micro to milli Ohms with current ratings to the tune of a few hundreds of A. When switch S_{w1} is gated on for the next switching cycle, the synchronous rectifier S_{w2} is turned off. Lower power loss in the synchronous rectifier, compared to the losses in the diode, increases the efficiency. Figures have been quoted that such use of synchronous rectifiers leads to 2% efficiency improvement in low-voltage output applications.

3.3.1.1.4 Synchronous Forward Converter

Consider a buck circuit with transformer isolation, known as a forward converter, shown in Figure 3.7a, with its rectifier diodes on the secondary side replaced with synchronous rectifiers using MOSFET switches S_{w1} and S_{w2} and shown in Figure 3.10.

When the primary-side transistor is turned on, note that the secondary voltage becomes positive, and for this condition, normally diode of S_{w1} conducts. That is taken over by MOSFET of S_{w1} by proper gating. The gating signal itself is provided by the transformer terminal that will give a positive voltage to this switch for its conduction as shown by the dot or polarity symbol of the transformer. When the primary side of the switch is turned off, the magnetizing current is taken over by primary winding with N_3 turns, while the voltage across the secondary winding reverses. At this time S_{w1} turns off and switch S_{w2} is enabled as its gate is positive with respect to its source terminal. That enables the bypassing of the diode in that switch to carry the inductor current. It is seen that the synchronous rectifiers are self-controlled as their gating signals directly come from the secondary side of the transformer itself and do not require separate gating power supplies and circuits thus saving considerable parts, losses in them, and cost. For low-voltage outputs <12 V, this arrangement of self-control works and for higher output voltages, gating circuits have to be different. Note that there is a need for output voltage and/or current control and they determine the switching of T_r on the primary side.

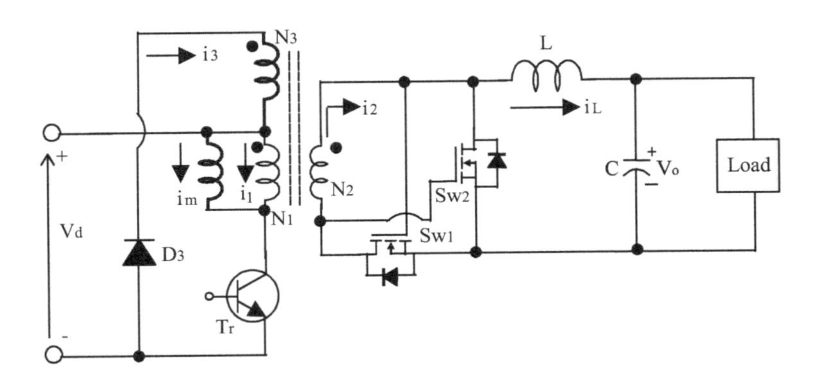

FIGURE 3.10 Synchronous forward converter.

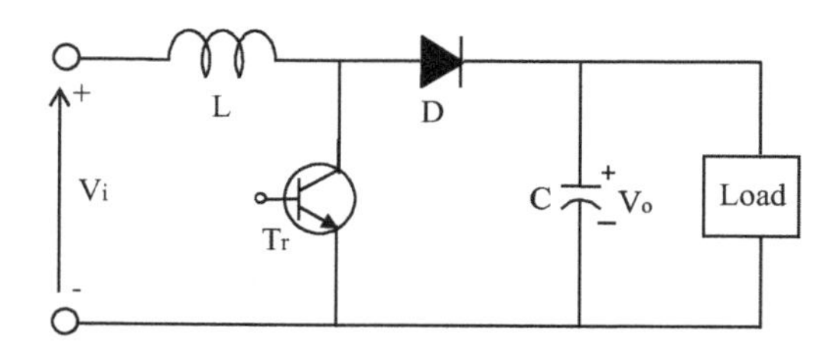

FIGURE 3.11 Boost converter.

3.3.1.2 Step-Up (Boost) Converter

It is shown in Figure 3.11. Its operation follows: Turn-on of transistor T_r applies the input dc source voltage of V_i on the inductor resulting in increasing current for a time, dT in a switching cycle of T. When the transistor is turned off, diode D takes over the current of the transistor and the current flows from input dc source, inductor, diode and capacitor as well as the load. A continuous current in the inductor is assumed for the entire switching period of T. The voltage applied across the inductor is equal to (V_i-V_o) for the off-time given by (T−dT). The volt-sec across the inductor for one cycle equals zero and can be written as,

$$V_i dT + (V_i - V_o)(T - dT) = 0 \tag{3.5}$$

On simplification, it leads to the ratio between the output and input dc source voltage as,

$$\frac{V_o}{V_i} = \frac{1}{1-d} \tag{3.6}$$

As the duty cycle is less than one, the range of output to input voltage ratio can only be greater than one leading to boost the operation of the converter. Practical values of boost required in industrial operations may be in the range of two to three, whereas in select nonindustrial applications they may exceed that number.

3.3.1.3 Step-Down/-UP (Buck Boost) Converter

A standard version called hereafter as buck-boost converter and one with transformer isolation called as flyback converter are presented in this section. Interested readers can research on synchronous rectifier based versions.

3.3.1.3.1 Buck Boost Converter

It is shown in Figure 3.12. Similar to other dc to dc converter treatment, it is assumed to have a continuous current in the inductor. The operation of this converter is as follows: When Tr is turned on for a duration of dT, a current is established in the inductor and it rises with time until the transistor is turned off and during this time the voltage across the inductor is equal to the dc source voltage, V_i. When the transistor is turned off, the current from the transistor is transferred

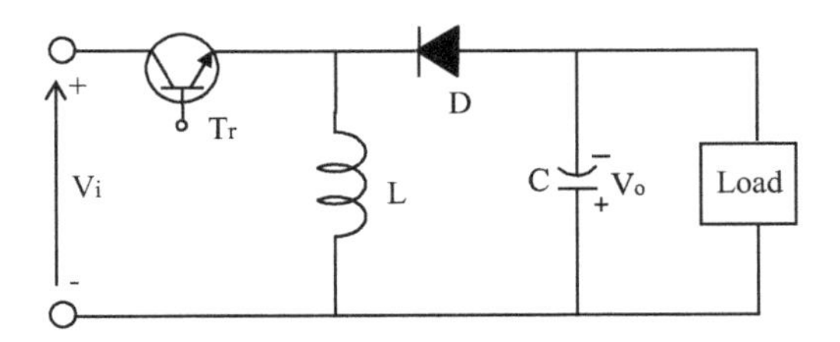

FIGURE 3.12 Buck-boost converter.

to the diode as the inductor current charges the capacitor in the polarity shown. The voltage across the capacitor then is $-V_o$ and this is for the remaining period of off time for the transistor given by (T–dT) where T is the switching period in a cycle. The volt-sec in the inductor is written for one cycle as,

$$V_i dT - V_o (T - dT) = 0 \tag{3.7}$$

This yields the ratio of the output-to-input dc source voltage as a function of duty cycle,

$$\frac{V_o}{V_i} = \frac{d}{1-d} \tag{3.8}$$

For duty cycles <0.5, the ratio of the output voltage to input voltage value becomes less than one giving rise to buck converter operation and for duty cycles >0.5, the ratio of the voltages becomes more than one indicating the circuit functions as a boost converter. For discontinuous inductor current operation, similar reasoning can be applied for analysis.

3.3.1.3.2 Flyback Converter

A transformer isolated step-down/-up converter is shown in Figure 3.13 which is very similar to the buck boost converter discussed and shown in Figure 3.12.

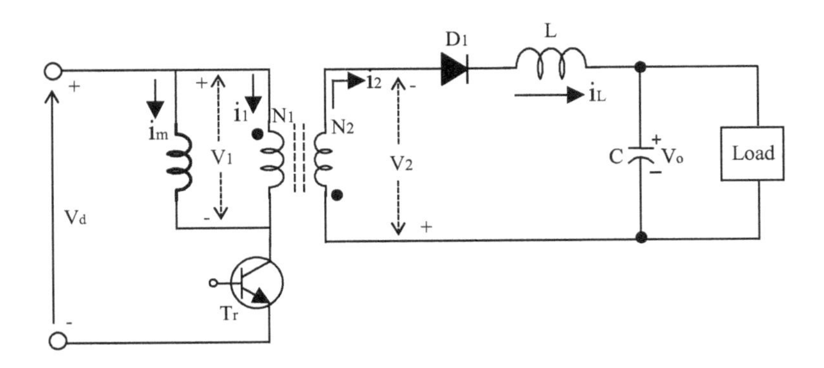

FIGURE 3.13 Flyback converter.

The difference is that the switching element Tr has been moved to the primary side of the transformer and everything else is left on the secondary side of the transformer. Because of the transformer, the ratio between the output and input voltage of the flyback converter is impacted by its turns ratio given by N_2/N_1 on that of the buck-boost converter and it is given by,

$$\frac{V_o}{V_d} = \frac{N_2}{N_1}\frac{d}{1-d} \tag{3.9}$$

where d is the duty cycle. Its operation is very similar to that of the standard buck-boost converter. The energy stored in the leakage inductance on the primary side has to be handled when Tr is turned off and that may be dissipated in a snubber, thus losing efficiency or modify the circuit which may require additional switching devices to handle this issue and to improve the overall performance as well.

3.3.2 Dc to ac Converters

The primary function of these converters is to generate an ac source from a dc source. Intended ac sources, for example, are to drive a dynamic load such as an electric motor where not only the magnitude of the voltages but also the frequency of the ac output is required to be varied over its operating region. Also, current control or current source ac is also required in some applications. Accordingly, there are two classes of dc to ac converters available and they are: (i) variable voltage variable frequency (VVVF) types and (ii) current source (or current controlled) variable frequency (CSVF) types. These converters are single phase, three-phase and multiphase (i.e., greater than three phases also for special applications). It is common to encounter the term inverters for these converters but many of them are also capable of rectification from ac to provide dc source. But this section addresses only the inversion functionality and the next section is reserved for the rectification aspects of these converters. Because of the maximum voltage limitation of the devices and high power requirement in applications, many devices are used in interesting arrangements to provide higher voltages in multiple step fashion and they are known as multilevel converters. Under the inverter section, single-phase and three-phase types with multilevel aspects are described.

3.3.2.1 Single-Phase Half-Wave Inverter

This is the building block of the inverters and is shown in Figure 3.14. It consists of two transistors T_1 and T_4 along with their antiparallel diodes D_1 and D_4. The emitter of T_1 and collector of T_4 are connected which is then connected to one end of a load with its other end connected to the midpoint of dc source. The collector of T_1 is connected to the positive of the dc rail and the emitter of T_4 is tied to the negative of the dc rail. Assume a load that is passive and having an inductor. The switching states are limited and they are: When T_1 is turned on, $V_d/2$ is applied to the load with positive as shown in the picture and the current, i_o enters from the

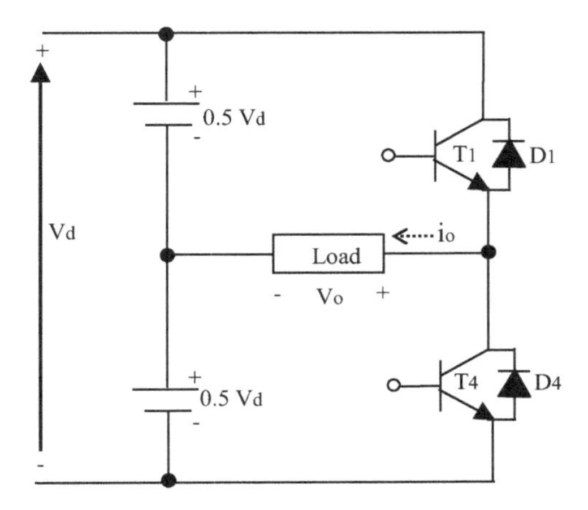

FIGURE 3.14 Half-wave inverter.

positive side of the load. In this mode, note that the load power is positive and that means the upper half of the dc source is supplying that energy to the load. When T_1 is turned off, if the current is continuous, then D_4 takes over the load current with the path for the current being, D_4, load, and bottom dc source $V_d/2$ with the result that the load voltage is $-V_d/2$ with respect to the midpoint of the dc source which is the assumed convention for load polarity. The load power is negative at this time and that amounts to energy being transferred from the load to the lower half of the dc source. If the current is not continuous, turning off T_1 will result in zero voltage across the load. Likewise, if T_4 is turned on, the current in the load is reversed and the load voltage is $-V_d/2$. When T_4 is turned off and if the current is continuous in the load, the upper diode D_1 will take over the current and the path is D_1, upper dc source $V_d/2$ and the load. Note that the load voltage becomes $+V_d/2$ and having a negative current results in the load transferring energy to the upper half of the dc source. If the current is not continuous or becomes discontinuous, then the load voltage goes to zero. The load experiences an alternating voltage, i.e., positive and negative voltage half-cycles, and hence an alternating current through the control of transistors T_1 and T_4. There is no way a zero load voltage can be achieved with loads having inductance with this inverter and hence limits additional freedom of control of this inverter. This is called half-wave inverter as only half the dc source voltage is applied to the load and not fully utilizing the full dc source voltage.

A simulation of the half-wave inverter with a resistive and inductive load is shown in Figure 3.15. This particular paragraph and the following three figures may preferably be visited after going through single-phase and three-phase inverter sections. The inverter output voltage and current are plotted in normalized units with the base of the voltage being $0.5V_d$ and the base current is arbitrarily chosen for this figure. The command voltage frequency is 200 Hz and the carrier frequency is 4600 Hz with the modulation frequency ratio of 23 which is

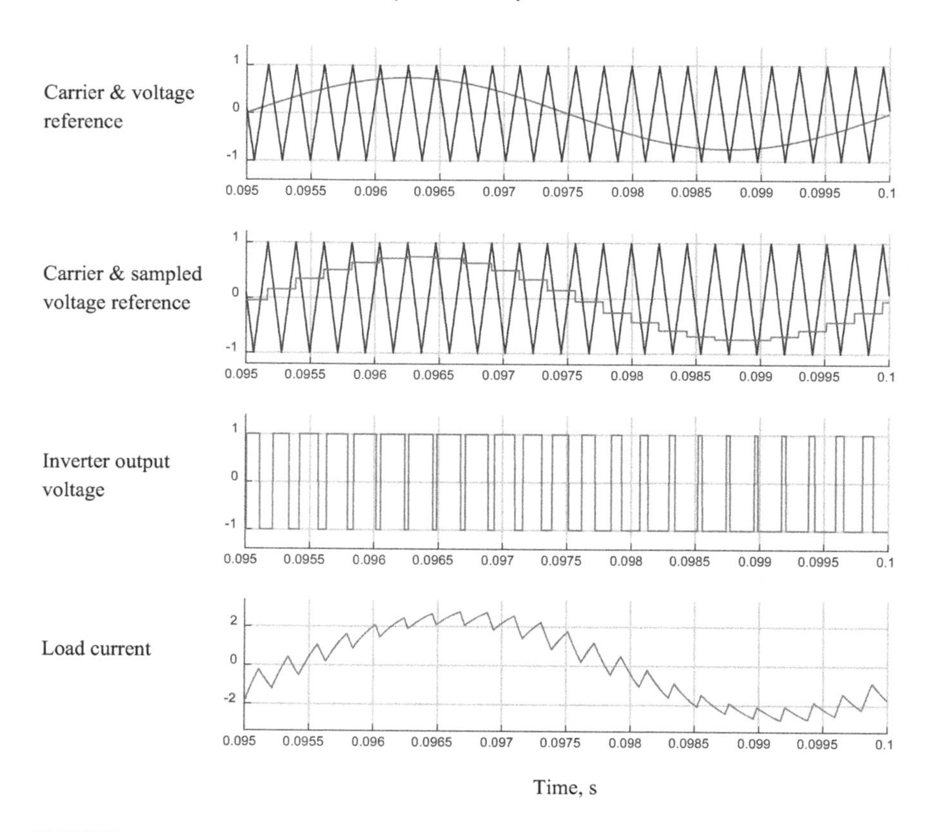

FIGURE 3.15 Simulation of the half-wave inverter operation with a passive load.

an odd number and preferred for pulse width modulation of which is explained in later sections. The load current is almost a sinusoid with current ripple at carrier frequency. The output voltage and current spectra are normalized to their respective fundamental spectra and given in Figures 3.16 and 3.17, respectively. The dominant harmonic is at the carrier frequency and the sidebands around it are very small.

3.3.2.1.1 Application in dc Power Supply

Consider the push-pull converter from Figure 3.8 discussed in the earlier section. That converter has two primary windings. For the present application, the primary winding is reduced to one and then that primary winding is energized and controlled by the half-bridge inverter with the secondary side of the push-pull remaining intact and such a converter is shown in Figure 3.18. It is known as half-bridge converter. One salient advantage of this circuit is that all the existing body of control methods including the basic method of control in the previous section are all applicable here at desired frequencies to obtain the required performance.

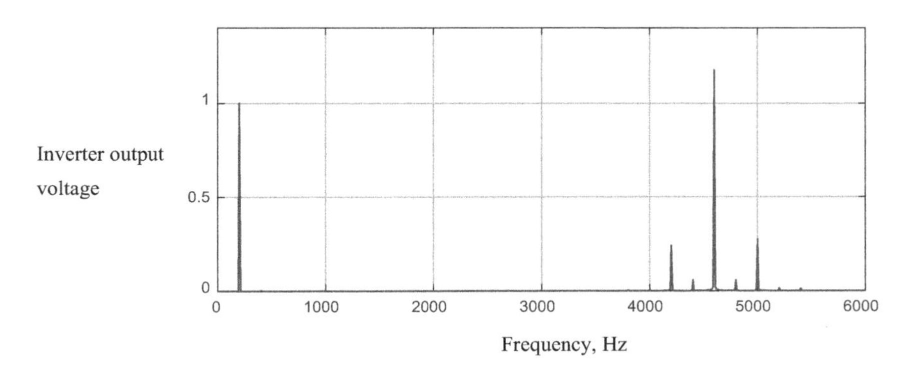

FIGURE 3.16 Spectra of the half-wave inverter output voltage.

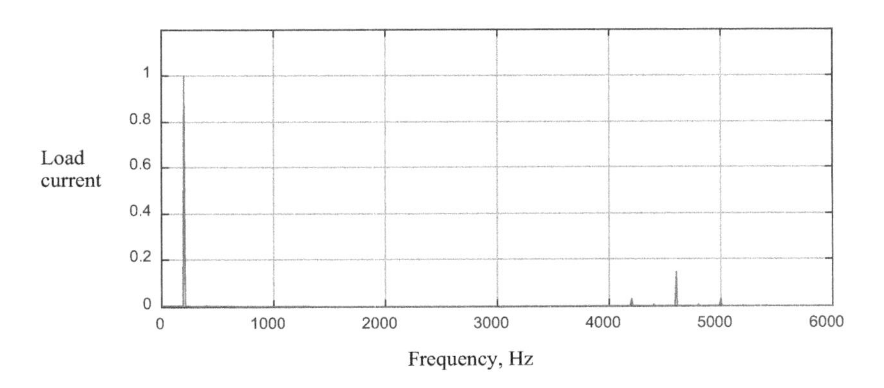

FIGURE 3.17 Spectra of load current in the half-wave inverter with a passive load.

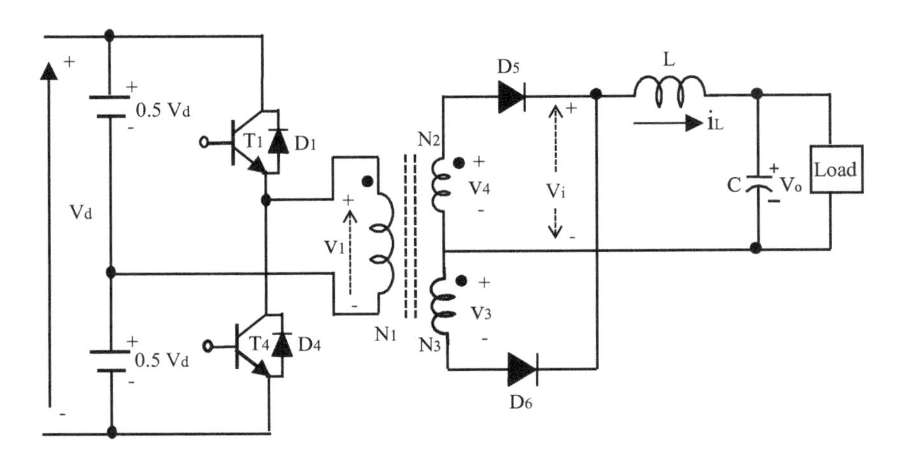

FIGURE 3.18 Half-bridge converter.

3.3.2.2 Single-Phase Full Wave Inverter

A single-phase full-wave inverter is shown in Figure 3.19. It consists of two-phase legs in parallel and each phase leg consists of two transistors in series with their antiparallel diodes and connected across the dc source. Phase leg one has transistors T_1 and T_4 in series with their respective antiparallel diodes of D_1 and D_4, respectively. Likewise phase leg 2 has identical arrangement. The centers of the phase legs provide the output terminals connected to a load. The load may consist of usually a resistance, an inductance and a voltage source and here only such a load is taken for consideration in the circuit operation. If the utility is the load, then the voltage source is nothing but the grid voltage source. This inverter is called a full-wave type because the magnitude of the maximum load voltage is as much as the dc source voltage. A circuit using a center-tapped dc voltage source of equal magnitude with only one phase leg with the load being between the center point of the phase leg and the dc source voltage is known as half-wave inverter as the maximum load voltage magnitude is equal to only equal to half of the total dc source voltage. The operation of the full-wave inverter is described below.

When an alternating voltage is applied to a load, an alternating current results with a phase shift determined by the load composition that may result in current having a phase lead, lag or zero with respect to the voltage. That means for any and all kinds of load, the inverter must be able to supply a voltage and a current of the same as well of opposite polarities to one another. Consider a lagging current and that means, from positive zero crossing of voltage to positive zero crossing of current, the output voltage has to be positive but the load current has to be negative. Then positive voltage with positive current, negative voltage with positive current and finally negative voltage with negative current for the remaining period of one ac cycle has to be supplied by the inverter. The operation can be visualized if the operational features are captured with voltage along the x-axis and current along the y-axis as shown in Figure 3.20. Using this, consider that

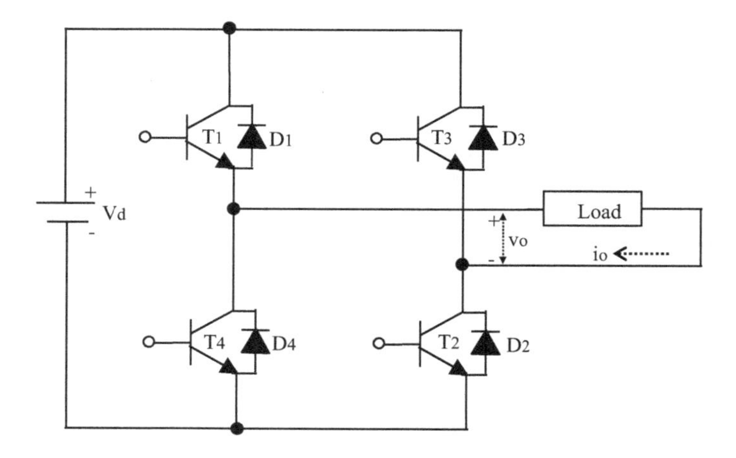

FIGURE 3.19 Single-phase full-wave inverter.

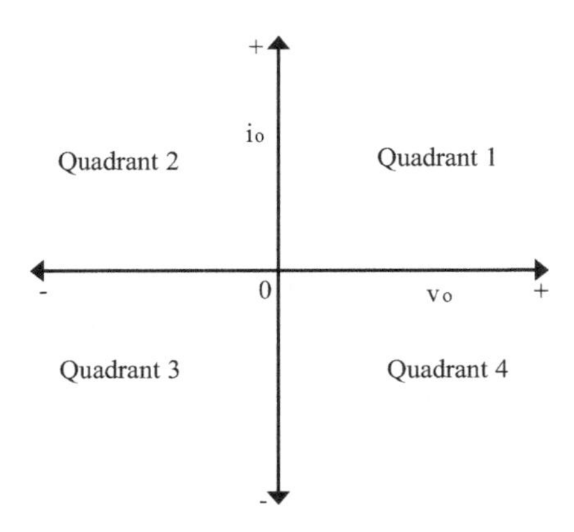

FIGURE 3.20 Operational four quadrants of the inverter.

the inverter has to operate in quadrant 1 with positive voltage and current which requires that transistors T_1 and T_2 are turned on that gives a load voltage equal to dc source voltage with the load current rising and positive (that is toward the load from T_1) (say, mode 3 shown in Figure 3.21). Now turning off both T_1 and T_2 means an alternate path for the load current has to be there and it is provided by D_4 and D_3 via the load and back to the dc source. Note then the voltage across the load is negative V_d, but the current is positive in the load (mode 5). It means the energy is being transferred from the load to the dc source in this mode and note that this corresponds to quadrant 2 operation. During this mode, the source is sinking the energy supplied by the load. Then when the load current becomes zero, turning on T_4 and T_3 applies the negative of the dc source voltage on the

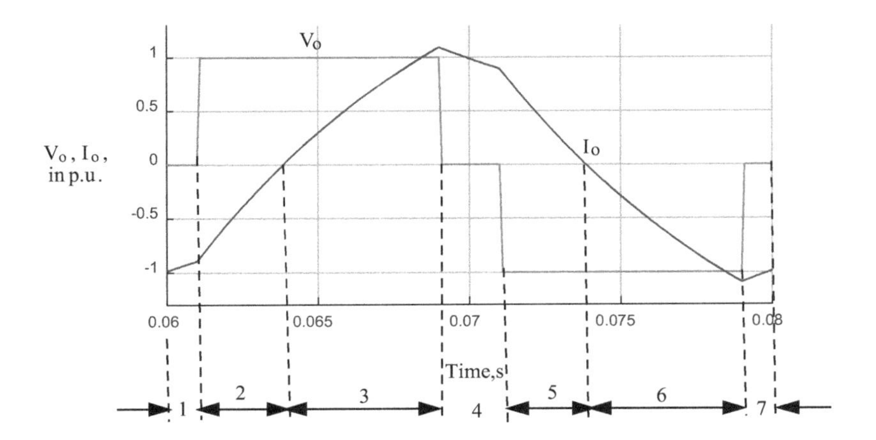

FIGURE 3.21 Programmed single pulse per half cycle inverter output voltage for passive load showing various modes of operation.

load and the load current is also negative with the mode in quadrant 3 giving a positive power to the load (mode 6). The current flows away from the dc source resulting in positive power transfer from the dc source to the load. If this current has to be extinguished in the load, then turning off both the conducting transistors enables the turn-on of diodes D_1 and D_2 resulting in a positive voltage across the load equal to V_d and that corresponds to quadrant 4. Now the load current is negative and load voltage is positive and therefore it supplies energy to the dc source (mode 2).

There is also an additional degree of freedom to operate by turning off one transistor only and keeping the other transistor off in each quadrant operation. Consider when in quadrant 1, the load current exceeds a set limit and it is required to control it within a limit. Then keep T_1 on but turn off T_2 which enables D_3 to conduction. Then the load current path is through load, D_3, and T_1 with the result that the load voltage is zero and the energy stored in the load inductor supports the current but that will keep declining (mode 4). If the current goes below a certain minimum set limit and the current demand is to get to the previous level, then turn on T_2 which will reapply the dc source voltage to the load and hence the current will increase. The principal advantage of this mode of operation is to limit the current within limits but without having to send the energy frequently from load to dc source to do so and thereby save switching losses in the transistor (as only one is turned on and turned off and not both at the same time), thus increasing its life, reducing the current ripple and leading to a slight reduction in its size and rating. Next time around, transistor T_2 can be kept on and T_1 is turned on and off to control the current in the load thereby increasing the cooling time for the devices and for having an equal thermal burden on each. A similar approach is applicable for other quadrants of operation where current can be controlled for allowable ripple magnitudes. Mode 1 shows the load current control when it is negative corresponding to its complement in mode 4 when the load current is positive. Various modes discussed are illustrated in Figure 3.21 for the case of a single-phase rectangular voltage output supplying a resistive and inductive load.

3.3.2.3 Inverter Control

Variable voltage control and variable frequency control of the inverter output are required for many dynamic loads such as ac motors for variable speed application. They are achieved in many ways and described in the following sections.

3.3.2.3.1 Programmed Pulses

Preprogramming the voltage vs. frequency and enforcing it on the inverter through pulse width modulation (PWM) of the gate signals to the switching devices is considered here. This is carried out in high-power inverters for motor drives where PWM switching frequency is required to be very small, say, in the order of three to ten times the fundamental frequency. Single-pulse per half cycle and multiple pulses per half cycle are possible options in this category of control. Both are dealt in the following:

i. Single-Pulse per Half Cycle: It has only one pulse which may be less than the half-cycle time and remains zero for the remainder of the half-cycle for both positive and negative half cycles, as shown in Figure 3.21. The load is passive and the dc link voltage (V_d) is considered as the base value and hence is equal to one per unit (p.u.) in normalized form. Likewise, various base values are assumed for other variables or related to some inherent relationships and can be normalized as given in many books and research publications. Neglecting voltage drops in the switching devices of the inverter and cable voltage drop, the plotted voltage is the inverter output voltage, V_o and that equals the programmed value or commanded value. The resulting load current, I_o is also plotted and the figure shows the steady-state operation as against the transient state. The steady-state is sufficient to look into the operation of the inverter circuit and based on the voltage and current at a given instant of time, it is shown to have seven modes indicated by numbers from 1 to 7 in the figure. Note that zero voltage output is not possible by simply not activating all the transistors when there is a load current and that is where the understanding of the modes of operation becomes essential. Each mode is discussed in the following:

Mode 1: The current is negative, i.e., current flow is from right to left in the figure, and voltage is required to be zero. For the negative current, note that transistors T_4 and T_3 must have been on to generate the current in the load. By turning off both these transistors while there is a current in the load will enable diodes D_1 and D_2 to take over the load current flow. But then the voltage will be equal to $+V_d$ (i.e., +1 p.u.) across the load which is not the commanded or preprogrammed value of the output voltage in this region of mode 1. Zero voltage is possible with shorting the load by keeping transistor T_4 on and turning off the transistor T_3 thus enabling diode D_2 to come on with the result that the load current has a closed path from D_2, load and T_4 which amounts to a shorting of the load for this region. An alternative also exists in the form of T_4 being turned off, but T_3 and diode D_1 being on to give the same result of zero voltage across the load. Note this corresponds to borderline between the quadrants 3 and 4 of Figure 3.20.

Mode 2: In this, the load current is negative, but the voltage has to be positive. Mode 1, say, has T_4 and D_2 conducting. To get positive voltage across the load, turn off T_4 with the result that now D_1 will come in and already D_2 is conducting with the result that the load voltage is $+V_d$ (1 p.u.). Likewise, the other option if T_3 and D_1 were conducting previously, turning off T_3 will enable D_2 to conduct with the result that load voltage is positive 1 p.u. If the current is discontinuous in this region, then to provide a positive voltage across the load, continuous gating of T_1 and T_2 will enable positive voltage across the load. As long as the current is negative, the devices in spite of being gated on will be reverse biased due to the conduction of their antiparallel diodes and no harm comes to the circuit by this action.

Mode 3: Both the load voltage and current are positive here, and they are obtained by keeping T_1 and T_2 conducting.

Mode 4: Current is positive and load voltage is zero which is very similar to mode 1 but with a negative current. Applying to similar logic of mode 1, the operation is obtained by keeping T_1 or T_2 on with the result that diode D_4 or D_3, respectively, coming in to support the load current and ensuring zero load voltage.

Mode 5: It is similar to mode 2 but with the polarities of the voltage and current being opposite. Similar reasoning can be applied to that of mode 2. As long as the current is continuous, turning off the conducting transistor which may be T_1 or T_2 will enable diodes D_3 or D_4, respectively, to come in with the result that only diodes D_3 and D_4 will be conducting thus giving a negative voltage across the load equal to –1 p.u. To guard against the discontinuous current in this mode's region, only transistors T_3 and T_4 may be turned on and they will not become active until their antiparallel diodes turn off, i.e., until the load current ceases to be positive.

Mode 6: Both the load current and voltage are negative which is exactly the opposite of mode 3. This mode is obtained by having transistors T_3 and T_4 in conduction (which is nothing but a continuation of mode 5 if the transistors had been on).

Mode 7: This is not an independent mode as this is exactly the same as mode 1 and hence the same switching logic applies here as well.

Introducing an alternating voltage source, say E, with the passive load elements of resistance and inductance results in an active load. The magnitude and frequency variations of E such as from generator or motor or utility grid provide the activeness to the load. With a single pulse per half cycle operation of the single-phase inverter catering to an active load in steady state is shown in Figure 3.22. Here the voltage source in the load, E is in phase with the inverter output voltage. The analysis with seven modes described earlier also applies here for this operation.

It can be seen that there are significant harmonics in the voltage and hence also in the current with a single pulse per half cycle voltage programming. The current has a better shape than the voltage due to inductance in the load but nevertheless has corresponding harmonics. It is desirable to reduce the lower order harmonics and their magnitudes so as to reduce the filter size in the system and hence lowering the overall cost of the power conversion system as well as the harmonic losses in the system resulting in better operational efficiency.

One way to improve the harmonic spectra is to increase the number of pulses in a half cycle as against one pulse per half-cycle as in the previous case. A three pulses per half cycle case in steady-state with passive load is shown in Figure 3.23a. The shape of the current has improved and has become more toward a sinusoid and it can be proven that the lower order harmonics have been reduced. When an

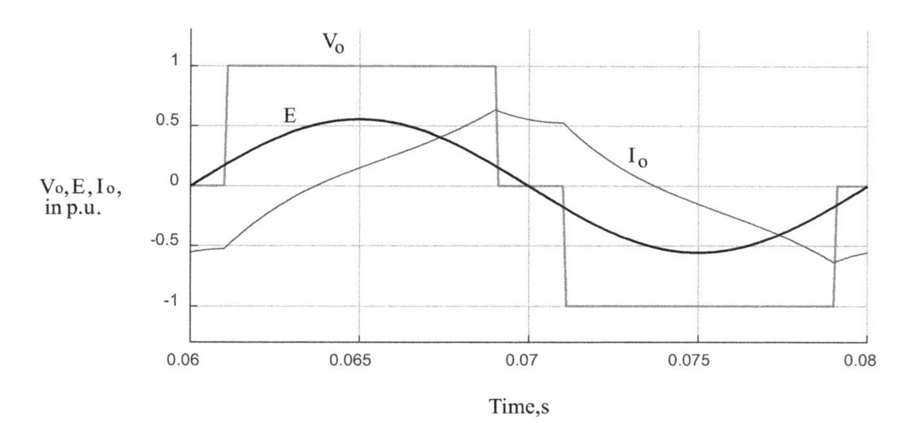

FIGURE 3.22 Programmed single pulse per half-cycle inverter output voltage for active load.

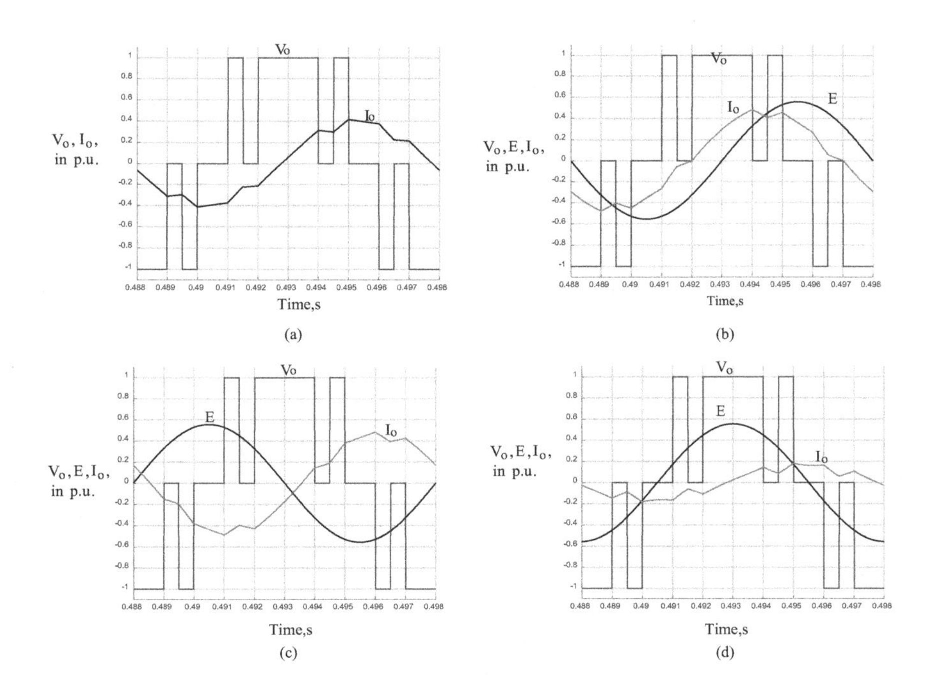

FIGURE 3.23 Three pulse per half cycle control of the inverter with passive load, and active loads with lagging, leading and in phase load emf source, E. (a) Passive load, (b) active load with emf, E lagging V_0 by 90°, (c) active load with emf, E leading V_0 by 90° and (d) active load with emf, E in phase with V_0.

active load is applied to such multiple voltage pulse-based inverter switching, the steady-state inverter output voltage, active emf source E and current waveforms are given in Figure 3.23b–d, for three cases of E lagging 90°, leading 90° and in phase with inverter output voltage, respectively. Note considerable improvement in current waveforms. In all cases, the load inductance is dominant. The phase of the load-side emf source E is not controllable as it is part of a load and then the only way to change the phase between E and inverter output voltage is to change the latter's phase as it can be shifted by the gating control and be achieved also instantaneously. Shifting the phase has a significant effect on the power factor and power flow of the system. The inverter voltage magnitude is controlled by changing the pulse widths of the multiple pulses, and its frequency is controlled by the varying the cycle time and the phase of the inverter output voltage with respect to the load's active emf source by shifting the start of the inverter output cycle at any commanded time. Therefore, the single-phase inverter is capable of voltage magnitude, frequency and phase controls. Hence, a full control of the load can be exercised, thus giving it a central place in power control applications now and in time to come.

3.3.2.3.2 Pulse Width Modulation (PWM) Control

A preprogrammed pulse width control for various frequencies is fixed and does not lend itself to changes in the system environment. Effecting that with considerable dexterity is made possible with PWM control by modulating a command voltage with a carrier signal such as a bidirectional and unipolar triangular signal. The magnitude, shape, frequency and phase of the inverter voltage together constitute the voltage command and these constituents can be varied at will from the command port of any system. Consider a control-level sinusoidal voltage is modulated with a high-frequency carrier signal of ramp or triangular shape and the result is a set of pulses that are fed to the gates of the switching devices in the inverter. These pulses of different widths are amplified by the inverter with a gain and the resulting output of the inverter is also a train of the pulses which contains a significant fundamental and a number of harmonics determined by the PWM switching frequency. When they are applied to a load with some inductance, the output current is quasi sinusoidal and is acceptable in practice. The control signal is a sinusoid whose magnitude and frequency determine the voltage magnitude and fundamental frequency of the inverter output. The advantage of the scheme is that the switching devices in the inverter are operated at a fixed carrier or PWM frequency for the most part and thus guaranteeing switching losses to be within the design limits.

Figure 3.24 illustrates the single-phase inverter operation with PWM control supplying a passive resistive and inductive load, and an active load emf source, E. The modulating signal is the command voltage, V_o^* for the inverter and in the present case is considered to be a sinusoid. The high-frequency carrier is compared with the command signal at all times. Whenever the command signal is greater than the carrier signal, the output is +1 and when it is lower than the carrier signal, the output is −1 thus producing a bidirectional signal. These signals

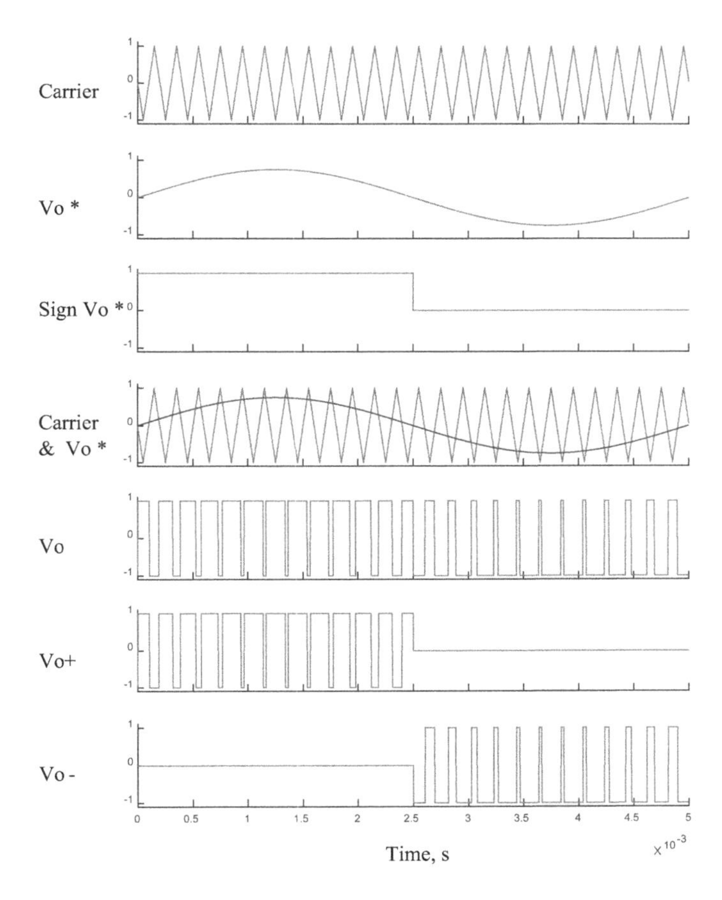

FIGURE 3.24 Sinusoidal modulation with a triangular carrier signal and resulting signals.

are amplified to provide the gate signals but with some modification to accommodate for switching delays of the power devices. The +1 signal when amplified is applied to the gates of the transistors T_1 and T_2 (refer to Figure 3.19) to produce a positive V_d across the load. The –1 signal is converted into a positive signal first and then amplified to go to the gates of T_3 and T_4 so as to provide a negative V_d across the load. The inverter amplifies these signals to track the command voltage signal, V_o^*. Note that the resulting signals from modulation produce inverter output voltage identical to them when the devices are considered ideal. The positive half cycle part of the resulting control signal, V_{o+} and negative half cycle part of the resulting control signal V_{o-} are obtained from V_o signals with the use of the positive and negative control signals derived from the command signal.

The Fast Fourier Transform (fft) of voltage command (modulation signal) at 200 Hz, PWM carrier signal at 5 KHz and resulting inverter output voltage are obtained and their power spectra are displayed in Figure 3.25. The sidebands in

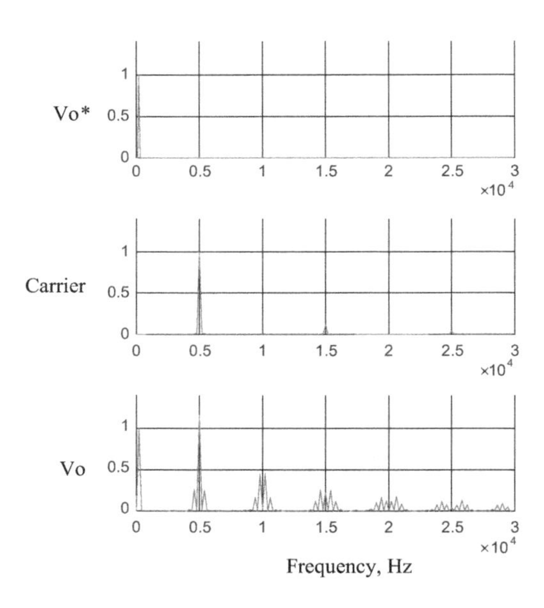

FIGURE 3.25 Spectra of the voltage command, carrier and resulting inverter output voltage.

the inverter output voltage are significant but nearer to the carrier frequency than to the fundamental. This shows that the higher the carrier frequency, the sidebands will be easier to filter with small filters. They are limited by the switching frequency of the inverter switching devices. This is one area where the newer devices such as SiC and GaN devices can make a difference in some power inverter applications.

3.3.2.3.3 Overmodulation

To maximize the fundamental of the output voltage, the modulation signal magnitude is increased to values greater than that of the carrier magnitude which results in longer duration pulses in the middle of the half cycles and that in the overall reduces the number of pulses in the inverter output voltage and in the switching frequency of the inverter even if the carrier frequency is maintained at a higher level. This is inferred from Figure 3.26 where the command or modulation signal and carrier are shown in the first part of the figure, and the middle part displays the control signal and hence the inverter output voltage also with the last part of the figure showing the fft with the power spectra of the inverter output voltage. The decrease in the inverter switching frequency (and note not the carrier frequency) with overmodulation is seen by comparing with Figure 3.24 where the switching frequency is the same as the carrier frequency. Likewise, the fundamental of the inverter output voltage has increased and can be seen by comparing the ffts in Figures 3.26 with 3.25. Overmodulation can also be seen as a tool to preprogram the inverter control signals at low switching frequency. But in the

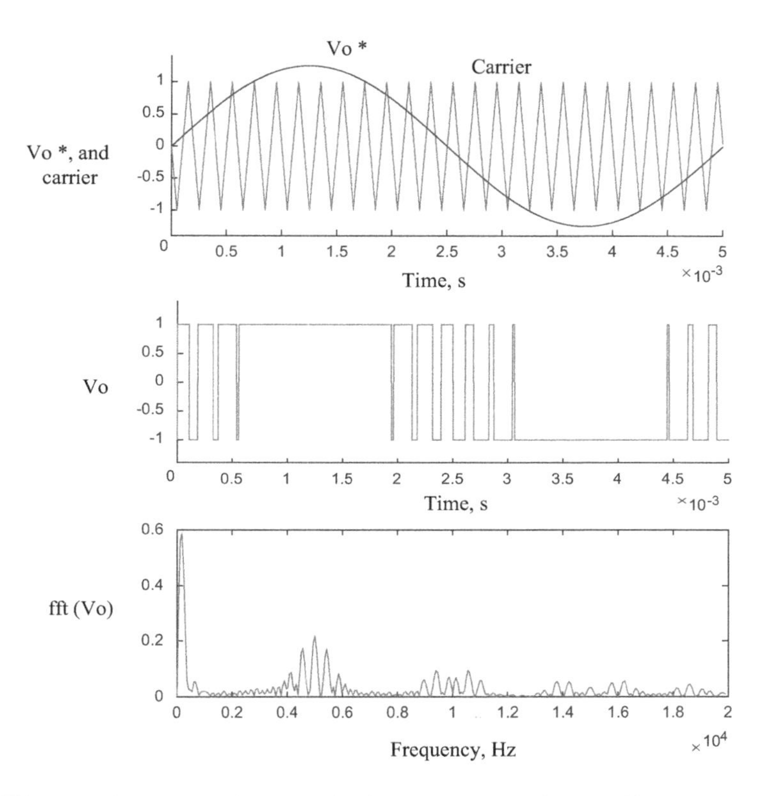

FIGURE 3.26 Overmodulation, resulting inverter output voltage and its power spectra.

case of high-power applications, various performance indices such as efficiency, harmonics and speed of operation in the case of a motor drive, etc., influence the derivation of gating signals at low switching frequency and may not be just derivable by using the overmodulation algorithm.

Switching Aspects: From Figures 3.24 and 3.26, it is seen that the inverter output voltage contains positive and negative voltage pulses during the positive and negative cycles. Consider the positive half cycle switching part. To impose, positive output, say, two transistors T_1 and T_2 (refer to Figure 3.19) are switched for part of the carrier cycle, and when the negative voltage has to be applied, both the transistors are turned off assuming continuous load current. Then the diodes D_3 and D_4 take over the load current and impose a negative voltage. That involves two transistors being switched on and off resulting in two device switching losses. That could be halved, if only one device is turned off, say, for instance, T_1. Then, the diode D_2 comes on and the load current goes through D_2, load and T_2. That results in the inverter output voltage being zero when ideal devices are considered. Therefore, with this kind of switching, transistor device switching losses are halved and also the inverter output voltage shape improves considerably having $+V_d$ and 0 only during its positive half cycles and $-V_d$ and 0 during its negative

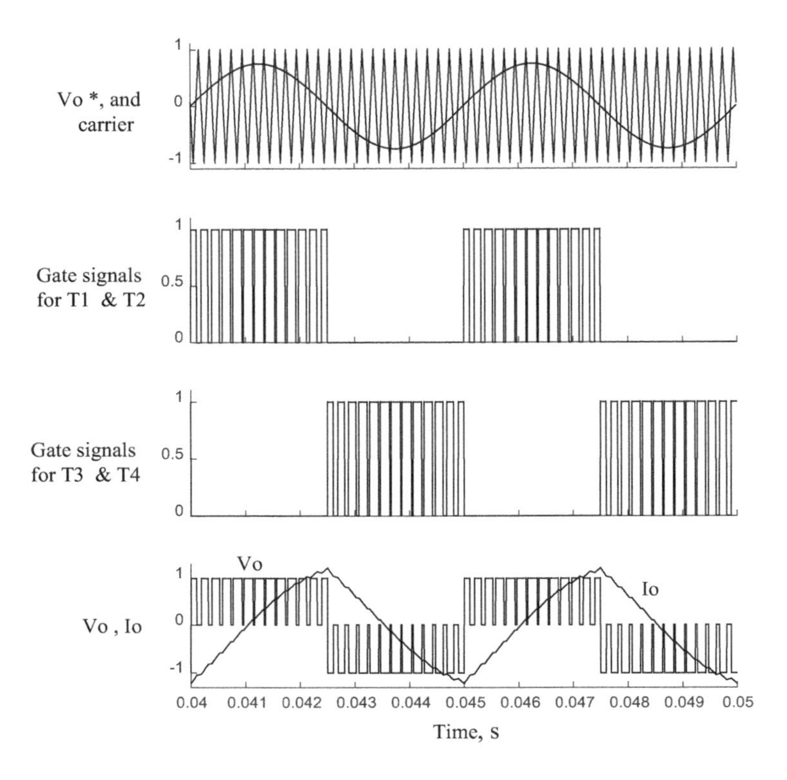

FIGURE 3.27 PWM with zero voltage enforced for a passive load.

half cycles, as shown in Figure 3.27 for a passive load. Note also, instead of T_1 being turned off, T_2 can be turned off while letting T_1 conducting producing an equivalent result of providing a zero voltage across the load. The control signals for this kind of switching may be produced by having a carrier signal that is unipolar, i.e., positive during positive and negative during negative half cycles of the modulation signal and limiting the output of the modulator from zero to a logic one. Also, notice the fact that the current has lower ripples during turn-off periods as the load experiences for positive half cycles only $+V_d$ to zero and not V_d to $-V_d$.

3.3.2.3.4 Modeling of the Inverter

The inverter has a gain, K_i that is given by the ratio between the output fundamental voltage and input command modulation signal. It has a time lag because the switching of the inverter phase leg from the top to a bottom transistor cannot be instantaneous because the devices have turn-on and turn-off times. If the turn-off time of the outgoing switching device is greater than the turn-on time of the incoming switching device, which usually is the case, then there will be a short circuit of the dc voltage source. Therefore, it is customary to introduce an intentional time lag in the inverter operation, i.e., from top to bottom or other way in the transistors for safe operation. That time lag plus a small safety margin is included in the time delay, T_i with the result that the transfer function of the inverter, G_i is modeled as,

$$G_i = \frac{K_i}{1+sT_i} \tag{3.10}$$

This model is useful as a part of the block diagram development of the practical system with inverter, its load, command values and voltage and/or current controllers and using it to compute the controller parameters such as that of the proportional and/or proportional and integral current controllers.

3.3.2.3.5 Current Control

The inverter with the voltage control is not sufficient in applications where current control is desired such as in electric motor drives. In that case, voltage is controlled continuously so as to provide the current control, say, to follow a given command or reference current. Many techniques are available and one such is hysteresis current control where the current is controlled within a specified band of deviations from the desired value. It is easy to implement but requires a high switching frequency to enforce the current control with the result that it is not preferred in applications given the limitations of the switching devices in the inverter. PWM control is another choice and its realization is considered here.

Command or reference current, i^* is enforced with a closed current loop control whose output is passed through a PWM controller as shown in Figure 3.28. The reference current is compared with the current signal i_{of} from the feedback circuit whose input is the load current, i_o after it is gain adjusted and filtered to remove the carrier noise. The current error between i^* and i_{of} is obtained from the summer and it is passed through a proportional plus integral (PI) controller whose output is limited to a certain maximum value. A PI controller is chosen so that the steady-state error between the command and its feedback is nulled which means that the output will follow the input with zero steady-state error. The output of

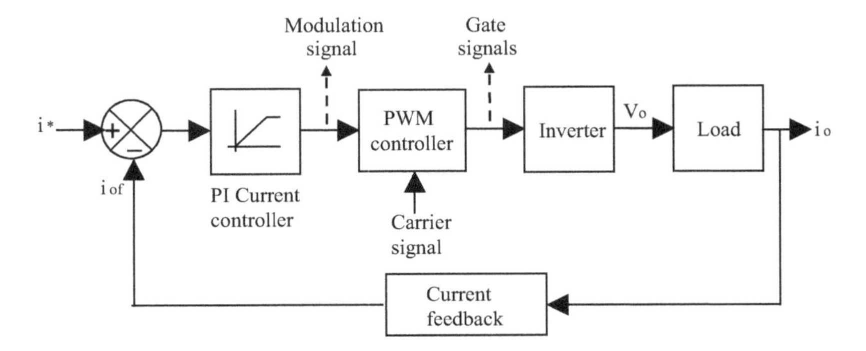

FIGURE 3.28 Inverter closed-loop current control schematic.

the PI current controller constitutes the modulation signal of the PWM controller whose other input is the triangular carrier signal of the desired switching frequency. The PWM logic is applied to these signals to produce the gate signals which then are input to the inverter device gates. The inverter output voltage, V_o which is applied to the load results in a current, i_o. It is usual to come across DSP-based implementations and that means the modulation signal cannot be tracked all the time and compared to the carrier signal as that would require a processor with infinite speed. Therefore, the modulation signal in practice is sampled once in a carrier cycle and held at that value throughout the cycle for comparison to the carrier which can be done, say, using a lookup table. Depending upon the application, the sampling time may be reduced or increased while sampling once in a carrier cycle is prevalent in practice. Note that this sampling function is not shown in the figure but it may be taken as a part of the PI controller block. Given a load, the transfer functions of the various blocks in the figure are derived. The only unknowns are the PI controller's gain and time constant and they can be calculated using linear feedback control theory. Electrical engineers and students are familiar with it and therefore no further development on that is given.

A simulation of the current-controlled inverter with PWM switching control is shown in Figure 3.29. The load voltage and current are all plotted in normalized units (per unit, p.u.). Here the current command and current output closely follow each other and only steady-state operation is shown except for the transition from a step change of 1–2 p.u. in the magnitude of the current command at 0.1 second. For the remaining graphs in the figure, note the time scale is changed so that the results can be viewed on a carrier cycle to cycle basis. The inverter voltage output is shown in the second graph also in normalized unit at a different time scale than the first graph. The carrier signal (one that is triangular with maxima of +1 and −1) and the output of the high-proportional current controller is shown in the third graph to correspond with the second graph. Note that this current controller output is not sampled yet as it is changing throughout the carrier cycles. When it is sampled at the positive peak of the carrier signal instant, the carrier and the sampled output of the current controller output are shown in the fourth graph of the figure. Note that the sampled current controller output is held for one carrier cycle and that in interaction with the carrier, produces the switching signals for the transistors, as shown in the fourth graph and hence the inverter output voltage is derived as shown in the second graph of the figure. The derivation of the turn-on and turn-off instances can be obtained not necessarily all the time of the carrier cycle but can be derived soon as sampling is over either by calculation or from a stored table between the values of the sampled current controller signals. The positive and negative on times of the inverter are derived from these on and off times and from that the gate signals can be developed. The performance of the current output to its command is highly satisfactory but contains ripple at the carrier frequency. They can be mitigated by further increasing the switching carrier frequency of the inverter if it is allowable and safe to operate with the employed devices in the inverter and then the filter at higher switching frequency

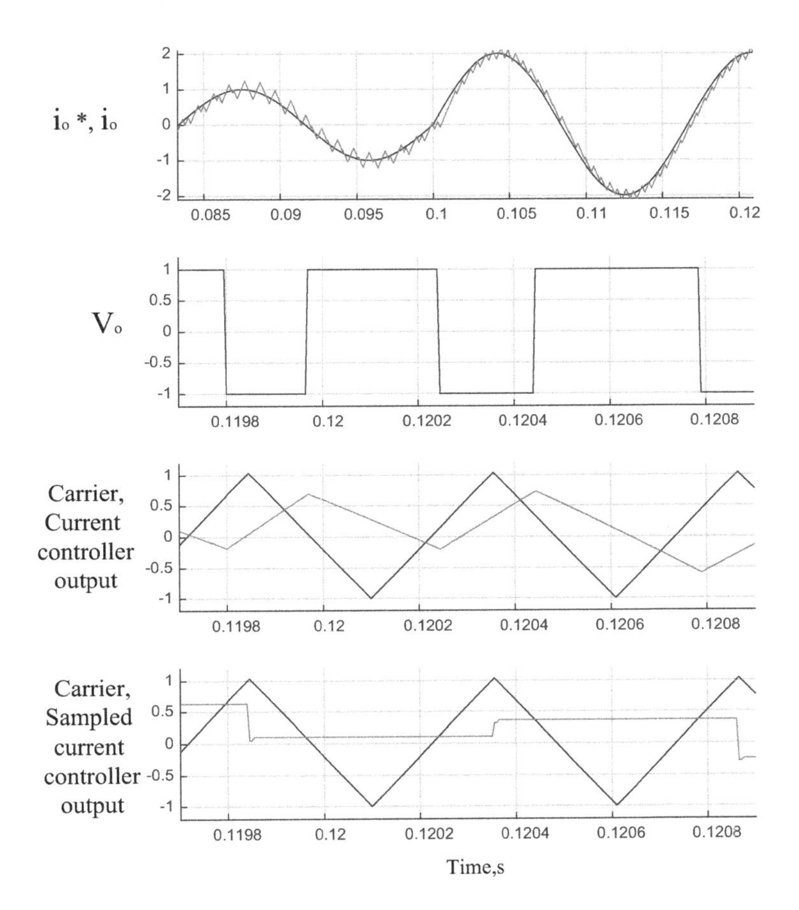

FIGURE 3.29 Current controller inverter operation with step change in current command magnitude.

becomes small. The simulation treats the devices as ideal and hence the switching times for the devices are considered to be zero, and hence the inverter lag also has been neglected. The readers are encouraged to include them in the simulation and see for themselves how acceptable are the changes for their own applications and it is always good to verify that before implementing them in a prototype and later in the product lines.

A step-change in the frequency of the current command from 60 to 120 Hz at 0.1 second is applied and the performance of the inverter is shown in Figure 3.30. All the waveforms are shown for steady-state operation and even with a step-change in the frequency, the inverter comes to steady-state very fast. Note that the time scales for the first and the remaining graphs are different to show some details with clarity like the previous figure.

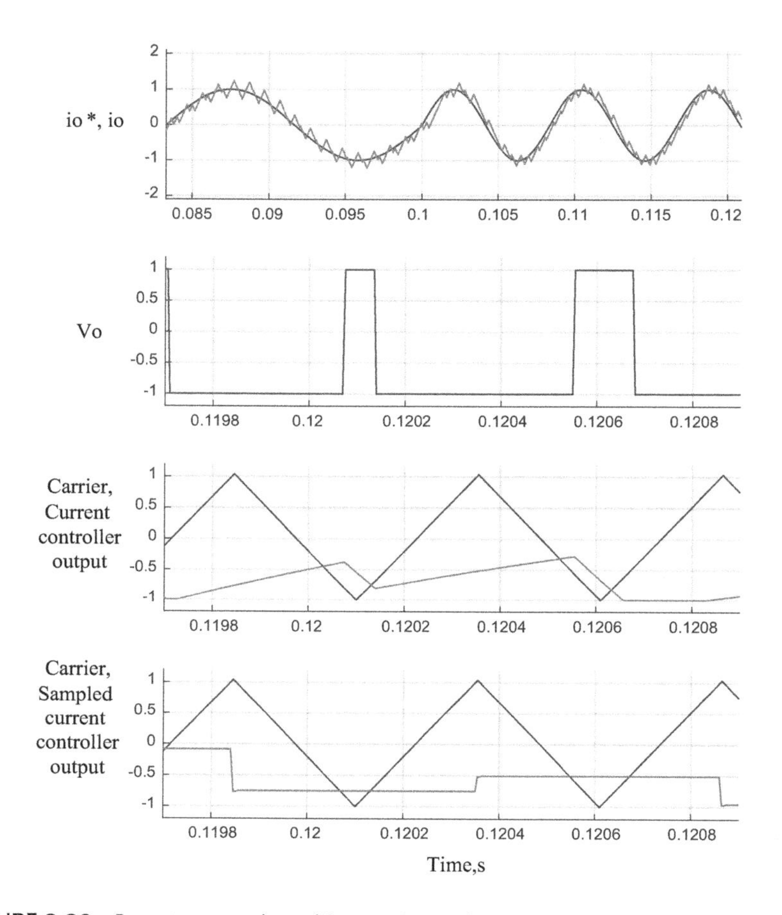

FIGURE 3.30 Inverter operation with step change in current command frequency.

A MATLAB program for the computation and the plotting of the PWM-controlled single-phase full-wave inverter is given below to relate to the write-up and derived figures in this section. Simulink programs can also be generated and it is left to the readers.

```
% PWM current controlled single phase full wave inverter
%  (Shows operation including change of current frequency at
0.1 second)
%  Program modification required to show current reference
change is also given

clear all;close all;

% Load Parameters
R=0.1;                          %Load Resistance
```

```
L = 0.07;                          %Load  inductance
% Dc link parameters
vdc=285;                           %DC Link voltage
wr=2*pi*60;                        %Rated frequency, rad/sec

%Converter and Current Controller (only P type here)
Parameters

fc=2000;                            %PWM switching frequency
Kpi=1;                              %Proportional gain of
                                    current controller

%Base values
Ib=4;                               %Base current, A
wb=628.6;                           %Base frequency in rad/sec
Vb=285;                             %Base voltage, V

% Initial values
ias=0;t1=0;                         % current and time initial
values
vax1=0;                             % PWM current controller
output-initial value
y=0;

%Initial Parameters

t=0;                                %Initial time
tfinal=.14;                         %Final time
ias=0;                              %Initialize  current
vas=0;                              %Initialize  voltages
signe=-1;                           % Carrier slope
ramp=+1;                            % Carrier max value
t1=0;
theta=0;                            % equal to wref*time at
                                    starting

n=1;
x=1;

% Command current peak value setting in A
iasc=4;

% sample at zero of the carrier when it goes from negative
to positive
% To make it happen flawless; make Tc/dt an integer, say,
equal to 100
Tc=1/fc;                      % Carrier cycle time, s
dt=Tc/100;                    % Integration time step; varies
                              with Tc and the integer
```

```
%Simulation Starts Here

while (t<tfinal),

   %Calculate reference currents
   %if t>0.1
      %iasc=8;
  % end
  % command frequency change at 0.1 s
  %
   if t>0.1;                % Changes the command frequency
                            from 60 to 120 Hz
       wr=120*2*pi;
   end

   theta=wr*t;
   iasr=iasc*sin(theta);

   %Calculate control voltages to be applied to PWM
controller

   vax=Kpi*(iasr-ias);      % Current controller output

   % Limiting of the current controller output between +1
and -1
   if vax>1,
      vax=1;
   elseif vax<-1,
      vax=-1;
   end

  %Sample and hold

  if ramp==1;
      vax1=vax;
      t1=0;
   end

   %PWM Controller generates the inverter output voltage
   if vax1>=ramp,
      vas=vdc;
   elseif vax1<ramp,
      vas=-vdc;
   end

   % Compute current from the load equation
   dias=(vas-R*ias)*dt/L;
   ias=ias+dias;

   %PWM RAMP
```

```
   ramp=signe*(2/(1/(2*fc)))*dt+ramp;
   if ramp>1,
      signe=-1;
   end
   if ramp<-1,
      signe=1;
   end

   t=t+dt;    %Increment time
   t1=t1+dt;

   %Plot variables
      tn(n)=t;                                 % Time
      iasrn(n)=iasr/Ib;                  % Normalized load current
                                         command (reference)

      iasn(n)=ias/Ib;                    % Normalized load current
      vasn(n)=vas/Vb;                    % Normalized load voltage
      carrier(n)=ramp;                   % Carrier signal
      pwmin(n)=vax;                      % Limited current
                                         controller output

      sampleout(n)=vax1;                 % Sampled current
                                         controller output

      n=n+1;
end

% Plotting
figure(1);orient tall

subplot 411
plot(tn,iasrn,'r',tn,iasn,'b');axis([0.0833 .1209 -2.1
2.1]);
box off;
grid;

subplot 412
plot(tn,vasn,'r');axis([0.1197 .1209 -1.2 1.2]);
box off;
grid;

subplot 413
plot(tn,carrier,'r',tn,pwmin,'b');axis([0.1197  0.1209  -1.2
1.2]);
box off;
grid;

subplot 414;
plot(tn,carrier,'r',tn,sampleout,'b');axis([0.1197  0.1209
-1.2 1.2]);
box off;
grid;
```

3.3.2.3.6 Space Vector Modulation (SVM)

This is one of the modulation techniques used in multiphase inverter control with advantages in minimizing the switching frequency and for high-performance control in the case of electric motor drives. Its treatment may be of interest to researchers and advanced product developers. PWM control in most cases can match its performance and sufficient for practical applications one normally encounters except in motor drives applications. Hence it is deferred in this text to the readers' discretion and interest by providing suitable references. Apart from original sources, two textbook references that deal extensively with SVM are given at the end of this chapter.

3.3.2.3.7 Application to dc Power Supply

Application of a half-bridge inverter in a push-pull converter has been described in the earlier Section (3.3.3.2.1). Instead of the half-bridge inverter, a full-bridge inverter has been substituted to drive and control the transformer's primary side of the converter without changing any other part of the circuit and it is shown in Figure 3.31. It is known as a full-bridge converter. It overcomes the disadvantages of the half-bridge inverter such as balancing requirement on the input capacitors and half dc input source voltage utilization at a given time whereas the full-bridge inverter utilizes the full dc input supply thereby transferring twice the power from input to output for a given current.

3.3.2.4 Three-Phase Inverter (2 Level)

This is made of three-phase legs and is shown in Figure 3.32. Note the numbering of the transistors in the inverter and that gives the sequence in which they are given the gating pulses to output a predetermined set of three-phase voltage waveforms such as the six-step voltages. This is very much similar to the rectangular voltage synthesis shown in Figure 3.23 for a single-phase case. The

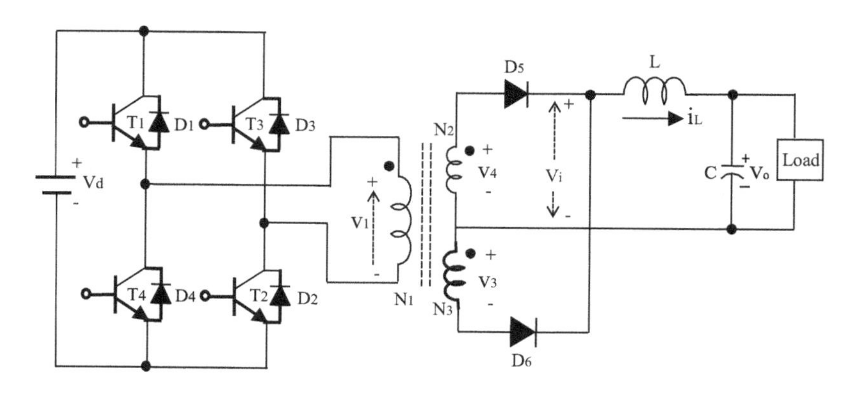

FIGURE 3.31 Full-bridge converter.

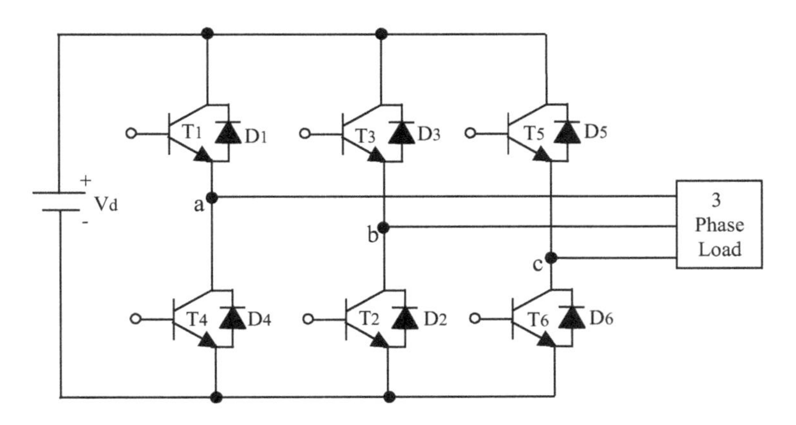

FIGURE 3.32 Three-phase inverter.

three-phase inverter is very popular in power electronic applications, the predominant being in the electric motor drives. As the three-phase inverter consists of three single half-wave inverter phase legs, its operation is very similar to that of the single-phase half as well as full-wave inverters as discussed earlier. Similar to single-phase inverters, the three-phase inverter also produces two levels of voltage at the output. The control techniques applied to the previous single-phase cases are equally applicable to the three-phase inverter case. The three-phase voltage magnitude, frequency and their phases are controlled in the three-phase inverter. The illustrations cover the inverter operation with PWM voltage control catering to a three-phase passive load of resistances and inductances, their closed-loop current control and also control of their 180° phase reversal.

Figure 3.33 contains the steady-state operational waveforms of the inverter under current control and initial startup transients are not plotted so as to focus only on the steady-state at this time. The various graphs are explained in the following. The first part on the left of the figure gives the current controller output and the carrier waveforms for phase A and the second gives the sampled controller output and the carrier for reference. The sampling is made when the carrier is at its positive maximum, i.e., +1 and held for one carrier cycle is evident from the waveform. The crossings of the carrier and sampled current controller output give way to switching of T_1 and T_4 for phase A and likewise for other phases, and they are nothing but the voltage across the phase terminal and the midpoint of the dc supply voltage, given as V_{ao} and shown on the right top of the figure. The voltages and currents are plotted in normalized values. Likewise, the B and C phase voltages between the respective terminals and midpoint of the dc supply source are shown as V_{bo} and V_{co} on the right side. From these phase voltages, the line to line voltages of the inverter are derived and shown in the bottom three waveforms on the right-hand side. From the line to line voltages the phase voltages of the load

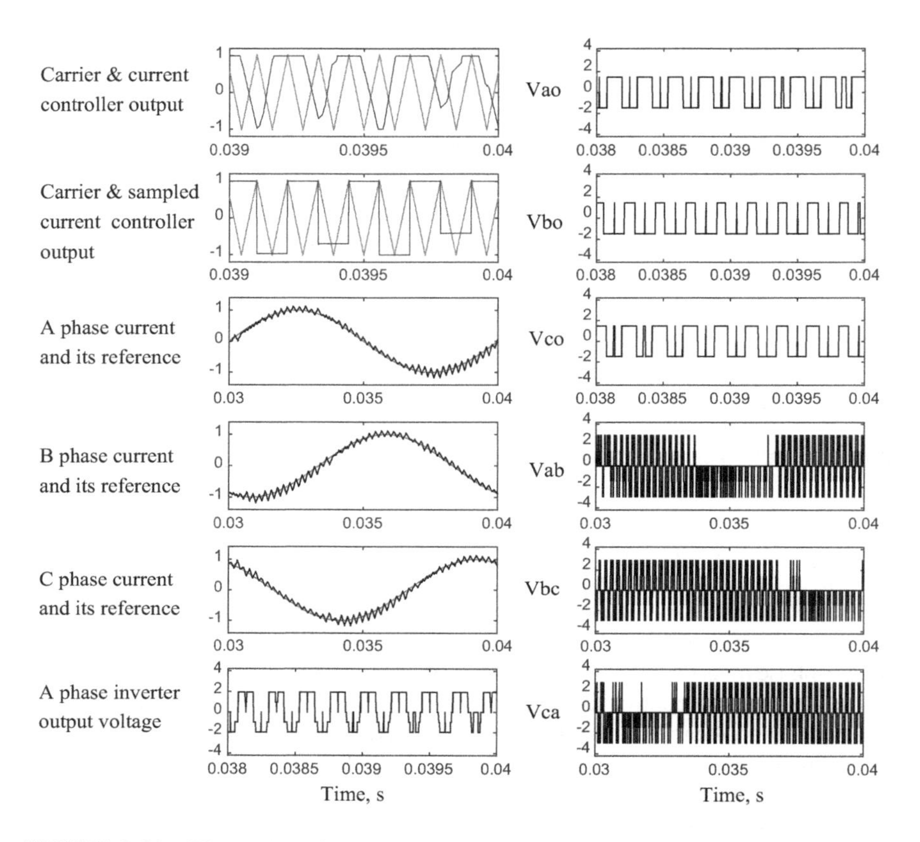

FIGURE 3.33 Three-phase inverter operational waveforms with current control with passive load.

are derived and phase A voltage (V_{as}) is shown in the bottom of the left-hand side of the figure. The resulting load currents and their references are shown in waveforms 3, 4 and 5 on the left-hand side of the figure. The actual currents vary from their sinusoidal references very slightly and only with small ripples around the references. Note that there are differences in the time scale of these waveforms so as to show the various variables in some details.

Figure 3.34 shows the operational waveforms when the inverter output currents are in steady-state for an active load with the load containing an alternating set of voltages or induced emfs, say in ac machines (E_{as}, E_{bs}, E_{cs}), of frequency to match that of the inverter. Closed-loop current control is enforced for the active load. Currents follow their commands very quickly from the startup, and a phase reversal of 180° of the current references is given at 0.03 second. The currents follow the references very fast and the line voltages are plotted also. Note that the time scales are not symmetric so as to bring out the details of some waveforms.

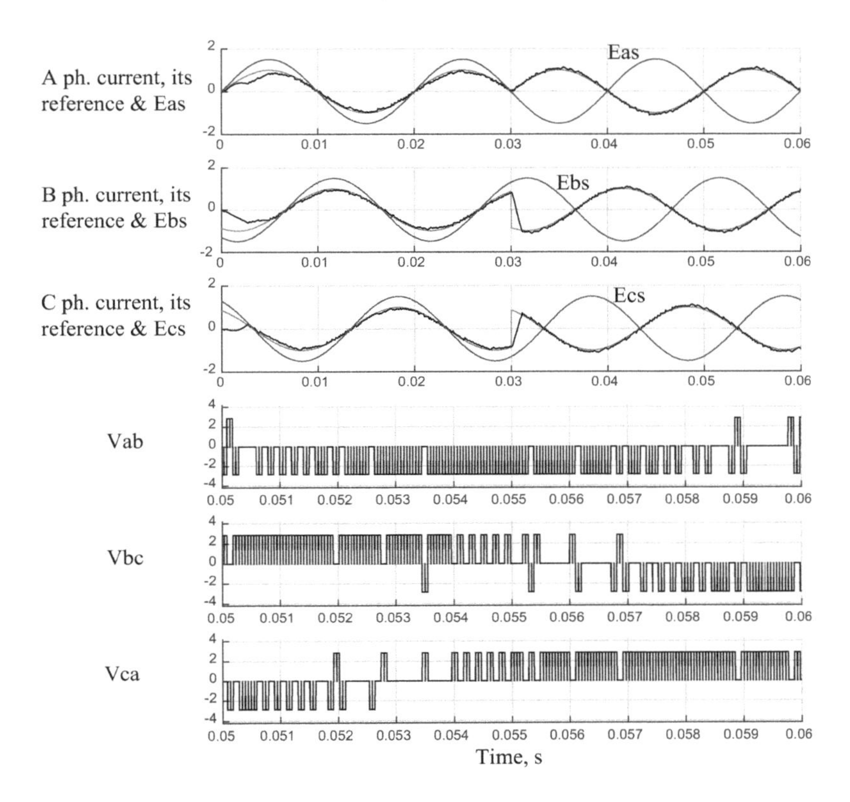

FIGURE 3.34 Waveforms for current phase reversal.

The phase reversal makes the power transfer from the inverter to load go from positive to negative indicating that the load receives power until 0.03 second from the dc bus and then it supplies the power to the dc bus as shown in Figure 3.35. The output power is taken as the product of the emf sources in the load and their respective phase currents. There is a current tracking error as the commanded current and references are not exactly matching here. The difference between them introduces current errors which contribute to the amount of power that is unrealized by this error. If such errors are not present, then the actual power output will match exactly the commanded value of the power. Both the unrealized power due to tracking error and the actual power output are shown in the bottom two waveforms in Figure 3.35.

MATLAB simulation of a three-phase inverter drive with an active load is given below for further self-study and understanding. Various aspects can be self-explored after understanding the code and many applications can be explored with the code forming the basis.

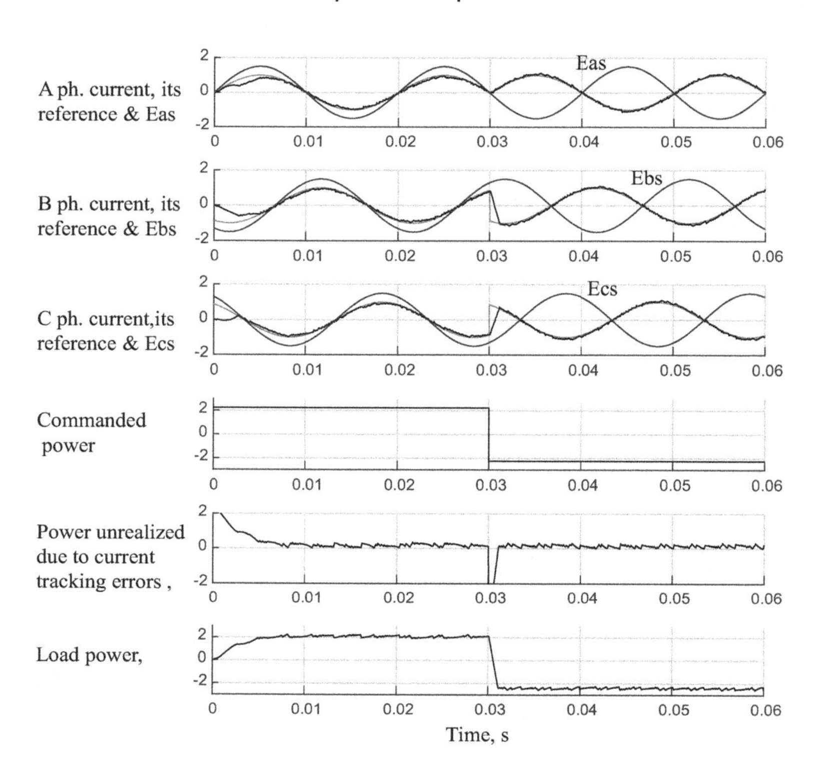

FIGURE 3.35 Current-controlled inverter with phase reversal command and power output study.

```
% PWM Current Controlled Three Phase Inverter
%    Inverter and controlled rectifier operations
% Active load with emf sources in the load (equivalent of an
         ac grid/utility)

clear all;close all;

%Load Parameters
R=0.1;                 % Resistance
L=0.008;               % Inductance
E=150;                 % Peak value of load ac source voltages
wre=2*pi*50;           % Ac source's frequency (on load side)

% Dc source voltage
vdc=285;               %DC Link voltage (say battery voltage
                           or rectified from
                       %  another ac source)

% Command or set values
```

```
wr=2*pi*50;                   %Reference frequency (or
                                 command) for currents
%   Inverter is at a fixed frequency to match
%      load source voltages frequency for grid applications

% PWM  Carrier frequency
fc=10000;                      % PWM Carrier frequency, Hz

%Converter and Controller Parameters
Kpi=100;                      % Proportional gain of current
                                 controller
Ki=15;                        % Integral gain of current
                                 controller

%Base values
Ib=20;                        %Base current value
Vb=100;                       %Base voltage value

%Initial values/inputs
theta_re=0                    % Initial position of load
                                 source voltages
theta_r=0;                    %Initial position of inverter
                              currents
t=0;                          %Initial time
dt=1e-6;                      %Integration Time step
tfinal=0.06;                  %Final time

n=1;
x=1;
t1=0;                         % Initial time
signe=-1;                     % Carrier slope
ramp=1;                       % Carrier peak value
ias=0;ibs=0;ics=0;            % Initialize phase currents
vax1=0;vbx1=0;vcx1=0;         % Initial sampled values of
                                 current controller outputs
y=0;
iasi=0; ibsi=0; icsi=0;       % Integrating current controller
                              initial values

% Reference currents peak values
   is_ref= 20;


%Simulation Starts Here

while (t<tfinal),
% The following changes current phase angle with respect to
  load source
 %    voltages, from 0 to pi at 0.03 second
```

```
angle=0;                  % Phase between currents and load
                            source voltages
if t1>=0.03
    angle = pi;
end

%Calculate reference phase currents

ias_ref=is_ref*sin(theta_r+angle);
ibs_ref=is_ref*sin(theta_r-2*pi/3+angle);
ics_ref=is_ref*sin(theta_r+2*pi/3+angle);

%Calculate control voltages to be applied to PWM
    controller
iasi=iasi+Ki*(ias_ref-ias)*dt;       %integral current
                                       error for phase a
vax=Kpi*(ias_ref-ias)+iasi;
if vax>=1,
   vax=1;
elseif vax<= -1,
   vax=-1;
end

ibsi=ibsi+Ki*(ibs_ref-ibs)*dt;       %integral current
                                       error for phase b
vbx=Kpi*(ibs_ref-ibs)+ibsi;
if vbx>=1,
   vbx=1;
elseif vbx<= -1,
   vbx=-1;
end

icsi=icsi+Ki*(ics_ref-ics)*dt;       %integral current
                                       error for phase c
vcx=Kpi*(ics_ref-ics)+icsi;
if vcx>=1,
   vcx=1;
elseif vcx<= -1,
   vcx=-1;
end

%Sample and hold the current controller outputs

if ramp==1;
   vax1=vax;
   vbx1=vbx;
   vcx1=vcx;
end
```

```
%PWM Controller

% vao is voltage between negative dc link and midpoint of
      phase leg
% vbo and vco -likewise for phases b and c.

if vax1>=ramp,
   vao=vdc/2;
elseif vax1<ramp,
   vao=-vdc/2;
end

if vbx1>=ramp,
   vbo=vdc/2;
elseif vbx1<ramp,
   vbo=-vdc/2;
end

if vcx1>=ramp,
   vco=vdc/2;
elseif vcx1<ramp,
   vco=-vdc/2;
end

%Compute line voltages

vab=vao-vbo;
vbc=vbo-vco;
vca=vco-vao;

%Compute phase voltages

vas=(vab-vca)/3;
vbs=(vbc-vab)/3;
vcs=(vca-vbc)/3;

% Load voltages

 eas=E*sin(theta_re);
 ebs=E*sin(theta_re-2*pi/3);
 ecs=E*sin(theta_re+2*pi/3);

% Integration of three phase voltage equations to solve
      for currents
% Runge Kutta method is not used but simple integration
      is performed.
% Interested readers may use Matlab's ode23 or ode45 for
      integration
```

```
d_ias=(vas-R*ias-eas)*dt/L;
ias=ias+d_ias;
d_ibs=(vbs-R*ibs-ebs)*dt/L;
ibs=ibs+d_ibs;
d_ics=(vcs-R*ics-ecs)*dt/L;
ics=ics+d_ics;

% position
d_theta_r=wr*dt;
theta_r=theta_r+d_theta_r;              % Current reference
                                        position
theta_re=theta_re+wre*dt;               % Load source
                                        voltage  position

%PWM RAMP

ramp=signe*(2/(1/(2*fc)))*dt+ramp;
if ramp>1,
   signe=-1;
end
if ramp<-1,
   signe=1;
end

t=t+dt;   %Increment time
t1=t1+dt;

%Plot variables

%if x>3,    This is used to store only one out of 3
                 computed values for plotting.
   t;
   tn(n)=t;
   vax1a(n)=vax1;
   vaxa(n)=vax;
   ias_refn(n)=ias_ref/Ib;
   ibs_refn(n)=ibs_ref/Ib;
   ics_refn(n)=ics_ref/Ib;
   iasn(n)=ias/Ib;
   ibsn(n)=ibs/Ib;
   icsn(n)=ics/Ib;
   vasn(n)=vas/Vb;
   vbsn(n)=vbs/Vb;
   vcsn(n)=vcs/Vb;
   carrier(n)=ramp;
   vaon(n)=vao/Vb;
   vbon(n)=vbo/Vb;
   vcon(n)=vco/Vb;
   vabn(n)=vab/Vb;
```

```
        vbcn(n)=vbc/Vb;
        vcan(n)=vca/Vb;
        easn(n)=eas/Vb;
        ebsn(n)=ebs/Vb;
        ecsn(n)=ecs/Vb;
        % Load power
        pin(n)=iasn(n)*easn(n)+ibsn(n)*ebsn(n)+icsn(n)
        *ecsn(n);
        % Inverter output power
        pgin(n)=iasn(n)*vasn(n)+ibsn(n)*vbsn(n)+icsn(n)
        *vcsn(n);
        % Unrealized power due to current tracking errors
           error(n)=(ias_refn(n)-iasn(n))*easn(n)+(ibs_refn(n)-
           ibsn(n))*ebsn(n)+(ics_refn(n)-icsn(n))*ecsn(n);
        % Commanded power
        pinerrn(n)=pin(n)+error(n);

        n=n+1;
        x=1;
     %end
     x=x+1;
end

% Plotting
%
figure(1);orient tall;
%
subplot(6,1,1)
plot(tn,ias_refn,'r',tn,iasn,'k',tn,easn,'b');axis([0.0 .06
-2 2]);
grid;box off
subplot(6,1,2)
plot(tn,ibs_refn,'r',tn,ibsn,'k',tn,ebsn,'b');axis([0.0 .06
-2 2]);
grid;box off
subplot(6,1,3)
plot(tn,ics_refn,'r',tn,icsn,'k',tn,ecsn,'b');axis([0.0 .06
-2 2]);
grid;box off
subplot(6,1,4);
plot(tn,vabn,'k');axis([0.05 0.06 -4.2 4.2]);
grid;box off
subplot(6,1,5)
plot(tn,vbcn,'k');axis([0.05 .06 -4.2 4.2])
grid;box off
subplot(6,1,6)
plot(tn,vcan,'k');axis([0.05 .06 -4.2 4.2])
grid;box off

%
```

```
figure(2);orient tall;
%
subplot(6,1,1)
plot(tn,ias_refn,'r',tn,iasn,'k',tn,easn,'b');axis([0.0 .06
-2 2]);
grid;box off
subplot(6,1,2)
plot(tn,ibs_refn,'r',tn,ibsn,'k',tn,ebsn,'b');axis([0.0 .06
-2 2]);
grid;box off
subplot(6,1,3)
plot(tn,ics_refn,'r',tn,icsn,'k',tn,ecsn,'b');axis([0.0 .06
-2 2]);
grid;box off
subplot(6,1,4)
plot(tn,pinerrn,'k');axis([0.0 .06 -3 3])
grid;box off
subplot(6,1,5)
plot(tn,error,'k');axis([0.0 .06 -2 2])
grid;box off
subplot(6,1,6);
plot(tn,pin,'k');axis([0.0 0.06 -3 3]);
grid;box off
```

Three-phase inverter studied is known as three-phase H-bridge inverter.

3.3.2.5 Multilevel Inverters

For high power in the range of 400 kW to multi MW range, the standard H-bridge inverter configuration is not sufficient to meet the requirement with the available ratings of the devices. In order to get up to that power, the dc source voltage has to be increased which necessitates multiple devices to be connected in series making the equal sharing of the voltage across them a daunting task. Or the devices can be paralleled without increasing the dc source voltage to increase the current-carrying capability which has the problem of equal sharing of currents among the devices because of slight variations in the device characteristics. While these techniques have been tried in the past, they are not prevalent currently because of the associated issues. A number of topologies have come into being to cater to high power and medium voltage conversion, and they fall under the category of multilevel inverters (MLI) as they provide more than 2 levels of voltage at the output. Very select but widely used multilevel inverter topologies are presented and that too at 3 levels. The same species can have higher levels of voltage at the output such as 5, 7, etc., and interested readers may further explore from the given references at the end of this chapter. These MLIs have been in industrial practice for more than 20 years. The three categories of prevalent and popular MLIs considered further are:

i. Diode clamped three-level inverter and its variation of active clamped version.
ii. Flying capacitor three-level inverter.
iii. Cascaded H-bridge (CHB) inverter with two units per phase.

3.3.2.5.1 Diode Clamped Three-Level Inverter

One phase leg of the diode clamped three-phase inverter is shown in Figure 3.36 with point o being between the capacitors and each of the capacitor having half the dc source voltage and the midpoint of the four serial connected transistors forming the phase output point, say a, for phase A's case and likewise for B and C phases. All the diodes and transistors have to support only half the dc source voltage and not the entire dc source voltage in this arrangement. Simultaneous conduction of T_1 and T_2 (with others being off) applies half the dc source voltage across phase A with respect to o. When T_1 and T_2 are on, diode D_6 ensures that the voltage across transistor T_4 is limited to half the dc source voltage with the other half dc voltage of C1 is blocked by T_3. Similarly, for T_3 and T_4 conduction, the output voltage becomes negative half of the dc source voltage and when they are turned off; diode D_5 ensures that the voltage across T_1 does not exceed half the dc source voltage. When T_2 and T_3 conduct, neutral point o is connected to the

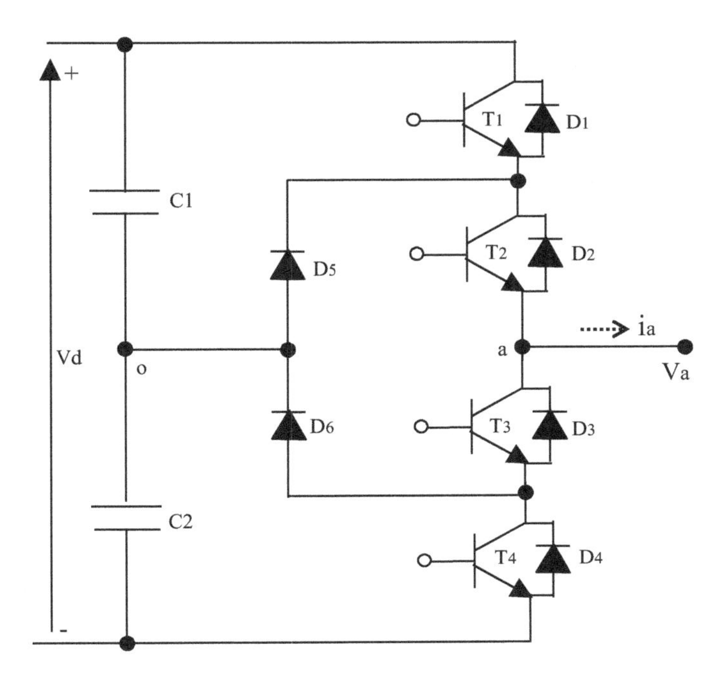

FIGURE 3.36 Three-level diode clamped neutral point inverter.

terminal a through the diodes D_5 and D_6 regardless of the direction of flow of the current in phase A thus clamping the voltage of the phase terminal a to zero with respect to neutral o. The switching states and the resulting phase voltage output of V_{ao} are given in Table 3.1. For the sake of simplifying the table, switch S_1 consists of T_1 and D_1 and likewise for S_2 to S_4. The same notation is used throughout for the multilevel inverters. For example, S_1 is on given as 1 in the table means either T_1 (positive load current) or D_1 (negative load current) is conducting depending on the direction of the current i_{ao}. From the operation and as captured in Table 3.1, it is seen that there are three levels of voltage available at the output and the diodes D_5 and D_6 clamp the neutral, i.e., point o, and hence the name goes by the level and diode clamp and sometimes referred to also as diode neutral point clamped (NPC) inverter.

Operational steady-state waveforms for this inverter with a passive load are shown in Figure 3.37. The voltage command, for phase A known as v_{ref}, carrier signal and its sampled values are shown in the first waveform. The resulting output voltage across a and o, v_{ao} along with its voltage reference is shown in the second waveform, and note that it contains three levels. The third part of the figure shows that for the b phase, and the fourth part shows the line-to-line voltage between phases a and b with its levels going to five. The fifth waveform corresponds to the phase A voltage across its own neutral, v_{as}, obtained by connecting the ending of all the three phases of the load and the final waveform shows the current in phase A, i_{as} for a predominantly resistive type load with a small inductance. Higher-level diode clamp NPC MLIs can be accessed from the references at the end of this chapter.

The advantages of this inverter are higher levels (three in this case) in voltage output resulting in lower harmonics in the voltage and current, devices having to withstand only half the dc source voltage in this case and effects of transient not appearing in the output and capability for delivering power levels up to 27 MW in practice. The major disadvantage is that they use many numbers of devices, though of lower voltage rating than that of the dc source voltage, as compared to a three-phase H-bridge inverter.

A variation of the diode NPC inverter clamping is through active means and such an MLI at three levels for one phase is shown in Figure 3.38. Its operational switching states are summarized in Table 3.2 with the convention that has been used for Table 3.1. Consider the first state where T_1 and T_2 are turned on to get

TABLE 3.1
Switching States and Phase Voltage for Diode Clamp NPC

S_1	S_2	S_3	S_4	V_{ao}
1	1	0	0	$V_{dc}/2$
0	1	1	0	0
0	0	1	1	$-V_{dc}/2$

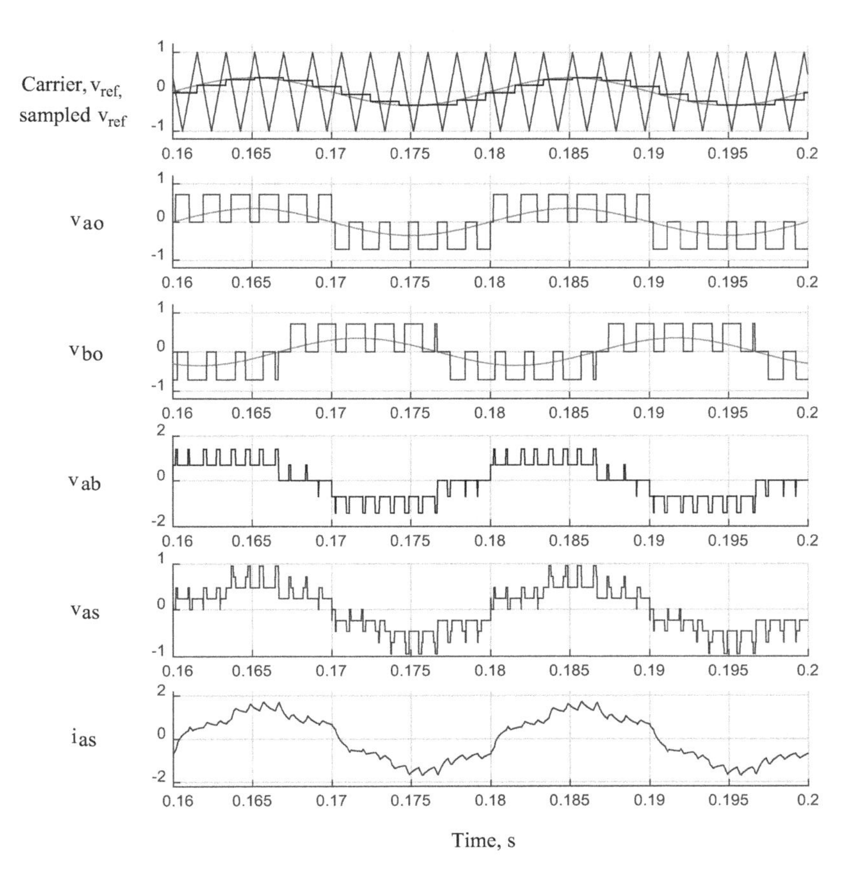

FIGURE 3.37 Operational waveforms of a three-level diode NPC inverter with a passive load.

a voltage output of $+V_d/2$ assuming that the voltage across the capacitors C_1 and C_2 are equal to half the dc source voltage. But in this state, T_6 is also turned on to ensure that the voltage across T_4 remains at $+V_d/2$ while T_3 blocks the same amount of voltage resulting in an equal voltage across the nonconducting devices. The active control ensures voltage balancing across the devices. Consider the second state where only T_2 and T_5 are on with the result that the output voltage is zero. Likewise, the remaining states can be derived. Complementary to the first state is the final state where T_3, T_4 and T_5 are all on with the result that the output voltage is $-V_d/2$. Active clamping ensures voltage sharing across the devices with the result that the switching losses also are minimized.

3.3.2.5.2 Flying Capacitor Three-Level Inverter

A three-level version of this inverter for one phase is shown in Figure 3.39. Switches S_1 and S_4 cannot be turned on at the same time as it will lead to the flying capacitor C_3 being in series with the source voltage resulting in almost short

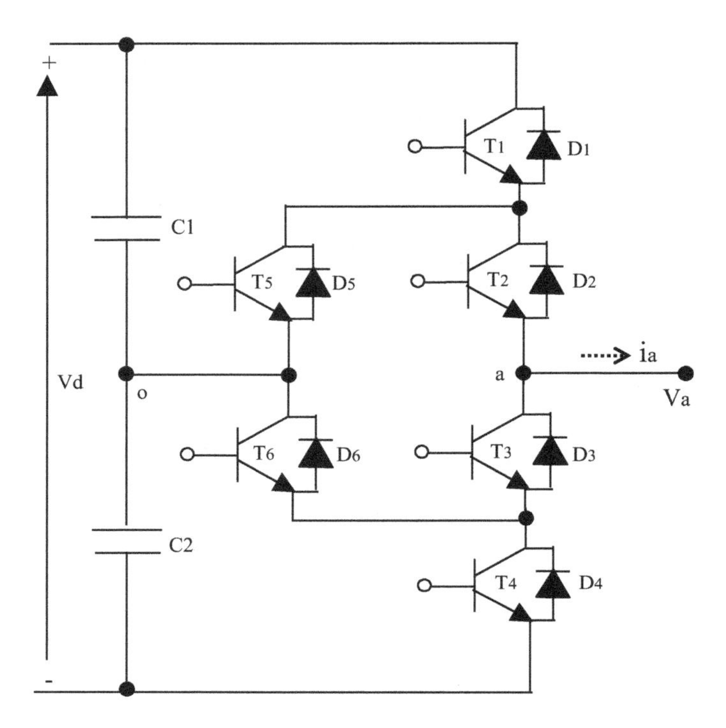

FIGURE 3.38 Active neutral point clamped three-level inverter.

TABLE 3.2
Switching States and Phase Voltage for Active Clamp NPC

S_1	S_2	S_3	S_4	S_5	S_6	V_{ao}
1	1	0	0	0	1	$V_{dc}/2$
0	1	0	0	1	0	0
0	1	0	1	1	0	0
1	0	1	0	0	1	0
0	0	1	0	0	1	0
0	0	1	1	1	0	$-V_{dc}/2$

circuit if the connecting cables have negligible resistance. Simultaneous conduction of switches S_2 and D_3 will result in a short circuit of the capacitor C_3 voltage. Turning on T_1 and T_2 gives the voltage of $V_{dc}/2$ across the output terminals given by phase terminal a and neutral o. Likewise, turning on T_3 and T_4 enables $-V_{dc}/2$ as the output. In order to charge the flying capacitor C_3, S_1 (i.e., consisting of T_1 and D_1) and S_3 (T_3 and D_3) are turned on with the result that the output phase voltage (same as load voltage) will be zero, and likewise discharge of C_3 is accomplished with S_2 and S_4 resulting also in load voltage of zero. With that,

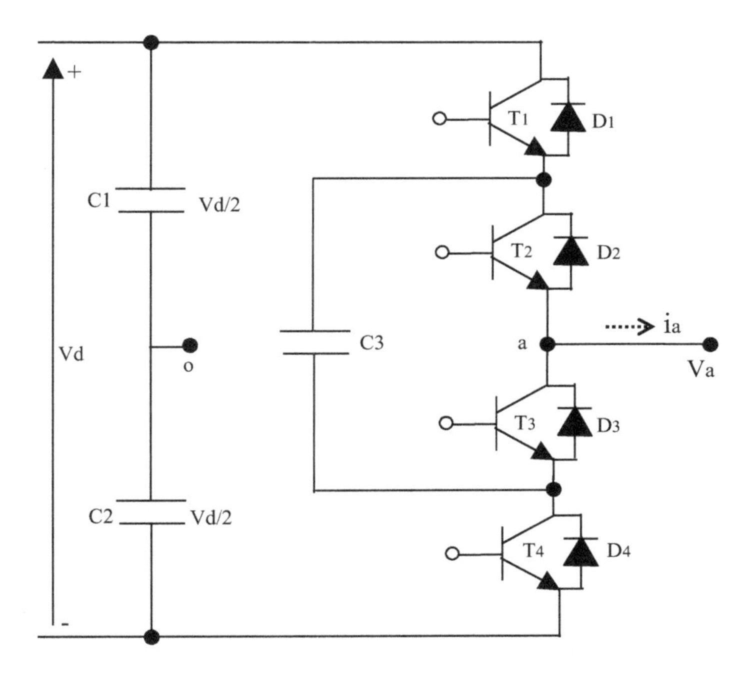

FIGURE 3.39 Flying capacitor three-level inverter.

this inverter provides three levels of voltage with the flying capacitor providing the voltage clamp resulting in devices evenly sharing the voltage. Its operational switching states are summarized in Table 3.3.

For three-phase inverters, three such phase units are assembled one for each phase and the neutral of the load need not be connected to the center of the dc source, o. That is the case for the three-level inverters including diode or active clamped NPC multilevel inverters. Also, it is to be noted that separate dc sources are not needed for each of the phases and that reduces the cost and complexity of providing more than one dc source. The dc source voltage sharing between capacitors C_1 and C_2 may become unbalanced and that requires attention so that the devices are not impacted with unequal voltages.

TABLE 3.3
Operational Switching States of the Flying Capacitor Three-Level Inverter

S_1	S_2	S_3	S_4	V_{ao}
1	1	0	0	$V_{dc}/2$
1	0	1	0	0
0	1	0	1	0
0	0	1	1	$-V_{dc}/2$

Higher level (more than three-level) flying capacitor MLIs can be accessed from the references at the end of this chapter.

3.3.2.5.3 Cascaded H-Bridge (CHB) Inverter

This inverter uses single-phase full-wave inverters as units to build on to the multilevel inverter. The individual units are connected in series with each other as shown in Figure 3.40 for one phase. As the input voltage to the unit full-wave inverter is added by the inverters, the dc source voltages have to be independent and therefore require a transformer isolation and rectification for each unit. The present case for illustration having two units of full-wave inverter then required two independent dc sources (with voltages V_{d1} and V_{d2}) and for a three-phase system the total requirement is then six independent dc sources or galvanically isolated dc sources. That requires a transformer with six secondary having six rectifiers overall for providing the dc sources. That invariably increases the cost and their cost alone is from 40% to 55% of the total cost of the inverter system. Note that such transformer-based dc sources are not required for the previously

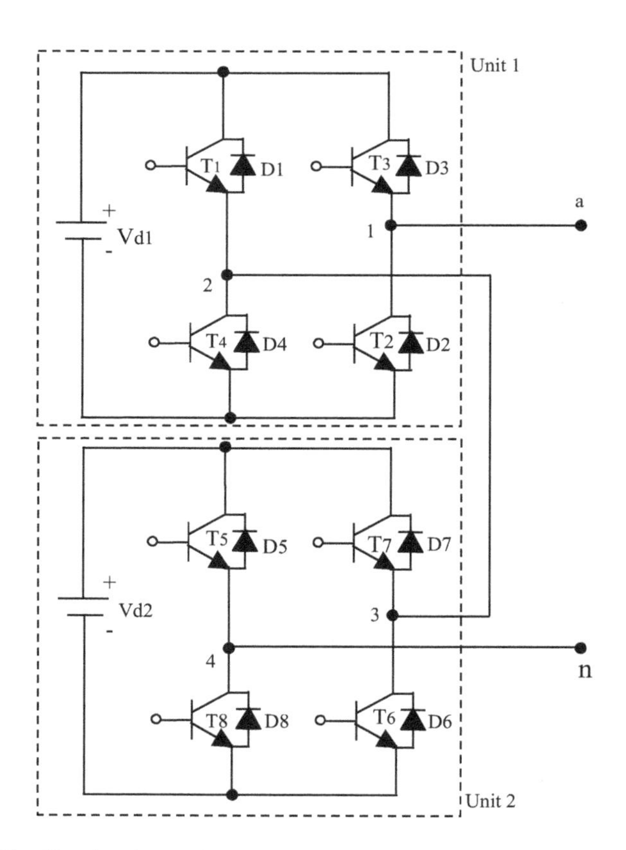

FIGURE 3.40 Five-level cascaded H-bridge inverter.

discussed diode clamped NPC or flying capacitor MLIs and hence their widespread use is up to 30 MW output levels. But if the power capacity has to be much greater, say, about 125 MW level (already implemented in practice), the CHB multilevel inverters seem to be one of the most attractive choices for the industry.

Consider the basic operation of this inverter shown in Figure 3.40. Switching is complementary for each phase leg, i.e., if T_1 is on, then T_4 is off and vice versa and that applies uniformly to all the phase legs in the CHB. For ease and compactness of representation, only four switching device states then are required to represent the inverter shown in Figure 3.40. The switching states and resulting output phase voltages are given in Table 3.4. For example, for the first item in the table, switching states are $T_1 = T_3 = T_5 = 0$, $T_7 = 1$ and therefore, by the convention explained earlier, what is left out and not shown in the table is that their complementary switch states are $T_4 = T_2 = T_8 = 1$ and $T_6 = 0$. Likewise, for all other switching states in the table. If $V_{d1} = V_{d2} = E$, then there are five levels produced in the inverter that are at voltages of 2E, E, 0, –E, –2E in the phase voltage and therefore it is a five-level inverter. Note that two independent dc sources per phase are required even if they are of the same magnitude. The major advantage of the CHB MLIs is that it is modular, simpler to control and it can be built to high-power levels.

Higher levels (greater than the discussed 5-level CHB inverter) can be easily derived from the understanding developed in this section and additional information can be found in the references given at the end of this chapter.

Many more topologies of multilevel inverters exist in the research and industrial product domains and interested readers may get additional information from

TABLE 3.4
Switching States and Phase Voltage of CHB Inverter

T_1	T_3	T_5	T_7	V_{34}	V_{21}	$V_{an}=V_{21}+V_{34}$
0	0	0	1	V_{d2}	0	V_{d2}
0	0	1	0	$-V_{d2}$	0	$-V_{d2}$
0	0	1	1	0	0	0
0	1	0	0	0	V_{d1}	V_{d1}
0	1	0	1	V_{d2}	V_{d1}	$V_{d1}+V_{d2}$
0	1	1	0	$-V_{d2}$	V_{d1}	$V_{d1}-V_{d2}$
0	1	1	1	0	V_{d1}	V_{d1}
1	0	0	0	0	$-V_{d1}$	$-V_{d1}$
1	0	0	1	V_{d2}	$-V_{d1}$	$-V_{d1}+V_{d2}$
1	0	1	0	$-V_{d2}$	$-V_{d1}$	$-(V_{d1}+V_{d2})$
1	0	1	1	0	$-V_{d1}$	$-V_{d1}$
1	1	0	0	0	0	0
1	1	0	1	$-V_{d2}$	0	$-V_{d2}$
1	1	1	0	$-V_{d2}$	0	$-V_{d2}$
1	1	1	1	0	0	0

the references given at the end of this chapter, initially from the survey papers and then from them going to the original sources.

3.3.2.6 Resonant Converter Circuits

Hereto the converters discussed in this chapter belong to the switch mode type. In them, the switches are turned on and off regardless of whatever the voltage across or current through them. That puts a stress in the form of rate of change of voltage and/or current (dv/dt or di/dt) and also significant switching losses in the devices. Consider switching the devices encounter zero voltage or current and then the switching comes stress-free practically. All of the circuits to achieve such a condition require an inductor and a capacitor either in series or parallel to the load in the circuit to resonate. All kinds of power conversion circuits have a resonant counterpart in them. As the switching stresses are removed or reduced with the resonant power circuits, the efficiency can improve in these circuits but practically that is not generally the case as there are losses associated with the resonant circuit elements as well. As the frequency is increased with the advantage of resonant circuits, that may introduce EMI unacceptable in some applications. Because of a number of these factors, there are selective applications for these power circuits compared to that of the switch-mode type of power converters. There are also some evolving applications of these converters in wireless charging and in high-efficiency power supplies, for example. Interested readers may access Refs. [2,6] to learn this topic.

3.3.3 Ac to Dc Power Converters

Dc source voltages, say for the inverter inputs, are generally obtained from the rectification of ac sources such as from the utilities. There are two options available for rectification, one that is uncontrolled using diodes only in the circuit and the other one with controlled devices. Both of them are considered in this section.

3.3.3.1 Uncontrolled Rectification

3.3.3.1.1 Single-Phase Rectification

A circuit for this purpose with a single-phase ac source input is shown in Figure 3.41. The circuit with diodes is in the form of full-wave single-phase H-bridge and the diodes are of the slow type as the device commutation occurs by the zero crossing of the current in the circuit. This type of dc rectification is prevalent. Because of familiarity of this circuit during the earlier curriculum (at sophomore level), only minimal treatment is given here. The inductance on the ac side is not necessarily an intended part of the circuit in low-power applications and it is shown here to represent the cable inductance that is connecting the utility terminal point to the converter. The capacitor at the load end is to smooth the variations of the rectified ac voltage and it is of considerable size usually. Diodes D_1 and D_6 conduct when the ac voltage is positive and only when the ac voltage is greater than that of the load voltage, V_d. And the conduction ends when the load

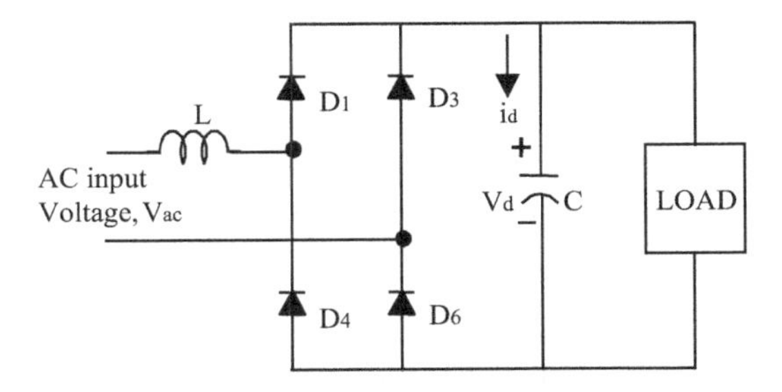

FIGURE 3.41 Single-phase uncontrolled rectifier.

current comes to zero and that is after the instant when the ac voltage becomes less than or equal to V_d. Similar is the operation for the negative half cycle of the ac voltage and where diodes D_3 and D_4 conduct. If the ac input is of frequency f_s, then the rectified voltage will have a ripple of $2f_s$ frequency. The operational waveforms are shown in Figure 3.42 for this circuit with no source inductance and with a resistive load, all of the variables given in normalized units and also the figure only displays the steady-state part of the operation. The ac input voltage is in the first part of the figure and the second contains the ideal rectified voltage which is the absolute value of the input voltage itself as well as the actual rectified voltage for a resistive load. Note that once the capacitor charges up to the peak of the ac input voltage, the capacitor current decays to zero. The load continues to draw the current which has to be supplied from the capacitor with the result being that its voltage decreases. The rectified ac voltage cannot charge the capacitor until the capacitor voltage equals the instantaneous value of the ideal

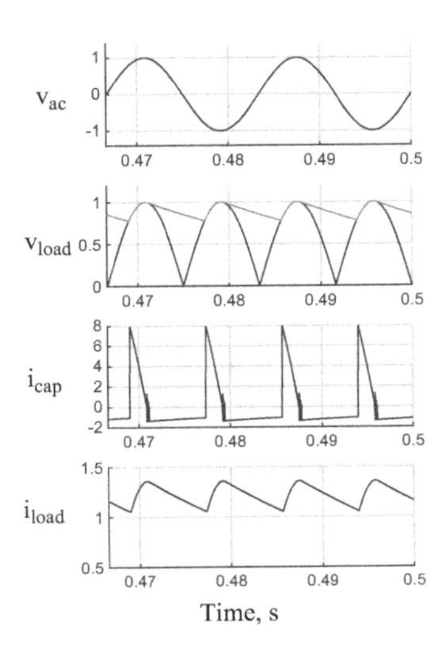

FIGURE 3.42 Operational waveforms of a single-phase bridge rectifier with resistive load.

rectified voltage at which point the capacitor gets charged from the rectified voltage. During this time, the resistive load will continue to be energized from the capacitor as seen from the capacitor current where the capacitor current equals the resistive load current. It is shown in the third graph of the figure. The last part shows the load current in the resistor which is identical to the dc load voltage but varies only in magnitude depending on the resistance.

3.3.3.1.2 Three-Phase Rectification

A three-phase H bridge with diodes for switches is used for rectification of three-phase ac source voltages and such a circuit is shown in Figure 3.43. The ac line voltages are shown in Figure 3.44 to assist the understanding of the rectification process. The highest of the line-to-line voltages will enable the corresponding diodes to conduct at any time in the H-bridge. To find the highest line-to-line voltages at any instant, the absolute values of them are shown in Figure 3.44 as well. The absolute values of the line voltages represent their rectification if each is distinct and separate from each other. When all the three phases are considered, the line voltages (and their rectified negative part of them) contribute only 60 electrical degrees of the maximum value for each of them. Line voltage between a and b, V_{ab}, is maximum from 60° to 120° during its positive half cycle. But in the negative half cycle, V_{ab} also has a negative maximum period from 240° to 300° and during this time the terminal a is negative compared to the terminal b. During rectification, this part becomes positive and since that requires inversion of the negative half cycle which means that the line voltage is V_{ba} indicating that the terminal b is positive compared to the terminal a. Likewise, when the other two line voltages are considered and drawn, the maximum of the line voltages is only for 60° duration and in one full cycle, there are six such periods as can be seen in Figure 3.44. Ideally, the rectification of three-phase voltages will be these 60° periods as shown in the first picture in Figure 3.45 that results from the diodes corresponding to the highest line-to-line voltages conducting for 60° electrical degrees in a cycle. For example, when the line voltage V_{ab} is the highest, diodes D_1 and D_6 conduct. The sequence of conduction is D_1D_2, D_2D_3, D_3D_4,

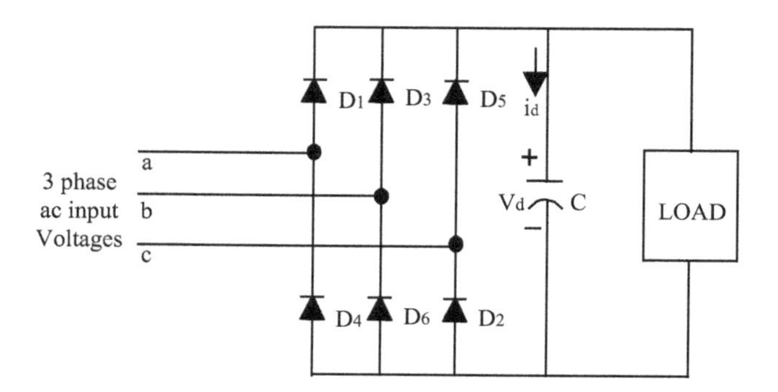

FIGURE 3.43 Three-phase uncontrolled rectifier.

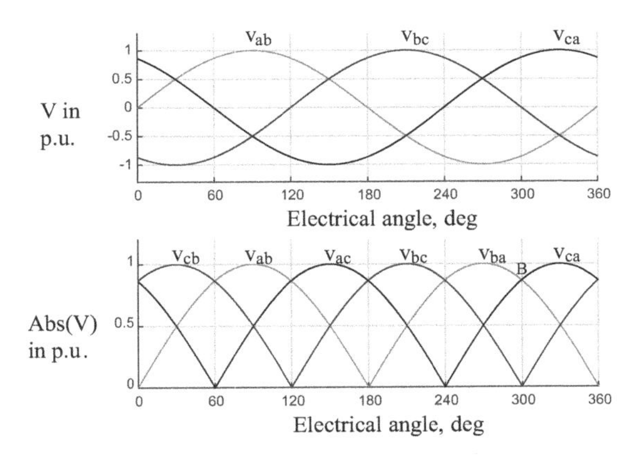

FIGURE 3.44 Line-to-line voltages and their absolute values of three-phase system.

D_4D_5, D_5D_6, D_6D_1 and goes back to the beginning of the sequence D_1D_2 to start the cycle. Neglecting the line inductances and having an output filter capacitor and resistive load is taken for illustrating the performance of this converter and given in Figure 3.45 and all variables are given in their normalized units, i.e., p.u. The simulation shows from the startup to a full-cycle operation. As the capacitor gets charged, the output voltage follows it and when the cap is charged to the peak of the line-to-line voltage as shown in the second part of the figure, and then the ac input current is cut off. Note that the output voltage from now on is influenced only by the capacitor and load until the next 60° subcycle. Also, the load is supplied only from the capacitor with the result that the voltage across the capacitor, which is the same as the load voltage decreases and when it comes to equal the maximum segment voltage in the next 60° maximum line voltage interval, the capacitor gets charged by the corresponding line-to-line voltage and so on. The output voltage is shown in the second part of the figure. The load current follows the load voltage as shown in the fourth part of the figure and since a resistive load is considered it exactly follows the load voltage in shape shown in the second part of the figure. For each line-to-line voltage, there are two 60° intervals of conduction in each positive and negative parts with the result that considering the phase voltage, say, phase a voltage which lags the line-to-line ab voltage, the ac current input to the phase is as shown in the last graph of Figure 3.45. In reality, the ac currents are not of four discrete pieces of current waveforms in a cycle but they will be currents of 120° interval with constant magnitude with a slight sine superimposed from the rectified voltage contributed by the line inductors as well as the filter inductances on the output side. The major issue with the diode rectification is the harmonic content in the ac input currents that impacts the power system utility. Uncontrolled rectification is widely used from low- to high-power inverter-driven electric motor applications.

Note that the output voltage both in the case of single-phase and three-phase rectifier circuits has a magnitude which is the average of the ac input voltage.

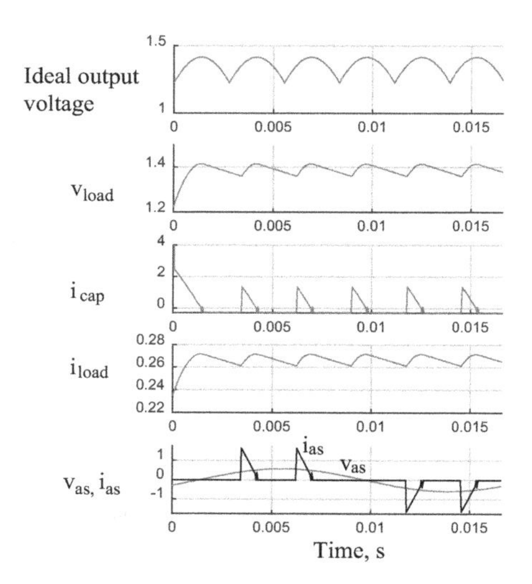

FIGURE 3.45 Operational waveforms of three-phase diode rectifier.

It is not possible to vary the output voltage and many applications require a different set of dc output voltages, for example, laptop computers requiring voltages at the level of 20 V (for example, Lenovo make). Hence, the need for controlled rectification as an alternative or using a dc to dc switchmode converter as discussed in an earlier section. The latter is widely used in appliances but the former is for high-power applications.

3.3.3.2 Controlled Rectification

When the diodes in the H-bridge rectifier are substituted with thyristors, shown in Figure 3.46, the converter is capable of controlled rectification as well as inversion. Thyristors lend themselves for controlled turn-on at a desired instant and until that time they are capable of blocking the positively biased anode against the cathode from conduction. Such a feature lends itself to incorporate a delay from

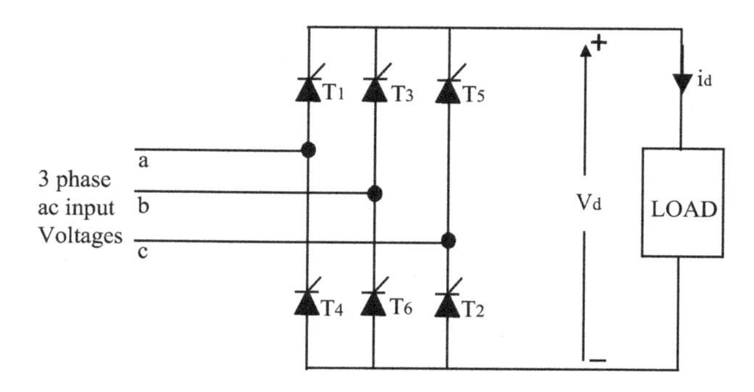

FIGURE 3.46 Phase-controlled converter for three-phase.

the natural conduction point which provides the maximum of the ac three-phase system's line input voltages. This delay angle, α is the variable that determines the dc output as shown later. The point of reference for the delay angle is where a particular input line-to-line voltage's magnitude is maximum compared to all other input line voltages at a given instant as discussed in the uncontrolled three-phase diode rectifier section.

The output voltage is obtained by the average integral of the 60° period of the line-to-line voltage after the delay angle, α. It can be derived as 1.35 V cos(α) where V is the rms line-to-line voltage in the three-phase system. When the delay angle is zero, it becomes the equivalent of the diode bridge rectifier with its output at 1.35 V. The operational waveforms of this converter in rectification mode are shown in Figure 3.47 having three sections for triggering angles of 0°, 30° and 90°. The first graph in each section shows the line-to-line voltages and their complements for rectification consideration, the second figure shows the waveform after rectification in situ with their corresponding line voltages, the third figure shows the output dc voltage and the load current and the final figure shows the phase ac input voltage and phase input current. The dc output voltage is the rectification waveform of the second figure shown in thickened lines and as the load is resistive in the case of triggering angle cases of 0° and 30°, their currents follow the output voltages. The phase a voltage, V_{as} lags V_{ab} by 30° and the ac input phase a current is from the segments related to terminal a. For the first two cases, only a pure resistive load has been considered. For the triggering angle of 90°, an inductive resistive load is introduced, and hence, the shape change in the currents. The current is not continuous for the 60° interval here due to the dominance of resistive load and when the current goes to zero, the thyristors see the reverse bias in the input voltage resulting in their nonconduction. If the inductance has been dominant, the current could have been continuous in that interval. The converter can also invert, from a suitable dc supply of reversed polarity to that of the rectification, to supply the ac source. The inverter operation is not prevalently used

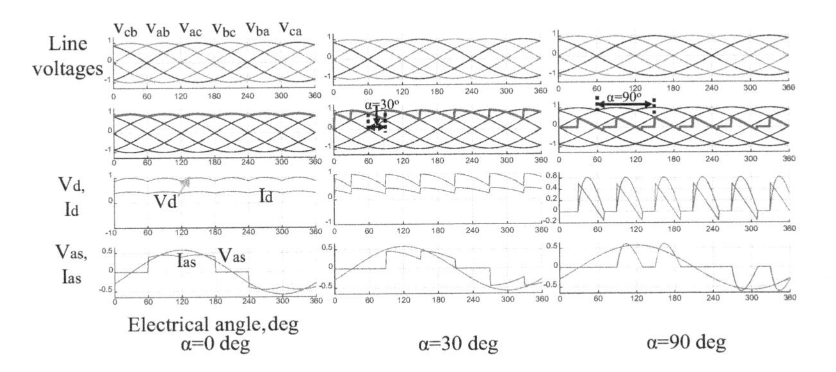

FIGURE 3.47 Operational waveforms of a phase-controlled converter for various triggering delay angles.

in many practical applications and hence not pursued any further here. One of the most highly used applications of the phase-controlled rectifier is in variable-speed control of large dc motors even at present and for providing front-end dc source for large inverter motor drives in the past.

3.3.4 Direct ac to ac Conversion

From the utility grid which is a fixed voltage and fixed frequency ac to a variable voltage and variable frequency supply is obtained by first rectification of the utility source voltages and then the inversion of the dc link voltage through an inverter. This method has been dealt in the inverters and their operation and it has limitations in that the energy has to be processed twice resulting in less than optimal efficiency and it further requires bulk dc link filters and ac input filters too. The latter increases packaging size and system cost. A direct ac to ac conversion without having to go through rectification and filtering has been in practice in limited applications for a long time. There are two kinds of converters for such conversions:

i. Cycloconverter: Uses the controlled H-bridge SCR converters with natural commutation for output frequencies lower than that of the ac supply input frequency and with forced commutation to obtain output frequencies higher than that of the ac supply frequency.
ii. Matrix Converter: Uses self-commutating devices that are bidirectional for each electronic switch in the converter configurations. There are many kinds:
    a. One uses matrix-like schematic appearance in converter configurations called direct matrix converter.
    b. Others that use the bidirectional switches also but having the standard ac supply to dc conversion and then dc to inversion stages similar to the conventional methods but without the dc link filter and they are referred to as indirect matrix converters. A subcategory of the indirect matrix converter uses some bidirectional devices but uses a large number of diodes to achieve the conversion and sometimes they are referred to as sparse matrix converters.

They are all briefly described in the following sections.

3.3.4.1 Cycloconverter

The controlled phase converter in numbers is utilized to synthesize direct ac to ac conversion. Such direct ac to ac conversion with variable voltage/variable current and frequency controls without a dc link consisting of a filter is the hallmark of cycloconverters. In addition, the cycloconverter is made with SCRs with natural commutation (i.e., without forced commutation circuits) resulting in high reliability and suitable for high-power applications constituting most of its applications. A full-wave single-phase cycloconverter is given in Figure 3.48. A three-phase

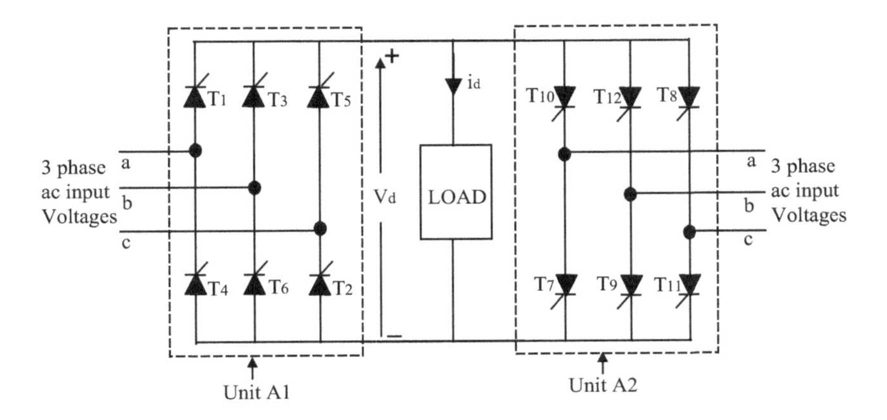

FIGURE 3.48 Single-phase cycloconverter.

system can be realized using three such single-phase cycloconverters. The single-phase cycloconverter has two units, named as Unit A1 and Unit A2, of three-phase controlled converters with SCRs for controllable devices with a total count of 12 SCRs for each phase with the result that the three-phase cycloconverter has a total count of 36 SCRs. Unit A1 provides a positive voltage and positive current to the single-phase load with its voltage v_d and current i_d as shown. This unit may be referred to as a positive converter unit. Unit A2 provides negative load voltage and load current and may be termed as the negative converter unit. Hence by switching between the units A1 and A2, an alternating voltage and current are obtained. The alternating output voltage is made sinusoidal in the average by suitable phase control of units A1 and A2. Note that the commutation of the SCRs in the units is achieved by the line input voltages requiring no extra circuits for that task. To obtain sinusoidal currents/voltages at the input and output, the output frequency is made a fraction of the input frequency, less than one-third with the result that there are many cycles of positive voltage and negative voltage are applied to the load. Even then, there will be considerable harmonics and that may not have a major effect in high-power motor drive applications. Because of the lowering of output frequency, this type of converters is referred to as step-down cycloconverters. The following are the main features of the cycloconverters:

i. Near sinusoidal shaped input and output currents and output voltages.
ii. Four quadrant operational capability.
iii. No dc link filer requirement and hence energy storage elements.
iv. Time-tested technology using phase-controlled converters with SCRs and hence great reliability in performance.
v. Ideal suitability for high-power variable-speed ac motor drives for low-speed applications.

A version with only half of the SCRs is available with half-wave converter configurations. There is also a single-phase ac input supplied single-phase cycloconverter with full-wave controlled converters having only four SCRs for each unit (i.e., for each positive and negative unit) with a total of 8 SCRs and they are usually intended for low-power applications.

Output voltage at higher frequencies than that of the ac source input is obtained in cycloconverters with force commutation of the SCRs in the controlled converter units. These are then referred to as step-up cycloconverters.

3.3.4.2 Matrix Converter

This converter uses bidirectional switches. Bidirectional switch, shown in Figure 3.49, is capable of carrying currents in both directions and consists of two self-commutating discrete devices, say, transistors with their antiparallel diodes and the switches are connected in series. Three such bidirectional switches are connected on the one end to the three lines of the three-phase ac supply and on the other end connected together to the single-phase ac output as shown in Figure 3.50. Therefore, for a three-phase ac output, three such phase units are required with a net 9 bidirectional devices giving a total count of 18 transistors and 18 diodes. Further, an ac input filter, say consisting of capacitors is required for the system to operate. The output lines are connected to a three-phase load. The operation is simple, and for example, consider that the phase A output is connected to a single-phase load which on the other end is tied to a neutral of the

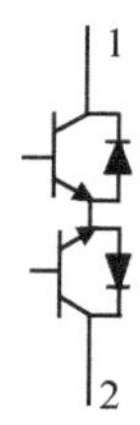

FIGURE 3.49 Bidirectional switch.

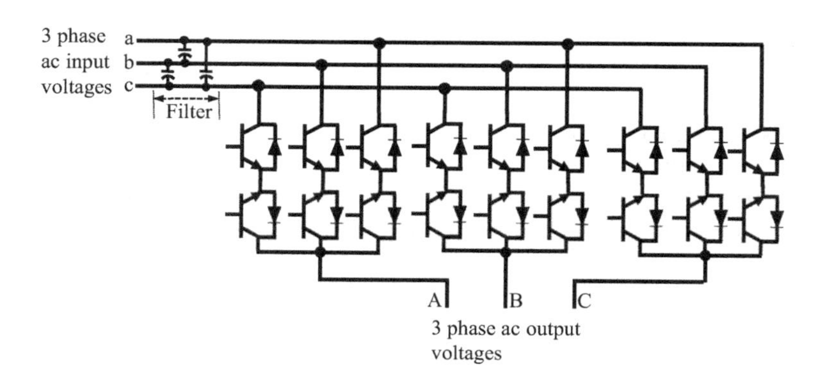

FIGURE 3.50 Three-phase direct matrix converter.

three-phase ac input. When input phase a is positive, the top transistor in the bidirectional switch is triggered and together with the bottom diode conduct a current to the load. A sudden turn-off of the transistor in that bidirectional diode can create a problem unless a path for the load current is provided either by switching the bidirectional switches connected to one of the other input phases. Likewise, when the negative current has to be injected to the load, say, when the input phase a goes negative, the bottom transistor in the bidirectional switch is triggered resulting in negative voltage and current in the load phase. Likewise, for the other phases, it results in three-phase ac output of variable frequency and variable voltage. Since in this converter there is no dc link at all, it is referred to as a direct matrix converter. The name matrix converter came about as the converter configuration with its bidirectional switches can be cast similar to a matrix as shown in Figure 3.51 and hence the name where the switches SaA, etc., are all bidirectional switches with lines a, b and c forming the input ac and lines A, B and C forming the output load end lines. The control details of the converter for further study can be obtained from the references.

Using 6 bidirectional switches and putting them in the form of H-bridge form for front-end converter and then having six devices based conventional H-bridge inverter for the load end results in the standard form of indirect ac to ac conversion with a dc link and hence the name indirect matrix converter, shown in Figure 3.52. But here such a converter with the bidirectional front-end converter does not have the dc link filter is to be noted. The function of the front-end converter is to provide ac to dc conversion during rectification mode and dc to ac conversion in the inversion mode. Its output voltage is always maintained to be dc whereas the current can be bidirectional and hence the need for bidirectional devices for it. The operation of this front-end converter is very much similar to the controlled rectifier in that the turn-on and turn-off of the devices happen at zero current. The function of the load-end inverter is to take the dc and convert it into three-phase ac of variable voltage and variable frequency. Note that this is similar to the standard inverter and hence likewise, the output frequency range

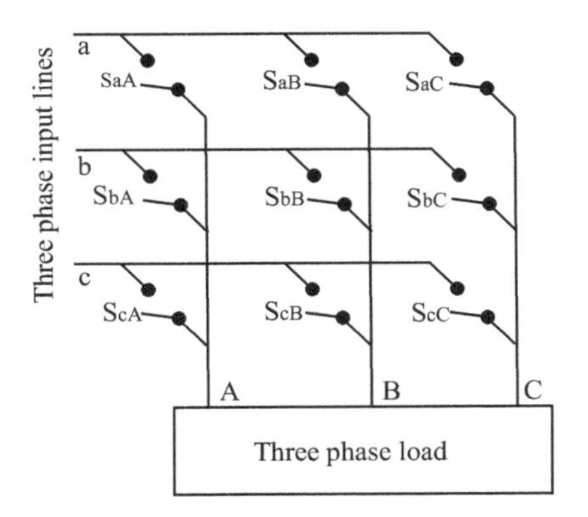

FIGURE 3.51 Direct matrix converter arranged in a matrix form.

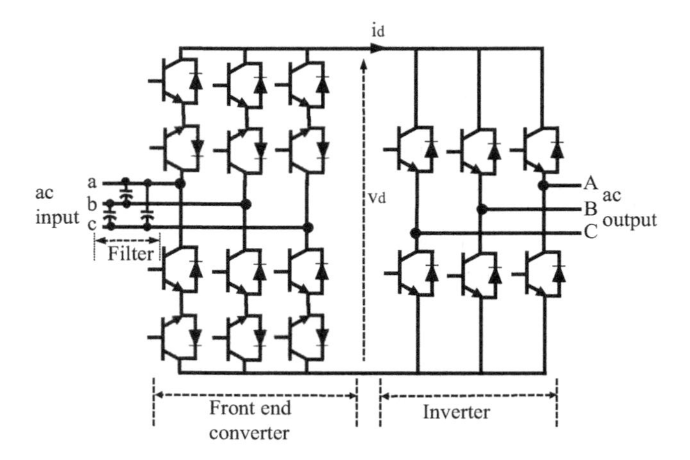

FIGURE 3.52 Indirect matrix converter.

is not constrained by the ac input frequency such as in the cycloconverters. This converter has the following advantages:

i. Pure sine output currents and voltages.
ii. Unity power factor with the control capability in the front end.
iii. Output frequency independent of input ac supply frequency.
iv. Four quadrant operation.
v. Standard PWM control techniques can be used with the inverter part of the converter system.
vi. Does not require a dc link filter.

The device count for this matrix converter is the same as that of the direct matrix converter is to be noted. The majority of the applications of the matrix converters are in aerospace applications and not usually in the MW range as in the case of cycloconverters.

3.4 APPLICATIONS

A large number of applications using power electronic converters have come into the market place over a period of nearly 60 years since the introduction of SCRs. They use standard converter circuits and topologies of which a fair amount of them has been discussed in the chapter. The control features are more specific to the applications and that is where the innovations continue to grow. But the development of new converter topologies continues to expand with heavy university-based research around the globe, but very few that emerge find applications has to be noted. Only a small number of high-volume applications are briefly described in the following sections and they may open up the possibilities for newer applications. Interested entrepreneurs are encouraged to make an extra effort to explore emerging applications for the power electronic converters and that may need newer topologies (much less) as well as suitable control techniques and features (very much).

3.4.1 Active Power Filter

The use of power electronic circuits generates nonlinearities with the result that harmonics are engendered in the power system loads, such as diode bridge or controlled rectifier fed loads, and draw source currents that are not pure sinusoids having a significant harmonic content. With ever-increasing demand on the utility such as the integration of renewable energy from wind and solar and catering to growth in customer base with differing demands has made the power system operation complex and susceptible to certain problems. Apart from the harmonics problem, there are frequency variations, reactive power demands, flickers, voltage unbalance, voltage distortion, voltage notching, current unbalance, etc., associated with the power system. Many of these problems are specifically made worse by the operation of the static power converters in the system. In the past, passive filters have been introduced to mitigate some of these problems. Compensation methods for these issues have been at the forefront for more than half a century.

The passive filter approach has the drawbacks of large size, sensitivity to impedance changes in the line due to load variations and resonance issues arising from them and difficulty in prefixing the filter frequencies in a system that is dynamically changing. The technical approach that is evolving, fully or partially in some cases, to avoid the passive power filters is through active power filter thanks to the emergence of power semiconductor devices and power converters with their demonstrated reliability in practice for a long time in various other applications. One of the most used power converters in active power filters is a three-phase H-bridge inverter.

There are four types of active filters as given in the following:

i. Shunt Active Filter: It is shown in Figure 3.53 with a load from a controlled rectifier connected to the utility sinusoidal source. It is highly desirable that the source current, i_{as}, is maintained close to a sinusoid. But the controlled rectification and resulting load current, i_{al}, is almost rectangular for 120° duration in each half cycle or even worse depending on the load as shown in Figure 3.47 in the operation of controlled rectifiers. Suppose the filter which is a three-phase inverter connected to a capacitor is connected to the ac lines as shown and it can be activated to draw the equivalent of load harmonic current, i_{af}. The result is that the source current is equal to the sum of the load and filter currents. From which the filter current is derived as the difference between the source and load currents and shown in Figure 3.54.

   The filter can be passive in which case it has to be tuned for various operating conditions without which the filter utilization can be neither optimal nor have a high quality of compensation. An active filter using a

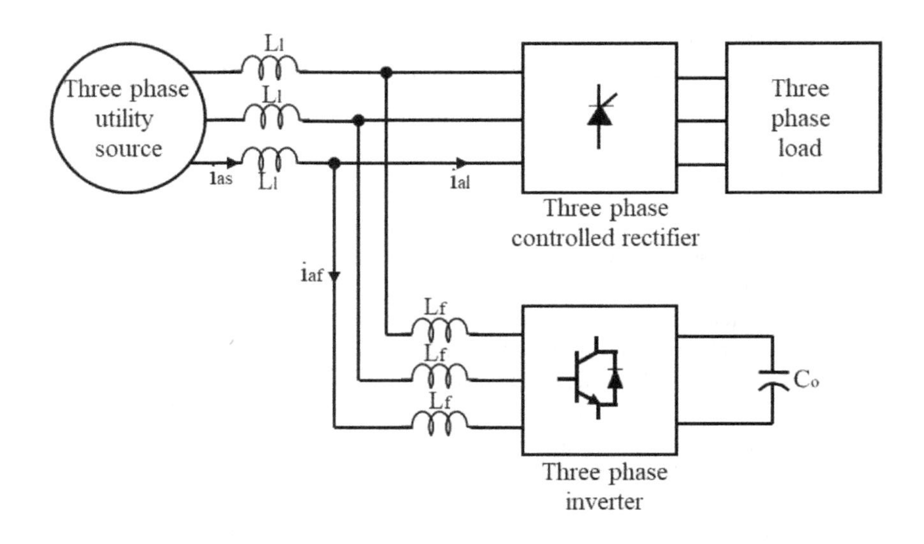

FIGURE 3.53 Shunt active power filter with a nonlinear load.

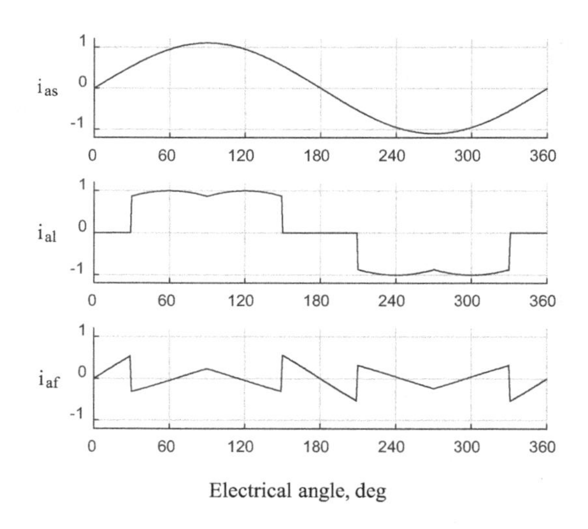

FIGURE 3.54 Active power filter illustration.

H-bridge three-phase converter can add or subtract the reactive current. The capacitor, C_o, of the inverter bridge has to have a higher voltage than the equivalent of the ac line voltages for it to send the current to the ac lines or load. That is where the filter inductances L_f in each line come into play as they can serve as a boost inductor with the inverter to get a higher desired voltage across the capacitor. As seen from the filter current requirement from Figure 3.54, the power requirement is zero for the inverter but requires only a small amount of power to meet the semiconductors' losses and losses in the passive elements, capacitors and inductors. But the reactive power requirement can be high depending on the level of required compensation and hence the volt-amp rating can be substantial is to be noted. A three-phase circuit implementation is shown in this illustration but a single-phase implementation is similar.

In the literature this type of active filter is also termed as a parallel active filter. Shunt active filters lend themselves to the paralleling of many and therefore, it is quite simple, elegant and practical to increase and upgrade the filter requirements at one location as the demand grows for them. Research and development have emerged to identify which active power filter is suitable for a particular compensation of the utility disturbances may be found in the references. Reactive power, current harmonic filtering, current unbalance and voltage flickers are the disturbances that shunt filters compensate in practice as they are inherently of current source types.

ii. Series Active Power Filter: It is shown in Figure 3.55 and it is directly in line with the utility and load lines, unlike the shunt configuration. This has the advantage of directly intervening in the line/load currents or voltages but also they have the disadvantage of having to stand the line values of the variables also. Note that the secondary phases of the transformer are embedded in series with the utility lines after the line filter inductors, L_l. They also require a transformer which increases the packaging size and cost too. This is preferred in certain line compensations similar to shunt active power filters. As the series power filter works like a voltage source, they lend themselves to compensating the voltage

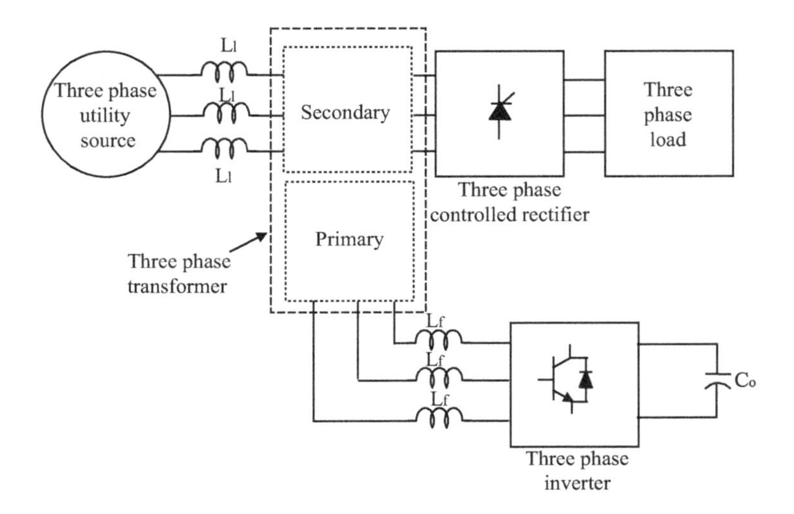

FIGURE 3.55 Series active power filter.

disturbances of any kind as well thus setting themselves apart from the shunt active power filters.

iii. Unified Power Quality Conditioner: This is a combination of both the shunt and series type to exploit the best of what both these types offer in compensating the disturbances in the power system. It gives the freedom to operate only one (shunt or series) or both active power filters in this arrangement, thus giving higher reliability compared to the system having only one of the active power filters.

iv. Hybrid Active Power Filter: It has either a shunt or a series active power filter together with a passive filter to enhance filtering capability at all times and to eliminate resonance in the system. Here again, the passive filter is always under operation whereas the active power filter can be made operational at will thus guaranteeing an additional degree of freedom in the control of the system.

3.4.2 Solar and Grid Interface

Photovoltaic (PV) panel output is connected to a regulating dc power supply so that the dc link voltage for inversion to line frequency ac via an inverter is available. In the case where it is not connected to the grid, the inverter stage is not necessary. When the solar power is intended for charging the battery in an uninterruptible power supply (UPS) system, then the first stage of dc voltage regulation stage alone is sufficient. The considered case is the solar panel-based energy extraction in individual installations such as in housetops, etc., where the power level is in hundreds of watts. In the case of solar farms where hundreds of acres are filled with solar panels, the system to extract energy and converting it to line frequency for supplying the utility grid involves similar blocks. But they have to be scaled to power levels of hundreds of MW also. In these cases, significant innovation to put in series the PV outputs to establish a dc link and then its inversion using multilevel inverter with harmonic control for tie-up with utility grid is achieved with most of the standard power converters but significant innovations happen in the stage of PV to regulated dc link. This particular field has opened significant opportunities for study and implementation.

3.4.3 Uninterruptible Power Supply (UPS)

UPS is a necessity wherever continuous power supply is required, whether it is ac or dc based. Many such applications are in hospitals, banks, commercial establishments, computers, public spaces such as libraries, railway stations, airports, etc., to name a few. The primary functions of the UPS are to provide a conditioned ac supply whenever utility power fails or even when it is functioning in some cases. The general topology for the UPS is as shown in Figure 3.56 that provides ac supply output with galvanic isolation using a transformer between the system and output, at line frequency. Batteries provide the stopgap power whenever the utility power fails and they need to be kept fully charged for that

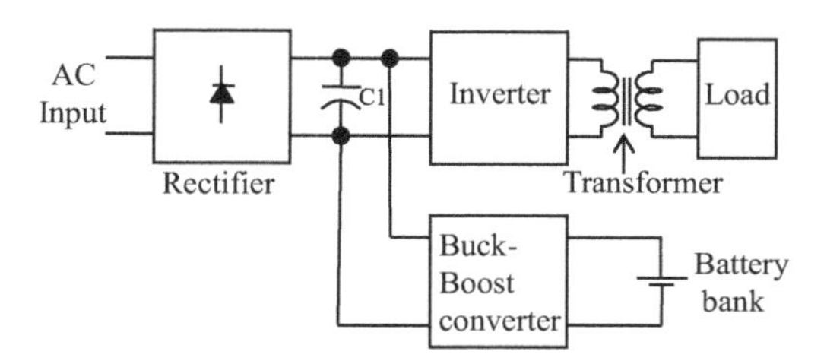

FIGURE 3.56 Uninterruptible power supply with line frequency isolation.

eventuality of power failure. That in turn requires a charging circuit from the utility which comprises of ac to dc rectification and conditioning of that output to match with the battery voltage during the charging mode using a buck type of switch-mode converter if the battery voltage is lower than that of the rectified dc voltage. When the battery steps in to supply power in the case of utility power failure, the converter is set to operate in the boost mode so that its output flowing from the battery to the dc link matches the previous dc link voltage. Then that dc link voltage is fed through an inverter (single or three phase as the case may be) to obtain the ac output at line frequency which is then isolated through a power transformer at line frequency. Therefore, this system uses utility power during the time it is on and uses battery power during the off-time of the utility supply. That means that this kind of UPS handles the entire power all the time when the utility power is on but with a decided advantage that there is no significant time to go from utility to battery power and vice versa. The power is uninterrupted in this case and hence the appropriate name UPS.

As the line frequency, transformer-based isolation is expensive and also takes a considerable packaging space a high-frequency isolation is deemed sufficient at least in low-power UPS. Such a system is shown in Figure 3.57. Between the buck-boost converter for the battery and the inverter, a high-frequency inverter, a high-frequency transformer and a high-frequency rectifier are introduced, which replaces the low- frequency isolation with a high-frequency isolation. The throughput efficiency is likely to be lower than that of the previous topology but the size of the package comes down, making it ideal for low-power applications where that is a dominant requirement.

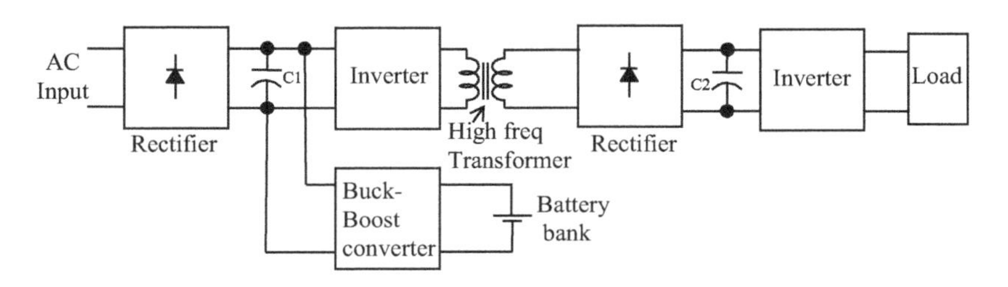

FIGURE 3.57 Uninterruptible power supply with high-frequency isolation.

3.4.4 DC Power Supplies

DC power supplies are in applications starting from a few watts to a few MW. The low-power applications are in consumer goods such as cellular phones, laptops, iPads and small battery chargers, etc., and are of high-volume category. They all use one or other forms of switch-mode converter circuit topologies and the innovations are more in making the circuits with newer devices and realizations at the chip level with the smallest packaging size and lowest cost. From a few tens of watts to MW dc power supplies, both uncontrolled and some controlled, are found in variable-speed motor drives. They use, say, diode rectifiers, or SCR converters or power factor correcting converters.

3.4.5 Motor Drives

Variable-speed dc and ac motor drives require various power converters most of which are standard types. The innovations have been mainly in the switched reluctance motor drives where the converter topologies have been tremendous and some may find applications in other fields as well. It is deferred to Chapter 6 on electric motor drives.

3.4.6 Future in Applications

Future power electronics applications can be obtained from various sources and they are:

i. University Research: To a smaller extent the insight as to future applications can be gained from this source as their emphasis is more involved with research than with applications per se. Further, the funding base (in USA) is mostly (nearly as much as 90%) federal agencies which require fundamental research focus than toward applications in general. This is very much in contrast to private company funding which requires current problems of interest to it be solved than on the futuristic applications.

ii. Emerging Applications in Certain Sectors: This is a better indicator of where the applications are of current interest but which are likely to go big in the market place. If that is the guiding tool to assess the future in applications, then grid-oriented applications including active power filters, grid interfaces to PVs, and most to do with the power system in overcoming the disturbances and unbalances in voltage and current and the emerging electric cars all point to a major market for power electronics applications.

iii. Entrepreneurship in Power Electronics: An additional insight into the near term and future market for applications to an extent is revealed by the entrepreneurship activities in this domain. Various techniques to handle issues related to present and futuristic systems are pursued in this arena but that does not mean that they all will be embraced by the market. As the success rate of entrepreneurship activities is low, the direction pointed out by them for future applications of power electronics has to be cautiously viewed. Not knowing which would succeed makes it important also to study these activities to be knowledgeable about the future applications that will go a long way to assist the engineers and their companies to position themselves in the exciting future and share in the prosperity. The current entrepreneurship effort in the form of startups is explored at length in the succeeding chapters.

3.5 EXERCISE PROBLEMS

1. Assess the qualitative and quantitative impacts of newer devices on dc power supplies and motor drives in terms of maximum converter output, voltage and switching frequency, losses, efficiency, packaging size and cost and in comparison to conventional MOSFET- and IGBT-based converters.
2. Discuss the input and output filter requirements for the following converter cases:
   i. Single-phase full-wave diode bridge rectifier
   ii. Three-phase diode rectifier
   iii. SCR controller rectifier
   iv. Switch mode dc power converters (choose one of them)

   Terms of study required to include all or any other specific parameters given below that are of interest to the class or individual projects:
   i. Harmonic contents
   ii. Cost implications
   iii. Size
   iv. Reliability
3. Compare the filter requirements of a three-phase inverter feeding the following loads at output power levels of 1, 10 and 100 KW:
   i. AC motor control
   ii. Uninterruptible power supply

   The parameters for comparison may be similar to that of Exercise Problem 1.
4. Develop the models for the simulation of the following using MATLAB:
   i. Buck converter with a passive load. Assume input voltage and desired output voltage for the simulation.
   ii. Buck converter with battery charging for a 48 V battery for an electric scooter from dc supply inputs of 160 and 350 V dc.
   iii. Buck-boost converter with a passive load where the input and output voltages for an application of your choice are considered.
   iv. Forward converter with an input voltage of 160 V dc and a desired output voltage of 3.3 V at 80 A. Select two sets of transformer ratios and evaluate the performance for both.

v. Push-pull converter with a passive load of your choice in input and output ratings. Assumptions to be stated clearly.
vi. Flyback converter with a passive load of your choice in input and output ratings.

5. Develop the models for simulation of the following (use may be made of the MATLAB programs given in the chapter):
   i. Single-phase inverter with passive load.
   ii. Three-phase inverter for utility/grid type load.
   iii. Three-phase controlled rectifier with input and output filters.
   iv. Half-bridge converter for a passive load with an input dc voltage of 350 V and desired output voltage of 12 V with a maximum current of 50 A. Select the transformer ratio and state all the assumptions made in the modeling and simulation.
   v. Full-bridge converter for a passive load with input voltage of 350 V and desired output voltage of 48 V and current of 200 A. Give the transformer ratio chosen and state all the assumptions made in the modeling and simulation.
6. Single-phase rectifier with power factor correction is to be achieved with a single switch. The circuit consists of a single-phase bridge diode rectifier connected to single-phase ac lines. The output of the rectifier is connected to a boost rectifier to provide a dc voltage feeding a load. (i) Draw the circuit in detail. (ii) Show how the power factor correction happens in this circuit and write a MATLAB simulation for the circuit operation. A resistive load for the output may be considered.
7. Research on the available standard packaged dc to dc power converters in the market. Choose one particular converter for the exercise. Compare similar products from leading companies and make your comments based on certain objective comparison criteria that you come up with.
8. Repeat the exercise for various other dc to dc converters studied in this chapter.
9. Repeat the same for various other inverters and multilevel converters at high power levels.
10. Make a study of multilevel topologies used in dc to dc power supplies by doing online research and submit a report with your comments.
11. Output filter designs and selection of capacitors at the input are very critical to contain the cost, maximize performance and packaging considerations. Given these factors, illustrate their designs for problems 4 and 5. Choose any one converter-related filter design in the problem set of 4 and 5.

3.6 CLASS PROJECTS

Project 1: Choose one of each that is used very much in your current work or planned work such as fast-acting diode, IGBT, MOSFET, SiC MOSFET, and GaN transistor from any of their manufacturers. Simulate them in a simple circuit such as a dc to dc converter of your choice in MATLAB or Simulink or any circuit simulation software you are familiar with to find the steady state operating waveforms of key variables such as device voltage, current, inductor current and voltage, output voltage, capacitor current and voltage and compute the system efficiency as a function of load. Individually list the computed device losses, inductor losses (note that the inductor has a coil which has resistance and accounts for that), and capacitor losses (due to its internal resistance).

Project 2: Let each student or a team choose one converter topology, investigate it and submit a report on various applications that use the specified converter topology and companies making products in that application market, their different approaches and their product specifications, and their significant differences and who dominates that market space and possible reasons for their dominance. The reports are to be graded.

Project 3: The students have to exchange their project reports so that other teams become conversant with the knowledge captured in that converter field and its industrial base and applications. They can be asked to improve the reports and demonstrate why their revision is significant and how insightful it is. The reports of the project are revised with the findings of project 2 and submitted for grading.

Project 4: After the process of exchanging projects of project 2 and their study is completed, the student teams have to choose one or two project reports only and then they can work in much detail on many aspects including manufacturability, marketability, IP base, etc., and then come up with how they can significantly improve these converter applications or converter topologies themselves or add or bring IP into this domain. The project study is then to be packaged and presented.

Project 5: Their improvements, if voted significant, are taken to the next stage with the teams contacting the companies in that domain and give them their study results. If the contributions are significant and self-standing such as in the business idea and IP creation, then the teams or interested members come together and can launch their own startup. The final project will capture the final report on how they will move forward and write a business plan learned in Chapter 2. This constitutes the final project report.

REFERENCES

General

1. R. Krishnan, *Electric Motor Drives*, Prentice Hall, Upper Saddle River, NJ, 2001 (also in Chinese, Greek translations, and various national editions).
2. N. Mohan, *Power Electronics: Converters, Applications and Design*, John Wiley & Sons, Hoboken, NJ, 1989.
3. M.H. Rashid (Editor in Chief), *Power Electronics Handbook*, Academic Publishing, 2001.

4. M.H. Rashid, *Power Electronic Circuits, Devices and Systems*, Prentice Hall, Upper Saddle River, NJ, 1993.
5. K. Heumann, *Basic Principles of Power Electronics*, Springer-Verlag, Berlin, 1985.
6. D.W. Hart, *Introduction to Power Electronics*, Prentice Hall, Upper Saddle River, NJ, 1997.
7. B. Wu, *High Power Converters and AC Drives*, Wiley InterScience, Hoboken, NJ, 2006.
8. D.G. Holmes and T.A. Lipo, *Pulse Width Modulation for Power Converters: Principles and Practice*, Wiley-Interscience, Hoboken, NJ, 2003.
9. P. Wood, *Switching Power Converters*, Van Nostrand Reinhold Company, New York, 1981.
10. R. Krishnan, *Permanent Magnet Synchronous and Brushless dc Motor Drives*, CRC Press, Baca Raton, FL, 2009 (also available in various national editions and Chinese edition).
11. R. Krishnan, *Switched Reluctance Motor Drives*, CRC Press, Baca Raton, FL, 2001 (available in various national editions).

Multilevel Converters-Survey Type Papers

1. H. Abu-Rub, J. Holtz and J. Rodriguez, "Medium-voltage multilevel converters: State of the art, challenges, and requirements in industrial applications", *IEEE Transactions on Industrial Electronics*, vol. 57, no. 8, pp. 2581–2596, 2010.
2. T.A. Meynard and H. Foch, "Multi-level conversion: High voltage choppers and voltage-source inverters", *Power Electronics Specialists Conference*, vol. 1, pp. 397–403, 1992.
3. T.A. Meynard and H. Foch, "Multilevel converters and derived topologies for high power conversion", *IEEE Industrial Electronics, Control, and Instrumentation*, vol. 1, pp. 21–26, 1995.
4. R. Teodorescu, F. Blaabjerg, J.K. Pedersen, E. Cengelci, S.U. Sulistijo, B.O. Woo and P. Enjeti, "Multilevel converters-a survey", *European Conference on Power Electronics and Applications*, 1999.
5. J. Rodriguez, J.S. Lai and F. Peng, "Multilevel inverters: A survey of topologies, controls and applications", *IEEE Transactions on Industry Applications*, vol. IA 49, pp. 724–738, 2002.

Wide Bandgap Devices

1. F. Medjdoub and K. Iniewkski (Editors), *Gallium Nitride (GaN): Physics, Devices and Technology*, CRC Press, Baca Raton, FL, 2016.
2. G. Meneghesso, M. Meneghini and E. Zanoni (Editors), *Gallium Nitride-Enabled High Frequency and High Efficiency Power Conversion*, Springer, Berlin, 2018.
3. H. Yu and T. Duan, *Gallium Nitride Power Devices*, Pan Stanford Publishing, Singapore, 2017.
4. J. Lutz, et al., *Semiconductor Power Devices*, Springer, Berlin, 2011.
5. B.J. Baliga, *Fundamentals of Power Semiconductor Devices*, Springer, Berlin, 2008.

Dc to dc Power Converters

1. M. Zehendner and M. Ulmann, *Power Topologies Handbook*, Texas Instruments, Dallas, TX, 2016.

2. F.L. Luo and H. Ye, *Advanced dc/dc Converters*, CRC Press, Baca Raton, FL, 2016.
3. M.K. Kazmierczuk, *Pulse-Width Modulated dc-dc Converters*, Wiley Edition, Hoboken, NJ, 2008.

Synchronous Rectifiers and dc to dc Converters

1. M.W. Smith and K. Owyang, "Improving the efficiency of low output voltage switchmode converters with synchronous rectification," *Proceeding of POWERCON* vol. 7, p. H-4, 1980.
2. C. Blake, D. Kinzer and P. Wood, "Synchronous rectifiers versus Schottky diodes: A comparison of the losses of a synchronous rectifier versus the losses of a Schottky diode rectifier," *Proceedings of* IEEE *Applied Power Electronics Conference*, 1994, pp. 17–23.
3. M.M. Jovanovic, M.T. Zhang and F.C. Lee, "Evaluation of synchronous-rectification efficiency improvement limits in forward converters," *IEEE Transactions on Industrial Electronics*, vol. 42, no. 4, pp. 387–395, 1995.
4. M.T. Zhang, "Electrical, thermal and EMI designs of high-density, low profile power supplies", Doctoral thesis, Virginia Tech, Blacksburg, VA, February 1997.
5. D.R. Sterk, "Compact isolated high frequency DC/DC converters using self-driven synchronous rectification", MS thesis, Virginia Tech., Blacksburg, VA, December 2003.
6. H. Jia, "Highly integrated dc-dc converters", Doctoral thesis, University of Central Florida, Orlando, FL, 2010.
7. R. Selders, Jr., "Synchronous rectification in high-performance power converter design", Application Notes, Texas Instruments, September 2016.

Neutral Point Clamped + Diode Clamped Multilevel Inverters

1. A. Nabae, I. Takahashi and H. Akagi, "A neutral-point clamped PWM inverter", *IEEE Transactions on Industry Applications*, vol. IA-17, pp. 518–523, September/October 1981.
2. P.M. Bhagwat and V.R. Stefanovic, "Generalized structure of a multilevel PWM inverter", *IEEE Transactions on Industry Applications*, vol. IA-19, pp. 1057–1069, 1983.
3. Z. Pan, F. Peng, V. Stefanovic and M. Leuthen, "A diode-clamped multilevel converter with reduced number of clamping diodes", *Applied Power Electronics Conference and Exposition, 2004. APEC '04. Nineteenth Annual IEEE*, Anaheim, CA, vol. 2, pp. 820–824, 2004.
4. A. Bendre, S. Krstic, J. Vander Meer and G. Venkataramanan, "Comparative evaluation of modulation algorithms for neutral-point-clamped converters", *IEEE Transactions on Industry Applications*, vol. 41, no. 2, pp. 634–643, 2005.
5. X. Yuan and I. Barbi, "Fundamentals of a new diode clamping multilevel inverter", *IEEE Transactions on Power Electronics*, vol. 15, no. 4, pp. 711–718, 2000.
6. N. Celanovic and D. Boroyevich, "A comprehensive study of neutral point voltage balancing problem in three level neutral-point-clamped voltage source PWM inverters", *IEEE Transactions on Power Electronics*, vol. 15, pp. 242–249, 2000.
7. S. Ogasawara and H. Akagi, "Analysis of variation of neutral point potential in neutral-point-clamped voltage source PWM inverters", *IEEE Industry Applications Conference*, vol. 2, pp. 965–970, 1993.

8. H.T. Mouton, "Natural balancing of three-level neutral-point-clamped PWM inverters", *IEEE Transactions on Industrial Electronics*, vol. 49, pp. 1017–1025, 2002.
9. M. Marchesoni and P. Tenca, "Diode-clamped multilevel converters: A practical way to balance DC-link voltages", *IEEE Transactions on Industrial Electronics*, vol. 49, pp. 752–765, 2002.
10. A.K. Gupta and A.M. Khambadkone, "A simple space vector PWM scheme to operate a three-level NPC inverter at high modulation index including over modulation region, with neutral point balancing", *IEEE Transactions on Industrial Electronics*, vol. 43, no. 3, pp. 751–760, 2007.

Flying Capacitor Multilevel Inverter:

1. B.P. McGrath and D.G. Holmes, "Analytical modeling of voltage balance dynamics for a flying capacitor multilevel converter", *Conference Record of Power Electronics Specialists*, Orlando, FL, pp. 1810–1816, June 2007.
2. O. Sivkov, " Five-level inverter with flying capacitors". Doctoral thesis, Czech Technical University, Prague, 2011.
3. C. Feng, J. Liang and V.G. Agelidis, "A novel voltage balancing control method for flying capacitor multilevel converters", *IEEE Industrial Electronics Society Conference*, vol. 2, no. 2–6, pp. 1179–1184, 2003.
4. L. Xu and V.G. Agelidis, "Active capacitor voltage control of flying capacitor multilevel converters", *IEE Proceedings-Electric Power Applications*, vol. 151, no. 3, pp. 313–320, 2004.
5. P. Kobrle, "Control Strategy of five level flying capacitor inverter", Doctoral thesis, Czech Technical University, Prague, February 2014.
6. G. Beinhold, R. Jacob and M. Nahrstaedt, "A new range of medium voltage multilevel inverter drives with floating capacitor technology", *Proceeding of EPE*, Graz, Austria, 2001.

Cascade H Bridge Multilevel Inverter

1. Y. Suresh, "Investigation on cascade multilevel inverter for medium and high-power applications", Doctoral thesis, NIT Rourkela, India, 2012.
2. J. Alan, "Modulation techniques for the cascaded H-bridge multi-level converter", Doctoral thesis, University of Nottingham, 2012.
3. R. Gupta, A. Ghosh and A. Joshi, "Switching characterization of cascaded multilevel-inverter-controlled systems", *IEEE Transactions on Industrial Electronics*, vol. 55, no. 3, pp. 1047–1058, 2008.
4. M. Malinowski, K. Gopakumar, J. Rodriguez and M.A. Perez, "A survey on cascaded multilevel inverters", *IEEE Transactions on Industrial Electronics*, vol. 57, no. 7, pp. 2197–2206, 2010.
5. A. Gupta and A. Khambadkone, "A space vector modulation scheme to reduce common mode voltage for cascaded multilevel inverters", *IEEE Transactions on Power Electronics*, vol. 22, no. 5, pp. 1672–1681, 2007.
6. R. Gupta, A. Ghosh, and A. Joshi, "Switching characterization of cascaded multilevel-inverter-controlled systems", *IEEE Transactions on Industrial Electronics*, vol. 55, no. 3, pp. 1047–1058, 2008.

Comparison of some MLC

1. S.S. Fazel, "Investigation and comparison of multi-level converters for medium voltage applications", Ph.D. dissertation, Technische Universität Berlin, Berlin, Germany, 2007.
2. C. Sourkounis and A. Al-Diab, "A comprehensive analysis and comparison between multilevel space-vector modulation and multilevel carrier-based PWM", *Proceedings of EPE-PEMC*, Poznan, Poland, pp. 1710–1715, 2008.

Modeling, Control and Applications of MLCs

1. T.A. Meynard, M. Fadel and N. Aouda, "Modeling of multilevel converters", *IEEE Transactions on Industrial Electronics*, vol. 44, no. 3, pp. 356–364, 1997.
2. T.V. Thai, "Multilevel inverters for high voltage drives", Doctoral thesis, Czech Technical University, Prague, 2003.
3. J.I.L. Galvan, "Multilevel converters: Topologies, modeling, space vector modulation techniques and optimizations", Doctoral thesis, University of Seville, Spain, 2006.
4. M. Fracchia, T. Ghiara, M. Marchesoni and M. Mazzucchelli, "Optimized modulation techniques for the generalized N-level converter", *Proceeding of IEEE Power Electronics Specialists Conference*, 1992, vol. 2, pp. 1205–1213.
5. J. Holtz and N. Oikonomou, "Neutral point potential balancing algorithm at low modulation index for three-level inverter medium voltage drives", *IEEE Transactions on Industrial Electronics*, vol. 43, no. 3, pp. 761–768, 2007.
6. A. BenAbdelghani, C.A. Martins, X. Roboam and T.A. Meynard, "Use of extra degrees of freedom in multilevel drives", *IEEE Transactions on Industrial Electronics*, vol. 49, no. 5, pp. 965–977, 2002.
7. J.K. Steinke, "Switching frequency optimal PWM control of a three level inverter", *IEEE Transactions on Power Electronics*, vol. 7, no. 3, pp. 487–496, 1992.
8. P.N. Enjeti and R. Jakkli, "Optimal power control strategies for neutral point clamped (NPC) inverter topology", *IEEE Transactions on Industrial Electronics*, vol. 28, no. 3, pp. 558–566, 1992.

Cycloconverter

1. B.R. Pelly, *Thyristor Phase-Controlled Converters and Cycloconverters*, Wiley, New York, 1971.
2. L. Gjugyi and B.R. Pelly, *Static Power Frequency Changers*, Wiley, New York, 1976.
3. M. Ishida, M. Iwasaki, S. Okuma and K. Iwata, "Waveform control of PWM cycloconverters with sinusoidal input current, sinusoidal output voltage and variable input displacement factor", *Electrical Engineering in Japan*, vol. 107, no. 3, pp. 95–103, 1987.

Matrix Converters

1. V. Jones and B. Bose, "A frequency step-up cycloconverter using power transistors in inverse-series mode", *International Journal of Electronics*, vol. 41, no.6, pp. 573–587, 1976.

2. M. Venturini, "A new sine in sine out, conversion technique which eliminates reactive components", *Proceedings Powercon 7*, San Diego, CA, pp. E3_1–E3_15, 1980.
3. A. Alesina and M. Venturini, "The generalised transformer: A new bidirectional sinusoidal waveform frequency converter with continuous variable adjustable input power factor", *Proceedings of PESC Conference Record,* pp. 242–252, 1980.
4. A. Alesina and M. Venturini, "Solid-state power conversion: A Fourier analysis approach to generalised transformer synthesis", *IEEE Transaction on Circuits and Systems*, vol. cas28, no. 4, pp. 319–330, 1981.
5. P.D. Ziogas, S.I. Khan and M.H. Rashid, "Some improved force commutated cycloconverter structures", *IEEE Transactions on Industry Applications*, vol. 1A-21, no. 5, pp. 1242–1253, 1985.
6. A. Alesina and M. Venturini, "Analysis and design of optimum-amplitude nine-switch direct ac-ac converters," *IEEE Transactions on Power Electronics*, vol. 4, no. 1, pp. 101–112, 1989.
7. L. Huber, D. Borojevic and N. Burany, "Analysis, design and implementation of the space vector modulation for forced commutated cycloconverters", *IEE Proceedings Part B*, vol. 139, no. 2, pp. 103–112, 1992.
8. D. Casadei, G. Grandi, G. Serra and A. Tani, "Space vector control of matrix converters with unity input power factor and sinusoidal input/output waveforms," *Fifth European Conference on Power Electronics and Applications*, vol. 7, pp. 170–175, September 1993.
9. P. Wheeler, "Matrix converter for variable speed ac motor drives", Ph.D. thesis, University of Bristol, September 1993.
10. M. Ziegler and W. Hofmann, "Semi natural two steps commutation strategy for matrix converters," *PESC 98 Record, 29th Annual IEEE Power Electronics Specialists Conference*, vol. 1, pp. 727–731, 1998.
11. J. Kolar, F. Schafmeister, S. Round and H. Ertl, "Novel three-phase ACAC sparse matrix converters," *IEEE Transactions on Power Electronics*, vol. 22, no. 5, pp. 1649–1661, September 2007.
12. D. Casadei, G. Serra, A. Tani and L. Zarri, "Matrix converter modulation strategies: A new general approach based on space-vector representation of the switch state," *IEEE Transactions on Industrial Electronics*, vol. 49, no. 2, pp. 370–381, 2002.
13. M.N. Popat, Simulation and analysis of matrix converter, M.Tech. thesis, IIT Roorkee, India, June 2006.
14. L. Zarri, Control of matrix converters, Ph.D. thesis, University of Bologna, 2007.
15. J. Wang, B. Wu, D. Xu and N. Zargari, "Multi-modular matrix converters with sinusoidal input and output waveforms," *IEEE Transactions on Industrial Electronics*, vol. PP, no. 99, p. 1, 2011.
16. K. Kobravi, "Modulation and control of matrix converter for aerospace application", Ph.D. thesis, University of Toronto, Canada, 2012.

Active Power Filters

1. L. Gyugi, and E.C. Strycula, "Active ac power filters", *Proceedings of IEEE IAS Annual Meeting*, pp. 529–535, 1976.
2. N. Mohan, H.A. Peterson, W.F. Long, G.R. Dreifuerst and J.J. Vithayathil, "Active filters for AC harmonic suppression," *In Proceedings of IEEE-PES Winter Meeting*, pp. 168–174, 1977.
3. J. Uceda, F. Aldana and P. Martinez, "Active filters for static power converters", *IEEE Proceedings B*, vol. 130, no. 5, pp. 347–354, 1983.

4. H. Akagi, A. Nabae and S. Atoh, "Control strategy of active power filters using multiple voltage-source PWM converters," *IEEE Transactions on Industrial Electronics*, vol. IA-22, pp. 460–465, 1986.
5. M. El-Habrouk, M.K. Darwish and P. Mehta, "Active power filters: A review", *IEE Proceedings Electric Power Applications*, vol. 147, no. 5, pp. 403–413, 2000.
6. L.A. Morán, J.W. Dixon, J.R. Espinoza, R.R. Wallace, "Using active power filters to improve power quality", *IEEE Transactions on Power Electronics*, vol. 40, pp. 232–241, 2004.
7. S.C. Prasad and D.K. Khatod, "A review of selection and usage of modern active power filter", *International Journal of Engineering Trends and Technology (IJETT)*, vol. 20, no.2, pp. 109–114, 2015.
8. A. Martins, J. Ferreira and H. Azevedo, "Power quality", Ch 9. In: *Active Power Filters for Harmonic Elimination and Power Quality Improvement*, A. Eberhard (Editor). InTech, London, 2011.
9. A. Kumar, H. Tiwari and P. Anjana, "Review of active power filters for improvement of power quality", *INROADS (International Conference IAET-2016 Special Issue)*, vol. 5, no. 1, pp. 135–144, 2016.
10. H. Akagi et al., *Instantaneous Power Theory and Applications to Power Conditioning*, Wiley-IEEE Press, Hoboken, NJ, 2017.

4 Entrepreneurship in Power Semiconductor Devices

4.1 INTRODUCTION

Company startups in gallium nitride (hereafter referred to as GaN) and silicon carbide (hereafter referred to as SiC) power semiconductor devices since the turn of this century are considered for the study. A description of GaN and SiC devices, their structures, operational characteristics, distinct features of various devices within these categories and targeted applications is given complementing their brief introduction in Chapter 3. Existing and evolving markets and state of the research efforts in these devices are given ahead of exploring the current startup activities in them. Then the study of startups in both the devices is presented with mostly examples from USA. Taken together these startups constitute the ecosystem in their domains and their distinct individuality comes and evolves with their selective pursuits in technology, products, markets, inventions and intellectual properties, all engineered by founders and the management teams, investment and funding sources, collaborations with other entities, successful exits, and/or continued growth. These factors and the role of the founders and their backgrounds together with their top management team constitute the heart, soul and body of the startups and they are all presented in relevant details. Then general observations based on the presented information offer the readers with additional points to reflect, review and make their own assessment in regard to the particular startup in the devices domain but also lessons that could be learned for application to entrepreneurship in other domains as well. Twenty startups are considered, thirteen of them in GaN devices, six in SiC devices and one in the device controller.

4.2 OBJECTIVES OF THE STUDY

A framework for the study of entrepreneurship is presented in Chapter 1. The same framework is applied for entrepreneurship activities as related to startups in the area of power semiconductor devices and further identify various key factors that may help in one's own understanding for possible future efforts in this or other areas. The startups studied are mostly USA based and together with a small number from countries in Europe and Asia. The general basis of the study can be applied to startups in a particular region and country of choice before embarking on entrepreneurial activities there. Expected and desired outcomes of such learning in addition to the methodology-based evaluation outcomes are:

i. Learn about the factors that are related to academic research, industrial thrust and emerging newer applications of the technologies, contributing to the rise of startups in power semiconductor devices in these times.
ii. Map and draw lessons from the educational, professional and business backgrounds of the founders of these startups and any commonality very specific to the startups' product domains.
iii. Identify, if any, specific funding patterns for these startups, and the role of investors and venture capital firms in this domain.
iv. Influence of location of the startups and the factors contributing to it.
v. Learn of their tactics and strategies for success in the usual sense, specifically how they address the markets and position their products, and level of their activity with regard to the current customer base and future projections.
vi. Understanding the vital role of their communication methods as they are at the heart of marketing and how the startups address by using very many ways of reaching engineers and business decision-makers through IEEE conferences, trade conferences, trade journals and their own websites.
vii. Looking into any special factors, apart from methodology-based factors, contributing to and/or influencing the success of the startups and likewise any significant changes in their organization and business.

A summary of the salient features of the entrepreneurship activities in GaN and SiC devices subdomains is presented. The summary is intended to open up discussion and reflection among the readers so that they can effectively apply some of the lessons to their own present and/or possible future entrepreneurial efforts.

4.3 SEMICONDUCTOR POWER DEVICES

Two distinct sets of semiconductor power devices in a wide-bandgap (WBG) sector, silicon carbide and gallium nitride based, have been in the study, development and limited applications arena for more than 20 years. A brief description of these devices has been given in Chapter 3 with their currently available highest voltage and current ratings for these devices and possible market opportunities. These devices, their structures, and targeted applications are given in more detail to get a broader engineering introduction. A brief narrative of ongoing efforts from academic research and industrial development is put in place to shed light on the basis for evolving entrepreneurship activities. Their applications and market size and value are presented that could be of vital importance to potential entrepreneurs and employees, present and future as well, in this area.

4.3.1 GaN Devices

This section has a brief introduction to WBG materials and their distinct features over silicon material to appreciate their strengths and for providing the basis for their invariable emergence in device domains. Then different types of GaN devices are described to understand the spectrum in which their play emerges. They are depletion mode (D mode) FETs, enhancement-mode FETs and Cascode with D mode GaN devices. Their operation and characteristics from existing GaN devices are presented as well as their presently targeted applications. Applications cannot be confined to what is described and emerged so far, but many more will evolve in the future with the new demands on existing or of newer systems in industrial and other domains.

4.3.1.1 Wide-Bandgap Materials

Silicon semiconductor material has been dominant in power devices for a long time. These devices are reaching their upper limits of operation and breakthrough on these limitations are yet to happen in research and development. One of the alternatives open is to look at different materials, and two materials namely, Silicon Carbide (SiC) and Gallium Nitride (GaN), have emerged with desirable characteristics in power devices. Their properties along with that of silicon are given in Table 4.1 for comparison.

SiC and GaN have higher bandgap and hence termed as WBG materials. They require higher energy for moving the electron from the valence band to the conduction band compared to silicon. This feature reduces leakage currents which is a benefit. Note that the bandgap is not large enough that it becomes a pure insulator also. WBGs have higher breakdown voltage as much as 7–12 times more than silicon enabling as many times reduction of the drift layer in devices making them very small compared to silicon. Because of higher breakdown voltage, WBG devices have lower on-state resistance with GaN having the lowest on-state resistance among them. SiC has the highest thermal conductivity whereas GaN is comparable to Si material. The GaN devices have been achieved since the early 1990s, but their research, product development and market realization are still ongoing. SiC devices have been pursued since the mid-1980s, and their applications in the higher voltage range have been slowly progressing.

TABLE 4.1
Material Characteristics

Characteristic	Silicon	4H-Silicon Carbide	Gallium Nitride
Bandgap (eV)	1.12	3.26	3.44
Electric breakdown field (MV/cm)	0.3	2.0	3.8
Electron mobility (cm^2/V-s)	1,500	1,190 (c axis)	2,000 (2DEG)
Saturated electron drift velocity × 10^7 (cm/s)	1.0	2.0	2.5
Thermal conductivity (W/cm-K)	1.5	4.9	1.3

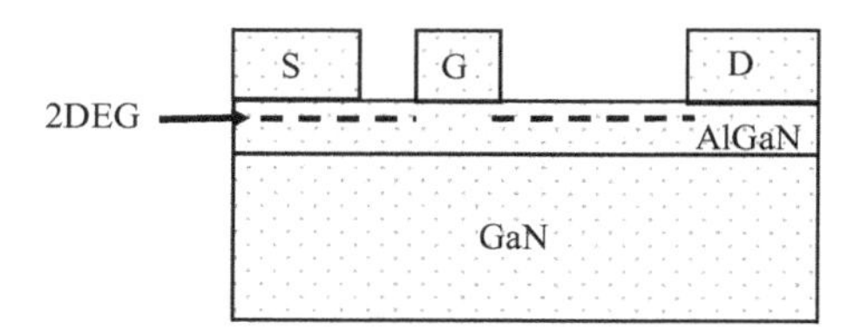

FIGURE 4.1 Basic GaN FET structure.

4.3.1.2 Depletion Mode GaN

A high electron mobility transistor (HEMT) and also known as GaN FET, shown in its basic lateral structure in Figure 4.1 is considered. A very thin layer of aluminum GaN (AlGaN) is grown on top of GaN material. The stress of AlGaN creates a path for high electron mobility between the source (S) and drain (D) when the gate (G) is at zero potential as the threshold of this material combination is zero or slightly negative voltage and a voltage is applied between the drain and source with the drain being positive than the source. The flow of electrons is on the surface and it is termed as two-dimensional electron gas (2DEG). As the source and drain are on the 2DEG itself, it serves to short them with the result that there is a current flow between them. In order to turn off the current from drain to source, the gate voltage is made negative in comparison to both source as well as drain with the result that electrons are depleted and conduction of drain to source current ceases. This type of HEMT is termed as Depletion mode GaN FET. This device is Normally ON at the time of the starting with no bias on the gate and that is not desirable in power conversion applications. In order to keep it nonconducting at the time of circuit startup, a negative voltage has to be applied to the gate and such a constraint in its use is not favorably looked upon also. Users of the power devices are quite used to operation with normally-off condition guaranteed by the silicon device at the time of starting of the circuit/application. Two approaches overcome this disadvantage in the D-mode GaN FET and they are described in the following sections.

4.3.1.3 Enhancement Mode GaN FET Structure and Its Operational Characteristics

There are many techniques to make the device to be normally off and one of them is to recess the gate terminal into the AlGaN layer, shown schematically in a lateral structure in Figure 4.2. The voltage generated by the junction between GaN and AlGaN is lower than the built-in voltage of the gate metal and that results in the depletion of electrons in 2DEG laterally resulting in nonconducting state with zero input gate voltage. Only when the gate is given a positive voltage over the threshold around +1.5 V or so, the device becomes conducting. Efficient Power Conversion Corporation's proprietary enhancement-mode GaN in lateral structure is shown in Figure 4.3 on silicon substrate resulting in low cost with the standard fabrication techniques currently in use and the references give

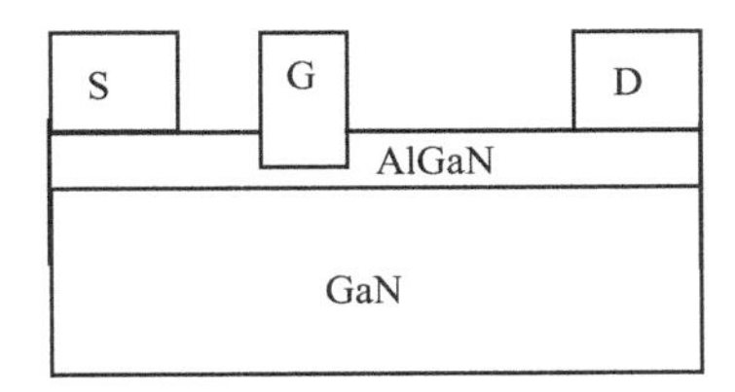

FIGURE 4.2 Enhancement (or E-mode) GaN FET structure.

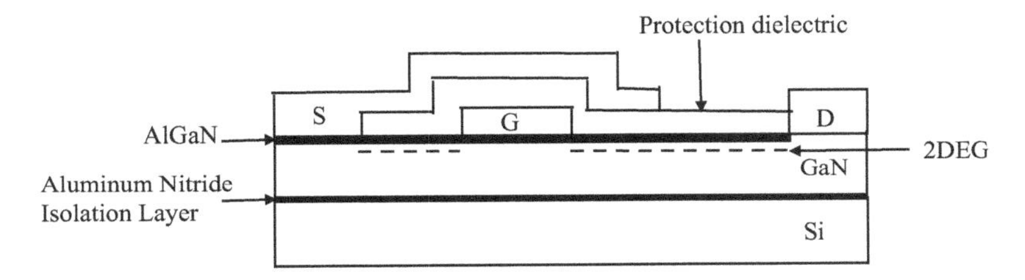

FIGURE 4.3 EPC's E-mode GaN FET lateral structure. (Courtesy of EPC Corp.)

further information. Vertical structure with its electron flow along the vertical direction with a higher drift region gives a higher voltage capability to the GaN devices and they are not common yet among most of the startups' efforts. It is construction-wise different from the lateral structure, but the principle of operation is the same.

Typical operating characteristics of e-mode devices may be garnered from datasheets provided by the manufacturers. One such from GAN Systems' device rated for 650 V, 22.5 A (max) with drain-source resistance of 67 mΩ is considered here. Drain to source current (i_{DS}) vs. drain to source voltage (v_{DS}) for various gate-source voltages for two junction temperatures of 25°C and 150°C are shown in Figure 4.4a and b, respectively. Note that the drain-source voltage drop during conduction becomes significant but less than that of the silicon devices in comparison. The reverse conduction characteristics are given in Figure 4.5 for the same device under discussion. Note that it is plotted as though it is in the first quadrant but notice that it is a plot of the source to drain current (i_{SD}) vs. source-drain voltage (v_{SD}) which is the negative of i_{DS} and v_{DS}, respectively. In the latter parameters, the graph will appear in the third quadrant and hence may be easier to comprehend the reverse conduction. Accordingly, the entire I_{DS} and V_{DS} characteristics for a typical e-mode GaN transistor are represented from these figures and shown in Figure 4.6. Note that, neither drain to source current nor drain to source voltages or threshold gate-source voltage (V_{sgt}) is given but the latter's value is around 1.3-1.5 V. The forward characteristics are straight forward whereas the reverse conduction has the feature that the source to drain voltage drop increases. It is due to (i) drain to source resistance is higher in reverse mode

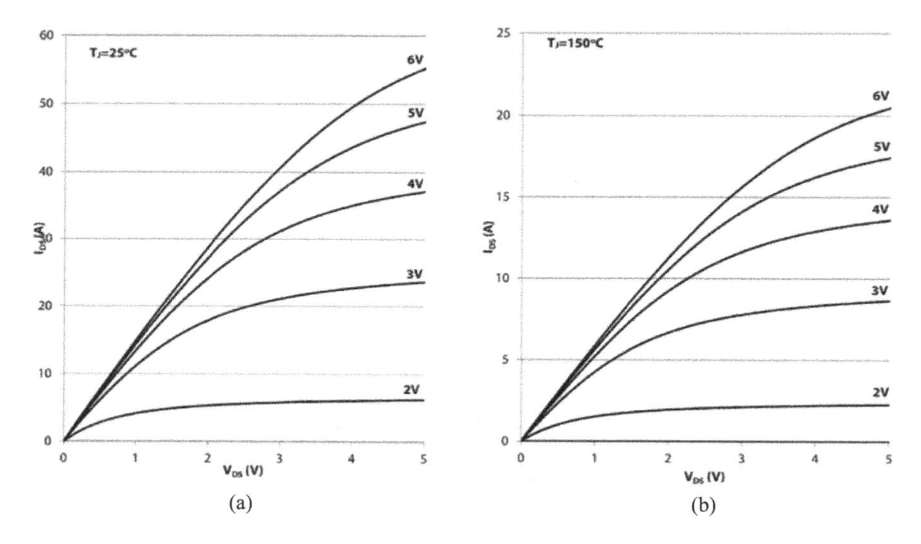

FIGURE 4.4 Typical i_{DS} vs. v_{DS} for two junction temperatures: (a) T_j = 25°C, and (b) T_j = 150°C - (G566506 T, 650 V, 22.5 A, 67 mΩ E-mode GaN transistor). (Courtesy of GAN Systems.)

than in the forward mode, and (ii) the difference between the gate to drain threshold voltage (which is closer to threshold gate to source voltage) and gate to source voltage adds up with the source to drain resistive voltage drop. This is not usually sustainable for more than a short time because of higher conduction losses than in the forward conduction mode.

4.3.1.4 Cascode (with the D-Mode GaN) Structure and Its Operational Characteristics

The typical characteristics of a depletion mode may be made similar to the enhancement mode GaN transistor but with the gate signal value different from it as shown in Figure 4.7. V_{gst} is the threshold voltage of the gate-source signal, which can normally be −12 V. For that i_{DS} vs. v_{DS} characteristics shown in the Figure 4.7 have the parameter x with a value of 2 V. For actual products in the market, they can be different.

In order to block the normally-on condition of the D-mode GaN, a low-voltage, say 30 V, silicon MOSFET is connected in series with the GaN FET. The connection for an NMOS device is the drain, D of the MOSFET connected to the source of the GaN FET, and the source of the MOSFET is connected to gate, G of the GaN FET and is available as terminal S of the complete cascaded device, as shown in Figure 4.8. The drain of the entire configuration is that of the GaN FET itself and named D. The final (total) device has terminals D, G and S as shown in Figure 4.8.

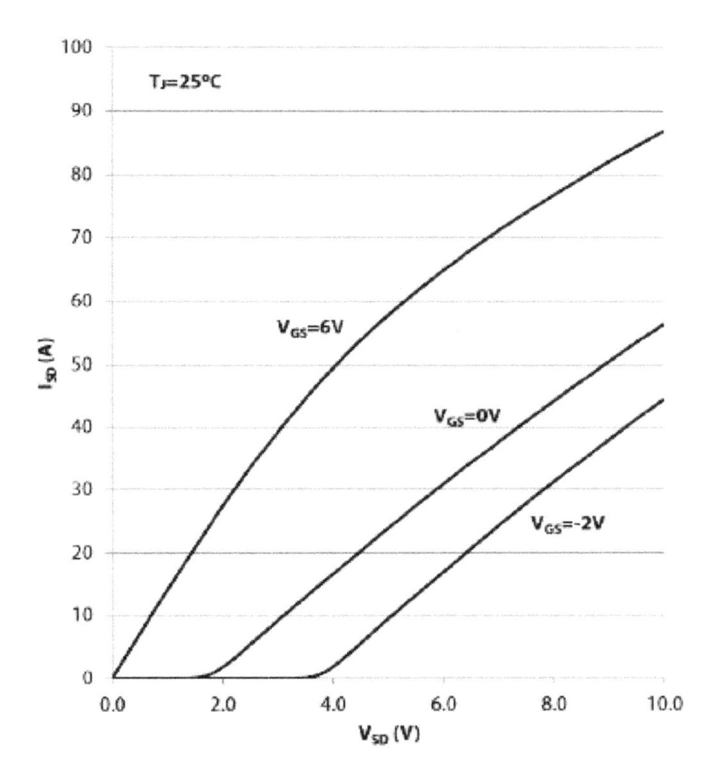

FIGURE 4.5 Reverse conduction characteristics at $T_j = 25°C$ (G566506 T, 650 V, 22.5 A, 67 mΩ E-mode GaN transistor). (Courtesy of GAN Systems.)

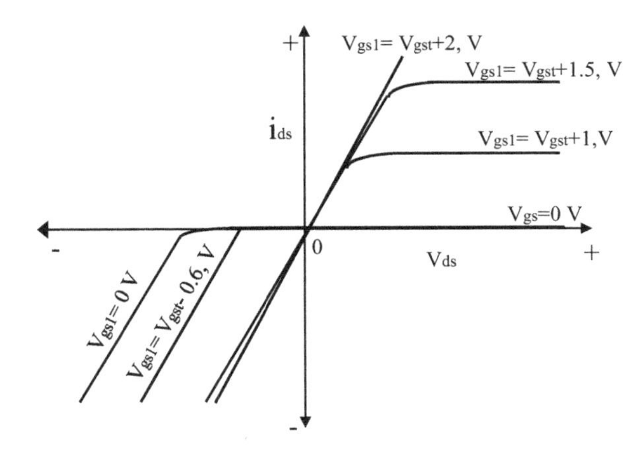

FIGURE 4.6 Typical conduction (both forward and reverse) characteristics of enhancement-mode GaN transistor.

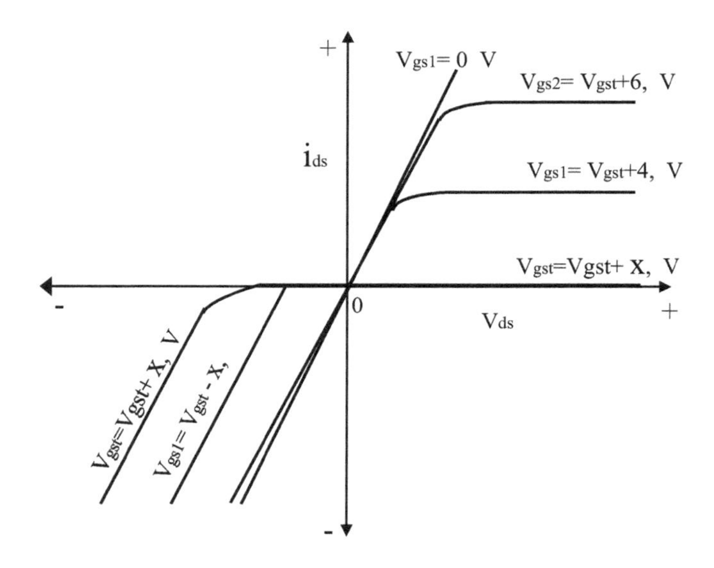

FIGURE 4.7 Typical i_{DS} vs. v_{DS} of depletion-mode GaN transistor.

The cascode's drain-source current vs. drain-source voltage characteristics, for a 650V, 15A cascade GaN from the startup, Transphorm are shown in Figure 4.9 for two different junction temperatures (T_j) of 25°C and 150°C. The reverse conduction characteristics of a generalized typical cascode GaN transistor are given in Figure 4.10. A noticeable difference is seen from earlier characteristics in the values of the gate-source voltages and also the reverse conduction characteristics. The voltage drop across the drain source is not high compared to an e-mode GaN device and that is a significant advantage in cascode devices. More of the cascade device operation is given in the following.

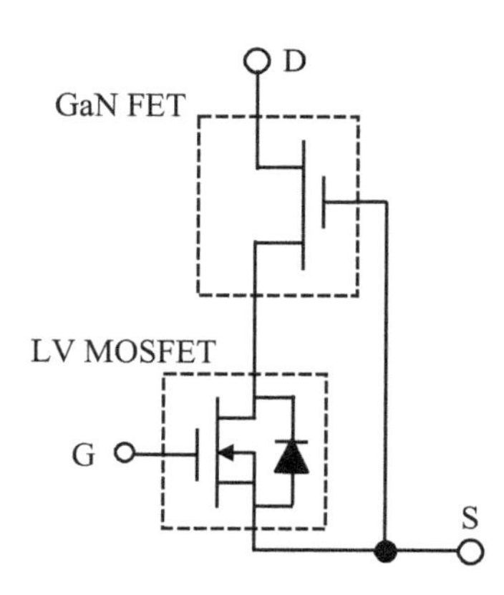

FIGURE 4.8 Cascode configuration of a D-mode GaN FET for normally-off operation.

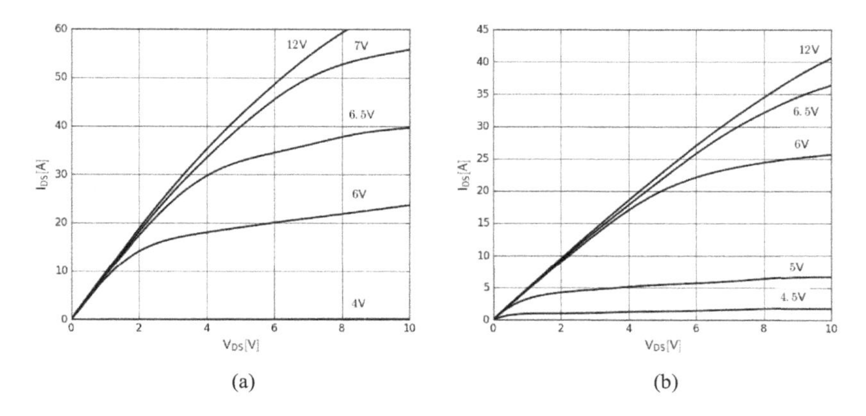

FIGURE 4.9 Typical drain-source current vs. drain-source voltage vs. gate-source voltages at two different junction temperatures for a 600 V, 15 A cascode device. (a) $T_j = 25°C$ and (b) $T_j = 150°C$. (Courtesy of Transphorm Company.)

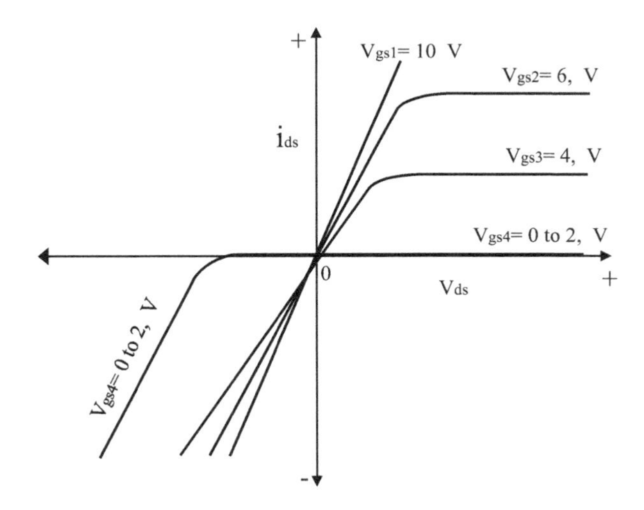

FIGURE 4.10 Typical i_{DS} vs. v_{DS} characteristics of a cascade GaN transistor.

It has three operational modes consisting of normally-off, normally-on and reverse conduction and they are described in the following:

a. Normally-Off Mode: The voltage across the GaN FET's gate and the source is equal to the voltage across the source and drain of the MOSFET. When a voltage is applied across D and S of the total device and at the time of the starting, let the gate G is at zero voltage. But the MOSFET will not conduct with the result that voltage across its drain and source increases and the negative of that voltage then is applied to the gate of the GaN FET resulting in blocking the conduction in GaN FET. This operational feature provides the normally-off characteristic to the D-mode GaN FET.

b. On Mode: For current conduction in the forward mode, i.e., from drain to source, a positive voltage of 10 V is applied to the gate of the MOSFET, as shown in Figure 4.11a, which drives its drain-source voltage to a small value equal to its conduction drop. Negative of that voltage appears across the gate source of the GaN device and note that a small negative gate voltage is sufficient to enable its conduction. During the conduction of the devices, the total device's drain-source voltage drop equals the sum of the drain-source voltage drops of GaN FET and MOSFET.
c. Reverse Current Conduction Modes: The GaN FET does not have a body diode since it does not have a parasitic bipolar junction. In the cascode structure, the device conducts in the reverse direction when the drain is negative with respect to the source in series with the body diode of the MOSFET in conduction. There are two modes for that to happen with the voltage across the gate source of the MOSFET at 10 and 0 V, let them be termed as mode (ii) and mode (iii), respectively. When the gate-source voltage of MOSFET is at 10 V, the device is in the region mirroring the first quadrant or positive drain-source current vs. drain-source voltage operational region into its third quadrant, shown in Figure 4.11b. When the gate-source voltage is zero and the drain is negative compared to the gate, the MOSFET is off, but the body diode of the MOSFET takes over the conduction of the reverse current and the GaN part will have the reverse current, as shown in Figure 4.11c. The voltage drop from the complete drain to source in the device is the sum of the body

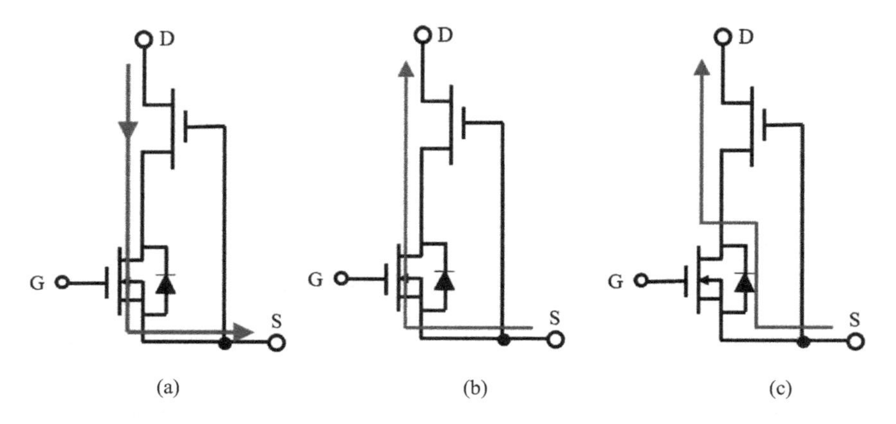

FIGURE 4.11 Cascode GaN FET conduction modes. (a) V_{gs} = 10 V, V_{DS} = positive, (b) V_{gs} = 10 V, V_{DS} = negative, and (c) V_{gs} = 0 V, V_{DS} = negative.

diode voltage drop and the voltage drop of the GaN FET. The body diode voltage is much higher than the resistive voltage drop of the devices, thus incurring higher losses during reverse conduction in this mode as against the previous mode.

4.3.1.5 Targeted Applications

Applications satisfied by the current silicon device-based power converters are not highly probable targets for reasons of matching the cost and unfamiliarity with the WBG devices and their requirements in use. Emerging applications and existing applications that could use the benefits of WBG devices to stay competitive in the market place are many. They are all prominently highlighted by the device manufacturers in their application notes and presentations in the conferences with prototypes and demonstration kits for customers. The application materials are available from the device manufacturers' websites as well. Only some are briefly given in the following for GaN device-based emerging applications.

i. Power requirement of wireless power chargers is <10 W for smartphones, <15 W for tablets, and <25 W for small laptops [Ref: Michael de Rooij & Yuanzhe Zhang, Comparison of 6.78 MHz amplifier topologies for 33W, highly resonant wireless power transfer, *Efficient Power Conversion Corporation*]. All these applications, assuming that they are safe in the environment and get respective agencies' approval, require charger function but with no wired connection (wireless) at all to the ac power supply. In addition, they may also need a fast charging feature and must come in the most compact form for these application devices. The preference for small and compact devices and willingness to pay for them are very high in the community of users with high-income backgrounds. Note that this is a high-volume market not only in advanced nations' but also throughout the world resulting invariably with a high revenue potential running into billions of dollars.

ii. Synchronous Rectifiers: High efficiency compared to diode at low power levels up to 40 W.

iii. Light Detection and Ranging (LIDAR) Device: It is like radar but with laser and used to measure distances between the source, which is the LIDAR device, and any other object such as vehicles and road signs. That information is used in autonomous vehicles and aerospace applications requiring a longer range and high speed of detection. These devices require pulsed power in the range of 4–6 kW within 8 ns requiring a current of 160 A with DC voltage supplies of 100 and 200 V. Such levels of power within very short intervals of time are not possible with silicon device-based pulsed-power converters and GaN-based devices are the natural choices. Note, that this emerging market is a high-volume market and is expected to be global and not just confined to only advanced countries [Ref: John Glaser, "Kilowatt laser driver", EPC].

iv. AC to DC power supply with the DC charger application using Active Clamp Flyback (AFC) converter: Fast chargers with the highest power densities of as much as 20 W/in^3 and above are preferred at power levels of 25–65 W. AFC-based converters using GaN devices are being positioned for dc to dc conversion applications as they deliver two to three times the power density of the existing silicon-based converter solutions (Ref: Lingxiao Xue, "Design Considerations of Highly Efficient Active Clamp Flyback Converter using GaNFast™ Power ICS", Navitas presentation in APEC 2018). Note these applications are all for 115/230 V ac input and the GaN devices used in these converter applications are rated for 650 V.
v. Dc to dc power converters for charging of laptops amongst other applications.
vi. Inverters for electric motor drives that do not have the diode across the GaN devices for high-efficiency operation and compact packaging. They have, in addition, advantages of sinusoidal current generation minimizing EMI with a switching frequency of 100 kHz or higher. Also, micro-inverters for PV applications are one of the high-volume applications.
vii. Ac to dc power supplies with power factor correction intended for high-efficiency applications that invariably include synchronous rectifier action of the GaN devices in the circuit.
viii. Envelope tracking of radio frequency (RF) amplifiers and the application involves the variation of its supply voltage so that efficiency is at maximum for the transmission of signals.

4.3.2 Silicon Carbide Devices

SiC materials and its distinct advantages are presented as a prelude to describing various devices in SiC, and they are normally-on JFET, normally-off JFET, SiC MOSFET, and their operational features and characteristics with illustration from existing devices from some startups. Various targeted applications are described to appreciate where they are emerging in spite of the stiff cost competition from existing devices.

4.3.2.1 SiC Material and Its Advantages

Various silicon carbide materials are available, and one material's (4-H SiC) characteristics are given in the comparison provided in Table 4.1 at the start of the GaN devices section. SiC is better than Si material in the categories of the comparison, and its application in devices gives a number of advantages (extracted from Infineon company's publication {Choi} and quoted in the references):

i. Higher electrical field capability of as much as eight times of Si with the consequent advantage in using a thinner die for its making.
ii. Current densities can be two to three times of Si with the result that leads to cost reduction.

iii. Higher thermal conductivity, about three times that of Si resulting in compact size and packaging.
iv. Its operating temperature maximum is 400°C, while that of Si is 150°C. This gives the advantage to operate the device at an ambient temperature of 100°C.
v. Because of its high-voltage capability, the devices are aimed up to a maximum of 25 kV rating. This certainly bypasses the present GaN device voltage rating by a very big margin.

SiC has been used in various devices, and some of them are rectifier diodes, Schottky diodes, junction FETs (JFETs), MOSFETs, BJTs and IGBTs. Only SiC JFET and MOSFET, both of which are getting popular in many applications in the market place, are briefly described in the following. The rest of them, depending on interest, may be accessed via the quoted references.

4.3.2.2 Normally-On JFET

A vertical structure of the SiC JFET, where the electrons flow from source to drain in the vertical direction, is shown in Figure 4.12. Consider the gate is not applied a voltage that equals to zero and the drain source is applied a positive voltage. A channel is formed from the source and along the top of p (also called p well in literature) from which the electron flow from the source to n-type epitaxial layer, n+ type SiC substrate reaching the drain plate and hence resulting in a current flow from drain to source in this mode. Since the device is on without excitation to the gate source, it is known as depletion mode or normally-on JFET. Recall the corresponding normally-on similarity in the D-mode GaN FET. Like the GaN FET, for turning off the current in the normally-on JFET, a negative gate-source voltage is applied that pinches the channel for conduction. If the source and drain are on the level and the flow of electrons is along the horizontal direction, then that device is termed as lateral JFET. Vertical JFET is preferred for higher voltages and hence has become the standard structure in product development.

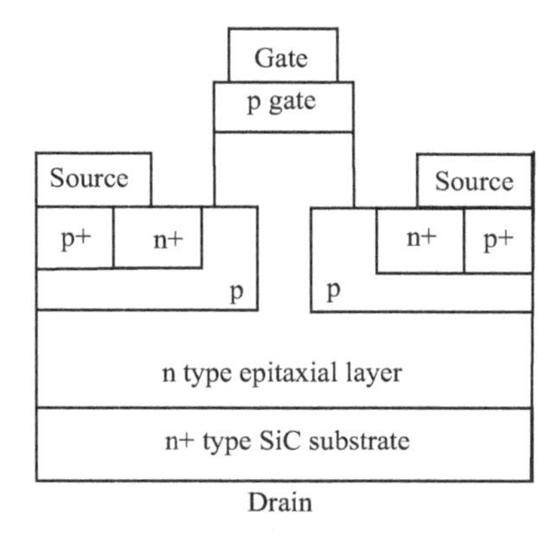

FIGURE 4.12 Lateral structure of SiC JFET.

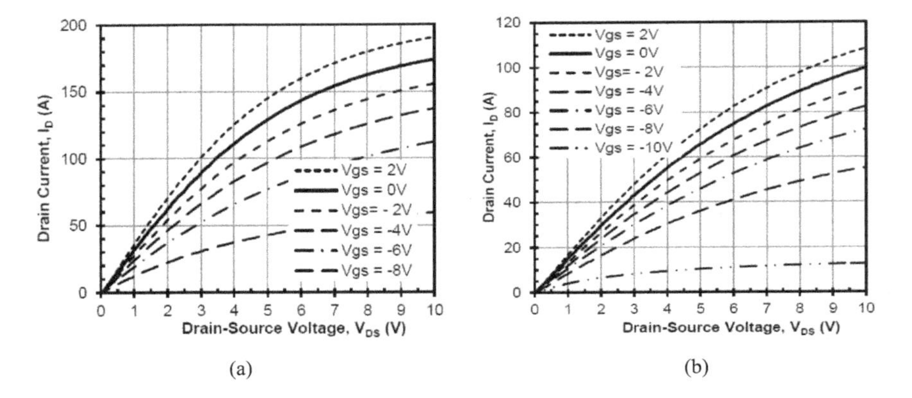

FIGURE 4.13 Drain current vs. drain-source voltage of a 35 mΩ, 1,200 V, 63 A, normally-on JFET device (UJ3N120035K35) at junction temperatures of (a) 25°C and (b) 175°C. (Courtesy of United SiC.)

Typical product characteristics are given from an available product in the market to appreciate the operational features of the SiC JFETs. The device has a rating of 35 mΩ, 1,200 V, and 63 A at 25°C from UnitedSiC which is one of the startups that is discussed later in the chapter. Its output of drain current vs. drain-source voltage drop for two junction temperatures of 25°C and 175°C are given in Figure 4.13a and b, respectively. Note that even for negative gate voltages of 10 V, the device is on. It requires a much higher negative voltage at its gate to turn off the device. As for forward conduction, the gate-source voltage is zero and at the rated current, the voltage across the drain source is only around 2 V or so which is less compared to Si devices of equivalent rating. A special gating circuit has to be designed for normally-on JFETs and existing gate drive circuits from Si devices do not function here as they are all normally-off devices. Further switching characteristics and drive requirements may be learned from the datasheets. The device manufacturers provide an excellent service to design and development engineers in the applications industry through their manuals, application notes, sample breadboards and direct consulting with the clients for no charge to learn more of the intricacies and trouble-shooting when their devices are being considered for application. Note that such service is a must for taking their devices successfully to market and a considerable amount of investment goes into this as well.

Normally-on JFET may not be preferable to many of the practitioners as it requires their time and efforts to convert it into a normally-off device.

4.3.2.3 Normally-Off JFET

Similar to cascade GaN FETs, the same approach of embedding a low-voltage (more like 30V) Si MOSFET in series with the JFET gives a very viable solution for obtaining a normally-off JFET. Usually, the Si MOSFET is made as a part of

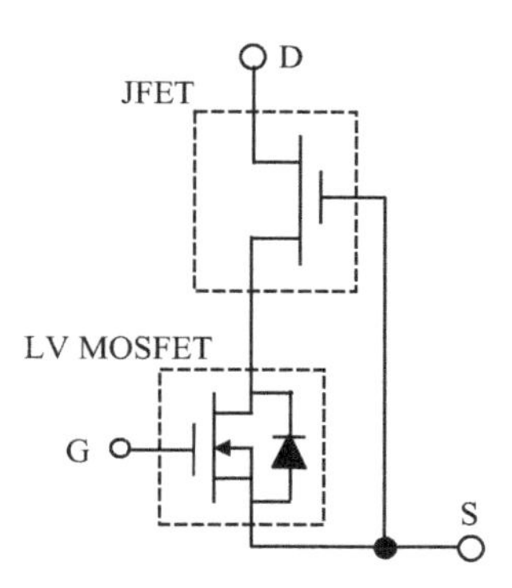

FIGURE 4.14 Cascode configuration of normally-on JFET.

the JFET structure itself. The gate of the JFET is connected to the source of the low-voltage Si MOSFET. The source of the JFET is connected to the drain of the MOSFET. Control is exercised through the gate signal of the MOSFET. The resulting device is as shown in Figure 4.14 with the final accessible three terminals, drain D, source S and gate G. Its operation is very much similar to cascade GaN FETs (given earlier in the text), and those who are familiar may skip the following:

i. Off State: At the time of starting, there is no gate signal and that turns off the MOSFET with the result that the voltage across its drain source starts to increase. Then the voltage across the gate source of the JFET is the negative of the drain-source voltage of the MOSFET. When that value reaches –15 V, JFET is put in an OFF position.
ii. On State: For conduction of current in the drain source (DS in Figure in 4.13), the MOSFET is turned on by applying a gate-source voltage. That reduces the voltage across its drain source, thus driving the gate source of the JFET below zero volt making it go to normally-on state and hence the cascade device will be conducting a current.
iii. Reverse Conduction Mode: The reverse current is handled by: (i) Stage 1: MOSFET's diode and the JFET when the gate source voltage is zero and at this time the voltage drop across the device is 3–4 V. (ii) Stage 2: On a regular positive gate signal, the reverse current goes from source to drain of the MOSFET and then on to JFET from its source to drain and the total voltage drop across the cascade device is very small compared to stage 1's voltage drop.

    But the reverse conduction has to go from stage 1 to stage 2 many times during circuit operation in an application. But stage 1 operation is not sustainable in practice for a longer time because of the resulting heavy losses in the devices, and therefore, a transition to stage 2 has to be with a minimum time delay between the stages of operation in the reverse conduction.

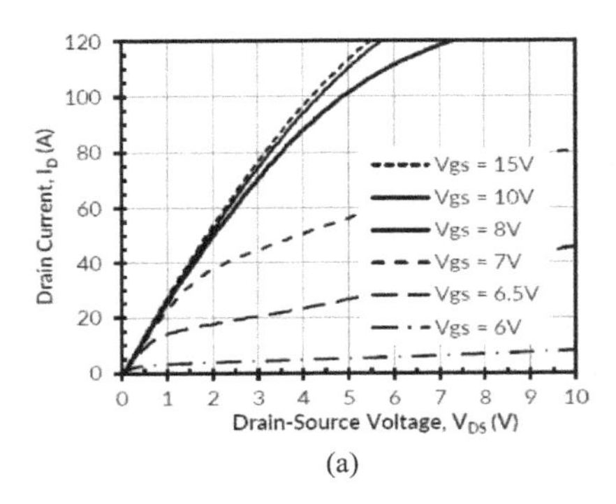

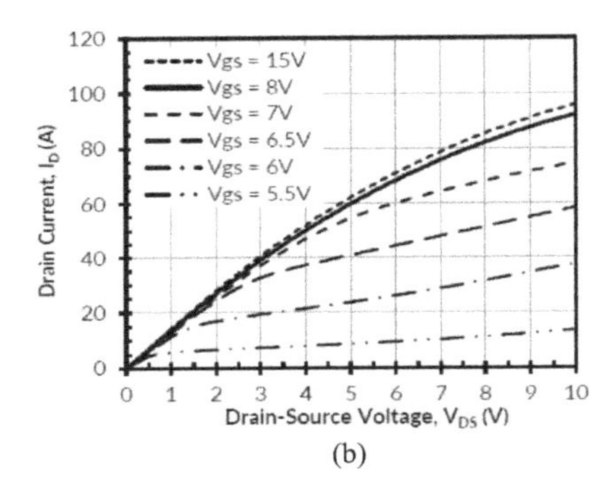

FIGURE 4.15 Drain current vs. drain-source voltage of a 35 mΩ, 1,200 V, 65 A cascode JFET (UF3C120040K4S) from UnitedSiC at two junction temperatures: (a) 25°C and (b) 175°C. (Courtesy of UnitedSiC.)

A commercial cascode JFET's performance characteristics are given in Figure 4.15 to get to appreciate its output aspects. The device information in brief: 35 mΩ, 1,200 V, 65 A cascode JFET (UF3C120040K4S) from UnitedSiC. Note that the drain-source current vs. drain-source voltage for various gate signals that are all positive (as against the normally-on JFET where all the gate signals are negative), and the cascode device gives the desirable characteristic of normally-off operation. At rated current, the device voltage drop is around 2.5 V and only when it is operated beyond the rated drain current, the conduction drop becomes excessive and is not sustainable over a long time. At a higher junction temperature of 175°C, the conduction voltage drop increases as the drain-source resistance increases. This particular feature has one advantage as the paralleling of cascade devices becomes easier as the resistances force current sharing amongst them evenly.

Reverse conduction characteristics for the same cascade device are shown in Figure 4.16. For fairly higher gate positive voltages up to 8 V, the reverse drain current resembles that of the linear part of the forward characteristics and goes through stage 2 described earlier. Only when the gate signal becomes zero and less, then the current goes to the diode, described earlier as stage 1. With junction temperature increase from 25°C to 175°C, a marked increase in the drain to source voltage occurs. That calls for a significant current derating of the device but that is common among all the electronic devices.

4.3.2.4 SiC MOSFET

Lateral and vertical SiC MOSFET structures are shown in Figure 4.17a and b, respectively. Consider the lateral device at zero gate voltage. There is no path or channel for the conduction of electrons from source to drain which are at the same level for this shown structure. This represents the normally-off operation of the device. When the gate source is given a positive voltage similar to its silicon

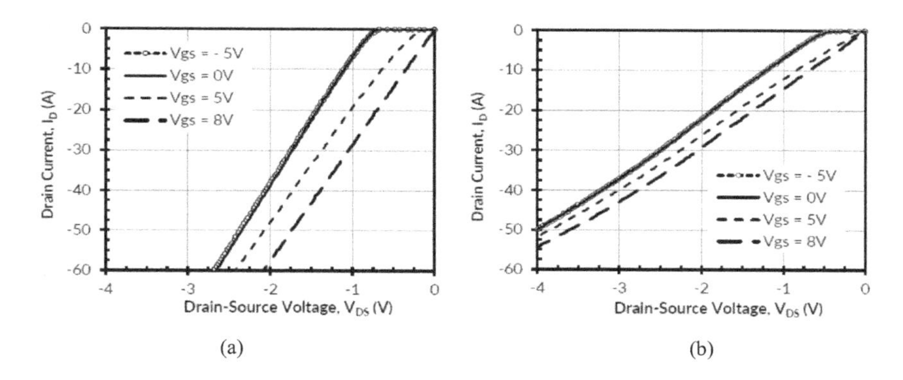

FIGURE 4.16 Reverse conduction characteristics of the cascade's drain current vs. drain-source voltage of a 35 mΩ, 1,200 V, 65 A cascode JFET (UF3C120040K4S) from UnitedSiC at two junction temperatures: (a) 25°C and (b) 175°C. (Courtesy of UnitedSiC.)

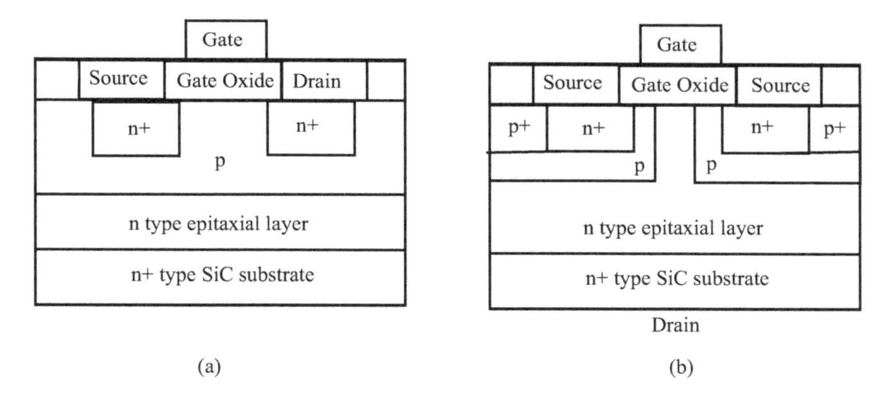

FIGURE 4.17 SiC MOSFET structures. (a) Lateral SiC MOSFET and (b) vertical SiC MOSFET.

complementary device, there is depletion in the p region below the gate resulting in an inversion layer or channel for the flow of electrons directly from the source to drain and by convention which means there is a current flow from drain to source. This is the normally-on operation. At this time, the drain-source voltage is equal to the resistive voltage drop across it. As the drain-source resistance is small (compared to its silicon complement), the voltage drop across its drain source is small thus giving it a higher efficiency in its operation. The device, similar to Si MOSFET, has a body diode which has a higher voltage drop in comparison to Si MOSFET when it conducts and carries a current from source to drain of the SiC MOSFET. Sustained operation in this mode for more than a μs or more has to be avoided. Like its Si complement, the SiC MOSFET can conduct reverse current so long the gate-source voltage is positive like other WBG devices. As the electron flow is along the horizontal direction since the source and drain are at the same level, this device is known as lateral SiC MOSFET. The symbol for this is the same as for its silicon complement of MOSFET and can be found in Chapter 2.

The vertical structure operation is similar to the lateral device operation except that the flow of electrons from source to drain is via the channel formed over the p well (that is the p material to the right side of the source n+) and the gate oxide, n-type epitaxial layer and n+ type substrate to the drain terminal along the vertical direction in the last two layers. Hence it is known as vertical SiC MOSFET. The advantage of this structure is that the voltage blocking capability is far higher than that of the lateral structure, a structure highly contributing to a growing interest in high-voltage applications targeted in the emerging market place.

The basic output characteristics are not very different and one such for vertical SiC MOSFET is shown in Figure 4.18 for a device from Infineon company offering with the nominal ratings of 30 mΩ, 1,200 V, 56 A SiC MOSFET (Part# IMZ120R030M1H) for two junction temperatures of 25°C and 175°C. It is usual and recommended that the gate voltage for standard operation is 15 V and not to exceed that under normal operating conditions. Similar to other devices, it has a linear, quasi and full saturation region and at elevated junction temperatures, the drain-source voltage drop increases more than two times as seen from the figure.

The reverse conduction (or also known as third-quadrant operation) characteristics are given for the same device in Figure 4.19 for two junction temperatures so as to learn the effect of temperature on the output. The MOSFET's body diode operation gives a higher voltage drop across the device and note that the gate-source voltage is zero or even negative. The reverse operation of the device (that

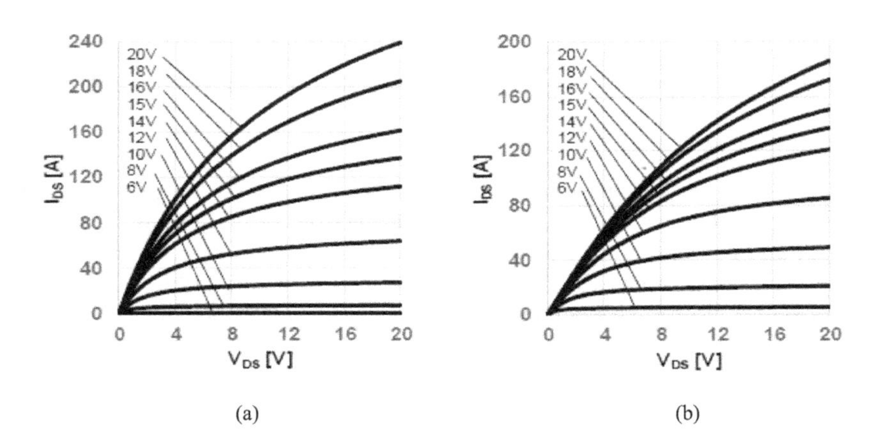

FIGURE 4.18 Drain current vs. drain-source voltage of a 30 mΩ, 1,200 V, 56 A SiC MOSFET (Part# IMZ120R030M1H) from Infineon, at two junction temperatures: (a) 25°C and (b) 175°C. (Courtesy of Infineon.)

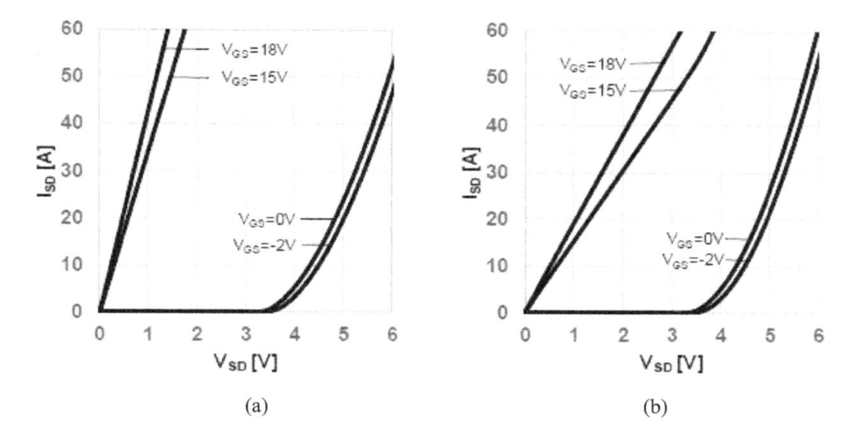

FIGURE 4.19 Reverse conduction characteristics of a 30 mΩ, 1,200 V, 56 A SiC MOSFET (Part# IMZ120R030M1H) from Infineon, at two junction temperatures: (a) 25°C and (b) 175°C. (Courtesy of Infineon SiC MOSFET.)

means not the body diode) leads to a smaller voltage drop when the gate-source voltage is positive, i.e., 15 V (which is normal gate excitation voltage) and higher than that which is 18 V. This property of the MOSFET to conduct reverse current is highly useful in many converter applications such as buck converter, and inverters by using these devices as synchronous rectifiers discussed in Chapter 2 delivering high efficiency.

SiC MOSFETs are available commercially up to 1,700 V, 100+ amperes, with low resistance between drain and source, and higher switching frequency as much as 50 kHz at higher power levels. They have made entries in applications ranging from electric vehicles and other power converter applications as may be seen from various SiC MOSFET companies' websites.

4.3.2.5 Targeted Applications

SiC devices and their incorporated equipment give not only high efficiency, higher operating frequency compared to their Si complements including IGBTs and BJTs with added reliability and lower packaging volume. Some 1–100 kW power applications are in the making and/or already embedded in solar inverters, power factor correction circuits, server centers, electric vehicle applications with successful implementation in Tesla's EV motor drives and battery chargers, and a host of applications in railways using SiC Schottky diodes, JFETs, Transistors and IGBTs.

Many in grid applications including a high-voltage solid-state transformer at 10 kV level, power electronic converter as edge equipment to correct voltage swings in the near vicinity, and power factor improvement or correction of the grid powered equipment are pursued in research and development domains. HV SiC devices have started to appear in the market and have been pursued in research for some time and very select items are also available from manufactures. Some examples of them are in the following:

i. SiC MOSFET and IGBT modules have been developed for 15kV and 40 A, and 24kV and 30 A, respectively, by Wolfspeed. These modules have demonstrated higher operational capabilities in efficiency and temperatures of 200°C. It is to be noted that the modules consist of at least 4 SiC MOSFETs/IGBTs and 4 SiC Schottky diodes across them constituting one switch position in the half-bridge inverter.
ii. SiC MOSFETs are in the 10kV rating but with a low current of 10 A and the maximum level of 15kV is being realized for solid-state transformer application. They bring with them advantages that are suitable for many of the grid-connected device applications.
iii. To enhance the power density, SiC JFETs are being promoted and 6.5kV devices with cascode version are available from UnitedSiC.
iv. SiC IGBTs at 15 and 22.6kV have been in R&D and possibly in the market in the near term.
v. Much activity has been reported in SiC thyristors and a commercially 6.5kV device has been released by GeneSiC for some time. Startups are working on these HV devices and applications are described in the entrepreneurial activity section.

4.3.3 Market

Industrial interest is driven primarily by the revenue potential when it comes to seeking and allocating resources for new product development and research. The semiconductor industry is not an exception to this practice. Therefore, a market forecast and/or determination regarding the size of the revenue compared to current revenue and product base will facilitate investors' interests as well as that of the engineering founders and entrepreneurs. The widely used semiconductor devices in the industry are silicon diodes, IGBTs and MOSFETS to name the dominant ones and other high-power devices such as SCRs and GTOs. The current total global market for the semiconductor devices is estimated to be $15B (that is the revenue per year) in 2018. It is expected to grow to $25B by 2025 and on this estimates vary from $28B to $39B.

GaN: The optimistic estimates for the GaN market are, by the year 2020 to be $1B and $4.3B by the year 2025. There are many market size projections without rigorous data backup for all of them in the open, and it is wise to treat them as best guesses/estimates only at this time. The applications for these devices are in power supplies, communication equipment, LEDs, automotive applications other than the main inverter drive for the electric vehicles, particularly cars, but may find applications in the main motor drive for low-powered two wheelers, such as scooters and motor bikes. Because of its limited voltage capability at present to 650V in practice, the applications are somewhat restricted when it comes to high-power converter applications. Voltage operation at 900 and 1,200V is claimed, but products have to follow these developments, and it seems to be a matter of time before they can be realized in applications also.

SiC: Available devices have maximum ratings, for example from Rohm are: (i) 1,700 V, 6 A, (ii) 1,200 V, 95 A, and (iii) 650 V, 118 A. Higher voltage modules for high-voltage applications in the range of 10–24 KV have come. The SiC devices market is slowly growing compared to early projections, but application market demands are more likely to match the forecasts.

4.4 GALLIUM NITRIDE DEVICE ENTREPRENEURSHIP

4.4.1 Avogy, Inc.

Avogy is a startup founded in 2010 by Isik Kiziyalli in the area of vertical GaN semiconductor technology.

Technology: The use of vertical GaN semiconductor means that the devices are based on vertical geometry as against the prevailing lateral geometry. It endows higher breakdown voltage and size reduction leading to cost and other advantages. It claims that its homoepitaxial (as against heteroepitaxial) GaN and vertical transistors are superior to many others in performance. Homoepitaxy is the process of growing crystal on a substrate of the same material, while heteroepitaxy is the process with the crystal growth material and substrate material of two different material types.

Team: Its founder has vast experience in technical and management positions in Bell Laboratories, Agere Systems and Nitornex Corp. and Alta Devices, Inc. all related to semiconductors and in their applications. The CEO is Dr. Dinesh Ramanathan who served as Executive VP at Cypress semiconductor for 9 years and previous to that in Raza Microelectronics, and his educational background is in computer science and information systems. The considerable experiences of the team members are all related to the technical scope of the company.

Intellectual Property: 80+ patents are granted and 40 more are in application, which constitutes a significant value. They are in the area of vertical GaN semiconductor technology, power topology and associated controllers and intended for power conversion.

Funding: Incubated and initially funded by the venture capital firm Khosla Ventures and later to the tune of $40M by Intel Capital in the year 2014 totaling nearly $73M during its existence.

Products: Diodes and junction gate field-effect transistors of GaN kind. They have not been available for sale to the public through electronics part sellers and agencies.

Company Status: It filed for bankruptcy (or went out of business) in 2017 and its assets have been acquired for $200 K by a subsequent startup NexGen Power Systems. The reasons for its failure are unknown at this time. After the sale of the assets, investor Khosla Ventures sued the CEO and the company which had an agreement (Soraa, another startup in a similar domain and funded by Kholsa Ventures) also sued Avogy. The court cases were dismissed.

Observations:

i. Avogy focused on the power sector for the GaN products, and it was one of the few companies in that domain at the time of its starting.
ii. Well-experienced founder and a CEO in relevant technologies gave credibility to the claims to start with.
iii. The intellectual property of Avogy is significant for a startup of this size within its short span of existence.
iv. Raised a fairly large sum and was in line with some of its competitors to begin with.
v. It was incubated and backed by a well-known venture capital firm, Khosla Ventures. This venture capital firm has been one of the earliest investors in the field of GaN startups and their knowledge of this industry and its business potential are great assets for any startup.
vi. In spite of all the right ingredients to make the startup, it failed. Failures are part of entrepreneurship giving very valuable lessons to the founders, investors, team members and potential future entrepreneurs in this domain and other domains as well.
vii. There could be many reasons for its failure and a few are surmised here:
   a. product realization and market entry times not acceptable to the investors or the norm of this industrial sector,
   b. not having an opportunity to work with a big user client for their devices in applications with a high potential for greater financial rewards, if not in the present but in the future,
   c. insufficient efforts to relate the technology and its benefits to the scientific community and applications engineers' purview,
   d. the market was slow to open in this sector and that affected its fate.

   Note that these are guesses and others may draw very different conclusions. Only the investors and management of the startup are privy to the reasons for the closure of the startup, and they may talk about them in private but not necessarily available to the public at this time.

4.4.2 Cambridge Electronics, Inc.

Cambridge Electronics, Inc. (hereafter referred to as CEI) is a GaN-based semiconductor startup founded in 2012 in Cambridge, Boston, MA, USA. It is a spinoff from the MIT campus and particularly from a group of faculty members and graduates in the electrical and computer engineering department. Founders and their current positions within CEI are:

i. Dr. Tomas Palacios (Board Member & Chief Advisor) – Associate Professor in ECE Department, MIT.
ii. Dr. Bin Lu (CEO) – Dr. Palacios' student and did his thesis relevant to CEI's core technology, MIT.

iii. Dr. Mohamed Azize (Director of Epitaxy) – Physics graduate and worked with others from 2009 onward at MIT.
iv. Dr. Anantha Chandrakasan (Chairman of Tech Advisory Board) – Professor, Electrical and Computer Engineering Department, MIT.

Technology: As claimed by CEI in various interviews and on their website are as follows:

i. Normally-off state devices from their GaN.
ii. Different material layer compositions in GaN transistors.
iii. Avoiding standard expensive processes and using materials that are compatible with silicon fabrication and hence going for cost reduction.
iv. Developing methods to use silicon foundries to deposit GaN on large wafers.
v. Developing power converter applications in laptop chargers to the most minimum size of packing it in the power plug itself.

Funding and Investors: $6.2 M so far by the investors and also through grants earlier at MIT to develop the technology. The investors are The Engine (VC), MIT Lincoln Labs, MIT, Mass Ventures Investment Bank, and others. The size of the funding as compared to companies of similar size in the GaN device domain is smaller.

Products: GaN devices for RF and power electronics applications and they do not give any details on the web and request direct contact. There is ambition to spread out to system products in data centers, power supplies, microinverters, power adapters, wireless communications, etc. It is claimed in 2015 that they developed the laptop charger with the smallest volume (1.5 in^3; approximating to 24.5 cm^3).

Intellectual Property: Not explicitly stated but it is reasonable to assume that there must be a sizeable amount as it comes out of the reputed academic laboratory with funding from many federal agencies.

Observations:

i. University spinoff and to start with a great name recognition because of MIT with investors, clients and potential collaborators that are in large established companies.
ii. Technology in development in a university laboratory and then in CEI for the last 9 years (and one of the best campuses that attracts large funding from defense and federal agencies).
iii. Great pipeline of graduate students from the founders' lab for their manpower with advanced skill and knowledge base – a big asset for the startups as seen from their team.
iv. Attractive claims regarding their technology and opportunities in their targeted product focus.
v. Big funding has not been raised (much to the surprise as they come from MIT) and therefore difficult to predict what their plans are to realize

their ambitious goals only with modest amount of capital as compared to startups in their space. They may remain fabless, i.e., without fabrication facilities, as they may reach an agreement to manufacture their products with some large semiconductor companies which may provide a kind of exit also for the company. Another alternative is to get more funding to put their own manufacturing facility which is not the usual way many semiconductor startups go.

vi. Products yet to be available to the wider public.
vii. CEI has all the desired ingredients, so far, for a startup to succeed.

Overall, success is staring at them from a glance at where the company is. Time will be the judge.

4.4.3 Efficient Power Conversion Corporation

Technology areas for Efficient Power Conversion Corporation (hereafter referred to as EPC) are in gallium nitride power devices and their application-oriented integrated circuits in power electronics and power conversion domains.

EPC was founded on November 1, 2007, in El Segundo, California, USA and it is in one of the suburbs of Los Angeles. It may be a coincidence that it is not located in Silicon Valley (even though a number of companies in their domain have made it their home) since its focus is to lead the semiconductor industry away from silicon to GaN! It is founded by:

i. Dr. Alexander Lidow (currently CEO),
ii. Dr. Jianjun Cao, and
iii. Dr. Robert Beach

Founders' Background: They have graduated from well-known Californian universities (Stanford, University of California, and California Institute of Technology, respectively) with doctoral degrees in applied physics or in engineering specializing in semiconductors and related devices. They have gained experience after that with research, development and working in the field of semiconductor power devices with a combined industrial experience of 60+years. Dr. Lidow worked for International Rectifier a well-known silicon MOSFET-based devices and controllers company in various capacities for 30 years and the last 12 years as its CEO making it a multibillion-dollar company during his leadership. He is also a co-inventor of the MOSFET device. He co-authored the first and well-known textbook on GaN transistors, "GaN transistors for efficient power conversion" in 2012. He is a popular speaker in technical and trade conferences. Dr. Cao is a prolific inventor with more than 30 patents and worked in R&D for International Rectifier also. Dr. Beach is also a former employee of the International Rectifier with work experience in device design and development. All three worked together in International Rectifier which contributed to getting together on EPC based on their capabilities and good working relationship. They have taken it to greater recognition.

Intellectual Property: Sixty issued patents in USA, China, Taiwan, Japan, and Hong Kong, which is a substantial accumulation within a short time of 12 years. Many more may be pending at the application stage and interested readers can access them from patent office websites.

Products: EPC is one of the very few companies having a large portfolio of products even though at low voltages (<350 V) and they are:

i. 51-plus enhancement-mode GaN FETs and ICs in many configurations (single, half-bridge, etc.) and in the 15–200 V range and the current range of 0.5–90 A.
ii. Demo boards for many applications and development board for a 48 V input to 12 V output buck synchronous rectifier and its performance in efficiency vs. output current are shown in Figure 4.20. The size of the board is impressive given the amount of power (300 W) being handled.
iii. Products in wafer form available for integration by customers and technical services to fit them to customers' requirements.
iv. Drivers and controllers for:
   Low side gate drivers,
   Half-bridge gate drivers,
   Controllers for synchronous rectifiers,
   Controllers for buck converters, and
   ICs for high-reliability applications.

Products by themselves do not sell in this market unless they are accompanied by technology description for easier understanding by application engineers in customer companies, application notes with technical information covering all

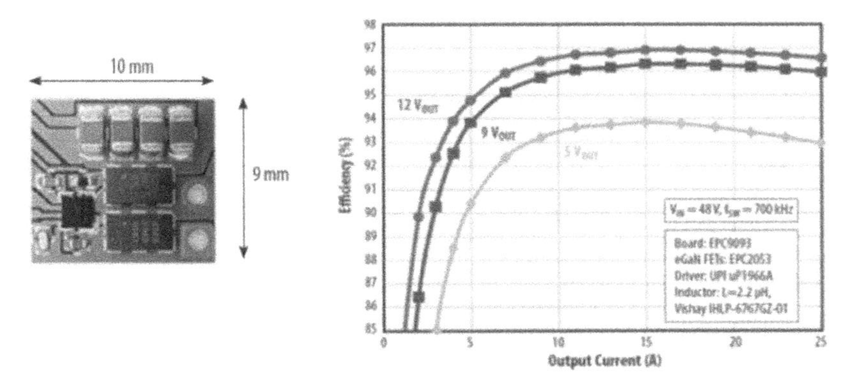

FIGURE 4.20 Development board of buck synchronous converter and its performance. (Courtesy of EPC Corporation.)

the concerns of the potential users, application-specific demonstration boards, handholding of large customers to work through their product adaptation by providing all kinds of in-house expertise and staying in front of customers' engineers in various trade and technical conferences. EPC is very leading in these efforts as seen from the available information on their website and materials made available to the public.

Observations:

i. Highly proven founders in research, development and finally product making and bringing them to market.
ii. Headed by Dr. Lidow who was in International Rectifier for 30 years, in engineering and management and for 12 years as its CEO making the company to attain a value of $4B – and this is a big plus for EPC
iii. Thanks to Lidow's reputation EPC has a billionaire partner in Archie Hwang, and together they both fund the company.
iv. Unrivaled literature creation and its unrestricted sharing in the form of applications of their products, data sheets and presentations in various trade and IEEE Applied Power Electronics Conference and Power Electronics Conference in the form of tutorials and product development efforts put EPC as one in the same level as established large semiconductor device companies. EPC's CEO (quite unusual in many startups and almost none in established companies) leads such efforts by his own participation.
v. Applications are in dc to dc conversion, automotive, envelope tracking, wireless power, space, lidar (light and radar sensor), class d audio, and inverter.
vi. It has gained a substantial revenue stream from its portfolio of a large number of products, which is what other startups yearn for.
vii. Very strong intellectual property position and probably one among the top few of the startups in this domain.
viii. They are limited by product voltage range at present, mostly up to 200 V, but they claim that they will get to 600 V range and a few competitors are claiming to be in that range already. They have the people with proven expertise to make it happen should their customers demand it.
ix. Wafer-level device offer for customer integration is a big advantage for them to work with large corporations in the applications sector.
x. The company has nothing but positives and accomplishments to make it to be an independent GaN technology company without having to exit in the distant future.
xi. Because of its president & CEO's expertise and background (who held a similar position for many years in International Rectifier), it is quite likely that it will dig in to become a major player with their manufacturing facilities and cultivating a larger customer base. It is worth following on this aspect as well to see whether established CEOs in their previous careers can again become as successful! If it is so, it becomes all the

more important to learn to bring on board the CEO with relevant expertise in your own startup.

xii. Company to watch for further learning by future entrepreneurs.

4.4.4 Exagan

This is a GaN startup from Grenoble, France. It was founded in 2014. Technology is on GaN power semiconductor devices but more to do at higher voltages of 650 and 1,200 V. The company's founders and their backgrounds are:

i. Mr. Frederic Dupont is currently its Chief Executive Officer (CEO). His professional background is: He was the general manager of the specialty electronics business unit at Soitec with exposure to semiconductors for and active development of technologies and business solutions at Soitec. He has graduated in materials engineering and also a business certificate in business.
ii. Mr. Fabrice Letertre is currently its Chief Operating Officer (COO). His background consists of: Long time exposure to GaN and SiC from 1985 and served as a director of corporate R&D at Soitec and has coauthored more than 60 patents. He graduated in engineering.

Technology: Given the founders' strong work experience at higher levels for some time at Soitec (www.soitec.com) which is a multinational French company in semiconductor materials and their applications, it has enabled them to see the potential of where the market is evolving. They obtained their technology of GaN devices from CEA-Leti (www.leti.fr/en) which is a French research and technology government organization with innovation and development resources in semiconductor devices and information technology with more than 1,700 patents by way of technology transfer program to private companies. Their technology, GaN on a silicon wafer, can be realized using a standard CMOS type fabrication facility and because of this, it has an excellent cost advantage. They achieved it with a partnership from X-FAB Silicon Foundries (www.xfab.com) in Germany.

Funding: Quantum of funding is not known in public but investment has been made by at least five parties.

Investors: Two types of investors are there for this startup and they are grouped under them as,

a. Technology partners are (i) Soitec and (ii) CEA-Leti both in France.
b. Investors (venture capital companies): (i) Innovacom, (ii) CM-CIC, and (iii) IRDInov.

Products: Released in November 2018 the following products: (i) FET transistors and fast switching drivers manufactured on their proprietary material technology of 200 mm wafer fabs and FETs are at 650 V also. Their technology is known as G-Stack™. (ii) Silicon driver ICs with GaN FETs packaging for low cost and compactness.

Exit: STMicroelectronics ($10B company) has signed an agreement to acquire a majority stake in Exagan in March 2020. Subject to necessary approval from the relevant government agency and the deal going through, then the remaining portion of Exagan will be acquired by the same in 24 months after the deal. The monetary value of the deal has not been revealed.

Observations:

i. Startup from extensive research and intellectual property from a public technology unit/agency (CEA-Leti) in France.
ii. Such kind of outcome is possible only in some European Union countries which is rarely a possibility in USA at all.
iii. The founders' background has been in CEA at a high operational level dealing with semiconductors mostly and that helps Exagan.
iv. They become partners and their reputation in regard to technology and IP gives very high credibility to Exagan's chances of success not easily available to other startups in this domain.
v. It gives enormous confidence to investors resulting in private investment capital flow to Exagan.
vi. As a startup in GaN, a fabless production route is least risky and most welcome to investors as well as to the founders as it frees them from not only raising a huge capital for the foundry but also from managing that facility instead of focusing on product development and marketing and sales. It has found in X-FAB (Germany) a great manufacturing partner to make their products.
vii. The products have been introduced at 650 V in late 2018 but not yet available to the public (may be small samples to big applications companies).
viii. It is one of the few startups having technology and products at 650 V level.
ix. That puts the company in an advantageous position to address high-volume applications such as electric motor drives, chargers, etc, in automotive, PCs and server applications.
x. They are behind in comparison to some other leading startup competitors in this domain in terms of greater product verification with multiple clients, fewer product availability and that too not to the general public.
xi. Worth studying about their model of genesis, fabless path, focus on higher voltage level products such as at 650 and 1,200 V and finally their market penetration.
xii. STMicroelectronics' signing to acquire Exagan shows the big corporations' interest in this technology. Many such deals are likely to happen in this domain given the scope of this technology in the emerging market.

4.4.5 Flosfia, Inc.

This is a gallium-based power semiconductor material and power devices company. It is founded in Kyoto, Kyoto Prefecture, Japan in 2011. There are not many startups in this domain in Japan as is the case outside of USA. It comes with certain characteristics that are different from the startups in USA as will be explained at the appropriate junctures.

Founders and their backgrounds: The founders are:

i. Mr. Toshimi Hitora (also its CEO at present): Graduate of Kyoto University and a serial and successful entrepreneur.
ii. Dr. Kentaro Kanek: An assistant professor from Kyoto University in its electrical and electronics department and well versed in devices and their applications.
iii. Prof. Shizuo Fujita: He is a materials engineering professor in Kyoto University who developed the breakthrough corundum structured gallium oxide (known as alpha gallium oxide, α-Ga_2O_3) for use in semiconductors forming the basis for this startup.

The founders have the common Kyoto university background either in study or in teaching and research and this startup is Kyoto university-based and launched thus having features that may be found amongst university incubated or developed startups.

Technology: Its manifold technologies and their implications are:

i. Developed technology for mist vapor deposition with Gallium Oxide for device development that helps to reduce the manufacturing cost.
ii. Alpha Ga Oxide in corundum structure which is its innovation (or acquired from Kyoto University and further developed in house later) has a wide bandgap of 5.3 eV as compared to 3.3 eV for 4H-SiC and 3.4 eV for GaN leading to higher breakdown voltage being bestowed upon on its devices. This is a very significant technical achievement that puts the company on the map. This structure endows high quality, reliability and comparable cost to silicon material.
iii. Development of normally-off FET using a newly developed p-type corundum structured material for integration with alpha Ga Oxide that puts this startup as one of the few having normally-off FETs. Most of the companies have normally-on FETs, and they have to be kept off by additional control means at the time of the start of the circuit and require extra means to achieve them, which is preferable to avoid in power electronics circuits and applications.

Funding: Flosfia has raised Japanese Yen 2.26 Billion (nearly $20.6 M) in three rounds until 2018. Funding is comparable to other startups of this size around the world but not in the top in funds as US startups.

Investors: It has a large number of investors and they are:

i. Mitsubishi Heavy Industries, Ltd.
ii. DENSO Corporation
iii. Yaskawa Electric Corporation
iv. Mitsui Kinzoku-SBI Material Innovation Fund,
v. Mirai Creation Investment Limited Partnership,
vi. Eight Roads Ventures Japan,
vii. Miyako Capital Ltd.,
viii. Energy &Environment Investment, Inc.
ix. Nissay Capital Co. Ltd.
x. Future Venture Capital Co. Ltd.
xi. Kyoto University Innovation Capital Co. Ltd.
xii. University of Tokyo Edge Capital Co.

It has three classes of investors and they are:

a. Heavy electrical industries (i to iii in the list): Mitsubishi Heavy Industries Ltd. has an annual revenue of $26.2 B in 2018, Denso Corporation has an annual revenue of $40.4B in 2017 and Yaskawa Electric Corporation has an annual revenue of $4.5B. Further Denso Corporation is also its industrial partner. They are all high users of power semiconductor devices in their power electronics circuits and as well as in other product applications. From this, it is easy to deduce the importance attached to the evolving gallium devices but specifically addressing their concerns of cost, reliability and quality.
b. Traditional venture capital firms (iv to x in the list) and all of them Japan-based.
c. University investments (xi and xii in the list) and one of the universities is a co-developer of Fosfia's technologies.

This unusual combination of investors is something unique to Japan and also to some degree in France as observed in the startup Exagan and almost nonexistent in the USA. The interest shown by the big industries is a powerful asset when it comes to convincing newer investors to join the select group of investors.

Intellectual Property: Flosfia has 14 Japanese patents granted and some more applications are pending. They encompass their key technologies and manufacturing methods. The number indicates a strong position given the short time of their existence.

Products: Flosfia has released in 2017 a Schottky barrier diode (SBD) with the lowest on-state resistance compared to any other device available in the market place and has come to achieve a normally-off FET with their patented technology using new p-type material with alpha gallium oxide in 2018. The products are yet to be available to general industrial users and they claim that they are gearing up production and product release with industrial partners.

Observations:

i. A Gallium-based semiconductor device startup in Japan where entrepreneurship space is very limited particularly in domains that large conglomerates dominate and it is a harsh/difficult environment for a startup to come up and kudos for this startup to emerge.
ii. Technology is from Kyoto University from materials engineering and from electrical and electronics departments and therefore can be claimed to have a strong science and technology base as is common for university-based startups.
iii. The major partner is Denso Corporation which is a very large company and likely to be a dominant user of devices for their product lines.
iv. The investment came from other Japanese multinationals companies such as Mitsubishi Heavy Industries and Yaskawa Electric corporation, and they are major users of power devices in their product lines. This demonstrates their interest from their product line point of view including Denso Corp.
v. Such a level of interest from similar large companies in USA in power device startups is absent and that indicates to an extent the lack of management's appreciation and understanding of technology and its impacts on their future revenues.
vi. The enthusiasm for this startup is due to the use of p-type material on alpha gallium oxide giving it a bandgap of 5.3 eV and the process of making the wafer at comparable silicon cost.
vii. It can be surmised that Flosfia has access to almost all large companies and confident about their help in Japan for their objectives as there are not many competing gallium device startups in this field for large companies to pick and choose to further their own interests. It is an ideally unique position for any startup to be in.
viii. One or more big companies may take over this startup when the products get a footing in some applications and that may be their sweet exit. It is only a guess and time will tell. This is one of the companies to watch for to learn lessons regarding the startup environment in this area and the peculiarities of its development in a complex environment such as in Japan.

4.4.6 GaN Power International, Inc.

GaN Power International, Inc. (GPI) is a GaN device and applications-based startup founded in 2015 in Vancouver, British Columbia, Canada. The cofounders and their backgrounds in brief are:

i. Dr. Zhanming (Simon) Li: Has a Ph.D. from the University of British Columbia, Vancouver, Canada in 1988. Worked with the National Research Council of Canada (NRCC) where he developed semiconductor device simulation software. In 1995, he founded Crosslight Software, Inc., Vancouver, Canada with simulation technology for semiconductor structures and devices, that goes under the name of Technology CAD (TCAD) on a technology transfer from NRCC. He is the chief designer of many semiconductor processes and device simulation software packages. He is a recognized researcher, product developer, coauthor of two books, and an entrepreneur. He also cofounded GanPower Semiconductor (Foshan) Ltd., Foshan, China in 2016, and more on this later in this section.

ii. Dr. Fred Y. Fu: He obtained Ph.D. and M.S degrees from the University of Central Florida and a Bachelor's degree from Zhejiang University, China. He has a working knowledge of both power semiconductor devices and power electronics systems. He worked for Crosslight, Freescale and Delta Electronics as a vice president, Member of Technical Staff and R&D engineer, respectively. Dr. Fu has co-authored two technical books: (i) "3D TCAD Simulation for Semiconductor Processes, Devices and Optoelectronics", Springer, 2012 and (ii) "Integrated Power Devices and TCAD Simulation (Devices, Circuits, and Systems)", CRC Press, 2017. Dr. Simon Li is also a coauthor of both and the second book has two other coauthors additionally. He holds US patents in power semiconductor devices and power electronics circuits.

Technology: Enhancement mode GaN HEMT devices and integrated GaN devices with control circuits, 650 and 1,200 V GaN HEMTs, a large variety of packages of these devices to suit many users and applications of these devices in power electronics circuits.

Intellectual Property: One issued patent until 2018.

Funding: Raised a reasonable sum in <2 years of its launch and they are:

i. Year 2015: Angel investment of $500,000
ii. Year 2016: (i) Foshan Science and Technology Innovation funding of CNY 12.5 Million (equal to $1.81M in 2019 conversion rate), (ii) Second angel investment of CNY 4.463 M ($0.65M).

Partner: GanPower Semiconductor (Foshan) Ltd., (GPSL) is a joint venture established by GanPower International, Inc., Foshan Jinze Investment Ltd., Foshan Huayu Investment LLC and Shenzhen Qianhai Wuxiandongneng Investment Ltd.

Division of labor between GPI and its partner GPSL: GPI develops and supplies e-mode GaN HEMTs, control circuits, applications and technical support to GPSL. GPSL is focused on system-level solutions to customers, and sales in China.

Collaborator: Air conditioner division of Midea Group which is part of the Midea Group of China. Midea Group is a Fortune 500 company with an annual revenue of $38 B in 2018 and a strong international presence. It has a significant share of the market in China and other countries in household electrical appliances and air conditioners among other things.

Products: Enhancement mode GaN HEMT devices with 650 and 1,200 V ratings, GaN HEMTs and integrated GaN devices with control circuits, a large variety of packages of these devices to suit many users, and specific applications of these devices in power electronics circuits such as notebook power adapter, solar panel microinverter, and EV charger in charging stations. Within a short time of 3 years since its inception, GPI has brought products for sale in the market. Their laptop charger and the conventional charger is shown in Figure 4.21 for comparison and note the size reduction. Note that many other companies are also in this product line, and one such is FINSix, a power electronics company that is discussed in the next chapter.

Observations:

i. This is a very recent entrant to the GaN startups group in North America and located in Canada where only another established GaN device company, GaN Systems, is in full swing.
ii. They have a unique product line with 1,200 V rated e-mode GaN HEMT which no other company at this time (2019) seems to have. With this, they have demonstrated that they are very innovative in technology and capable of product development in unchartered territory.
iii. Raised capital, mostly in China within a short time and that has been possible most probably by the CEO who is a founder of another company and connections to investors from China.
iv. The CEO and COO are both accomplished in their fields of expertise thus rallying the confidence of the investors. In addition, they have through their partner GPSL brought on board a power electronics expert, Prof. Yan-fei Liu, to assist them in power converter applications that can take advantage of the properties of their devices.

FIGURE 4.21 Conventional (left of the figure) and GPI's chargers. (Courtesy of GPI.)

v. The capital raised is small (around $3M) compared to many other startups in their domain and may move beyond this range soon enough to achieve their ambitious goals.

vi. The business strategy of starting GPSL in China and bringing it on as a partner to GPI is very laudable from the point of view that it allows them to one of the biggest markets in the world and connections to finance and companies. This strategy has no parallel to any other recent startup.

vii. In a fairly record time (<3 years), they have released, samples as well as for sale an impressive number of devices and integrated devices with control over the GPSL website. The startup through its partner has listed the price of the devices, unlike many others who do not release that information over the web. For many customers, this is a big help enabling them to make cost estimates for their new applications.

viii. They have a collaborator in Midea Group and particularly with their air conditioner division. It has the potential to open up an immense market for them in high-efficiency inverter drives for their speed control. That may lead them to other opportunities in home appliances such as washers and dryers, refrigerators, floor cleaners, and fans, to mention a few. Note all of these applications are in high volume not only in China but around the globe where Midea has a sizeable market share.

ix. Their work on products such as notebook power adapter, microinverter for solar panel, and EV chargers clearly indicate that they are not only interested in GaN power devices but in their applications so as to get into high-volume market with high-revenue potential.

x. GPI's intellectual property position is very small and in order to survive the potential threat from competitors, they may have to enlarge them very quickly which probably they are already at.

xi. Their success in making a 1,200 V rated e-mode HEMT puts them at a great advantage. Cascode HEMT device startups such as Transphorm, Inc. and VisIC products also match in voltage rating and that may cut into their technical edge in marketing and applications.

xii. At this time, there is no clear study to demonstrate the superiority of e-mode or cascode HEMT at 1,200 V rating, one way or other on the cost difference in making them and this affecting ultimately the price difference to the customers. Time alone will settle this!

xiii. It appears that this startup is the only one on GaN devices with China connections and China-based joint venture with GPSL. With political turmoil in terms of some restrictions on the USA-China trade deal in the recent political climate, China may have a substantial interest to see GPI and its partner GPSL to succeed so that China will have this asset in case business relationships become difficult with western governments and companies. GPI may gain strategic importance in such a situation that may work very well for its investors and the startup.

xiv. This is a company to watch for in the future to draw valuable lessons in many regards for budding entrepreneurs and learn from them, including for countries such as India.

4.4.7 GaN Systems

This is a startup in Canada and it is one of the two startups in the area of GaN power semiconductor devices. It is located in Ottawa, Province of Ontario and founded in 2008. The founders and their professional backgrounds are:

i. Girvan Patterson: He served also its founding chief executive officer (CEO) until his retirement in 2016. He previously founded graphics workstation producer Orcatech and co-founded Plaintree Systems, a pioneer in Ethernet switching.
ii. John Roberts: He served also its founding chief technology officer (CTO) until his retirement in 2016. He was the founder and president of two Ottawa-based semiconductor companies, Calmos (later Tundra) Semiconductor and SiGe Semiconductor.

Both Mr. Patterson and Mr. Roberts are serial entrepreneurs and during their professional careers have taken a dozen companies to successful exits with either initial public offerings (IPOs) or acquisitions (that is sale to other companies). Both are prolific inventors as seen from the granted patent portfolio to GaN Systems where they are listed as coinventors.

Technology: Normally-off type GaN FETs up to 650V product line and matched by very few in the game.

Funded: $60M (estimated) {(i) C$20M in 2015 raised with lead by Rock Capital, and (ii) BMW Ventures coming in bringing around C$40M (estimate) in 2017}. Funding magnitude at the high level of very few of its competitors shows that the company's management and its intellectual and product portfolio and plans are very attractive to the investment community.

Products: A class of 100 V up to 120 A and 650 V up to 150 A GaN FETs, and evaluation boards. Many companies have adopted their products to applications such as 1 kW energy storage system converter (GPhilos company), and 1.5 kW inverter for brushless motor drive system (Pi-Innovo company).

Intellectual Property: 18 patents are granted.

Partnerships with: (i) In wireless charging with another startup, PowerSphyr (founded in 2016, Danville, CA), (ii) Wireless power and charging with NuCurrent (founded in 2009, Chicago and Venture funded), and (iii) Master supply agreement with e2V for aerospace and defense applications. All of these partnerships have a significant impact on this company as discussed later.

Observations:

i. A Canadian startup and many do not come into that land easily, very similar to France and Japan.
ii. Startup from entrepreneurs well versed in inventions, startups and successful exits and great proven experience.
iii. This has no university/research institution background or tie-ups like others from Japan and France.
iv. Some similarity to them is seen in that a big company (BMW) venture funds is a leading participant in funding from 2017.

v. Because of (iv), there is good recognition that the GaN products from this startup have relevance to the motor vehicle sector and possibly gain the in-house market should BMW go the path of GaN devices in its subsystems.
vi. Its partnership with a sales and marketing firm (e2V) in aerospace and defense shows that the management is quite aware that cost insensitive (as compared to industrial sectors) outlet for their products is the way to go in the short run to gain revenue.
vii. Their partnerships with wireless power chargers-based startups in USA bodes well for GaN Systems since it is another sector that is bound to grow worldwide in near future as applications starting from cell phone charging to vehicle chargers and other wireless power transfer applications at home and offices. (More of this in the startups on this wireless power transfer in a separate section to follow.)
viii. GaN Systems has put its focus where it mattered most in making a line of normally-off 650 V GaN FETs and that puts it as one of the few companies with such a product line and because of it, a clear differentiation follows in the perception of the users of the GaN devices helping the company to claim a prestigious and noticeable spot.
ix. It has developed literature relevant to its products, evaluation boards and a strong presence in trade and technical conferences to meet with engineers from potential application customer companies.
x. The intellectual property position as seen from the number of patents indicates that GaN Systems is sufficiently secure covering most of the desired topics and grounds of GaN technology, reliability of the devices, packaging, control aspects in the gate drive circuit requirements, etc. in their patents.
xi. As the market is opening up, it is quite likely that they will grow to a $100M dollar revenue as the founders wished when they retired in 2016.
xii. May be disheartening for young entrepreneurs to see this model of founding but valuable lessons can be learned in how the experienced founders steered the company to such prominence in a very competitive field. Interested entrepreneurs must follow the future of this company to learn as much through the available public sources of information.

4.4.8 MicroGaN GmbH

MicroGaN GmbH (hereafter referred to as MicroGaN) is a limited liability company (LLC) with emphasis on GaN products as emphasized in its name itself. It was formed in August 2002, Ulm, Germany. The founders and their backgrounds are:

i. Dr. Ingo Daumiller: He did his Diploma thesis and Ph.D. from Ulm University. His work was in simulation, development, processing and characterization of GaN-based devices. He has 36 publications in high-temperature behavior of GaN-based devices. Several patents have resulted from his work. The results of his work enabled the foundation of the MicroGaN. He is currently a Managing Director of MicroGaN.

ii. Dr. Mike Kunze: He did his Diploma thesis and Ph.D. at the University of Ulm. He has 25 publications. His work is in the fields of semiconductor epitaxy and analysis that led to the foundation of the MicroGaN GmbH. Several patents from his work have been granted. He is also a Managing Director of MicroGaN.

iii. Two other founders are being acknowledged on their website but their names have not been identified or listed anywhere.

Technology: Design and manufacture of GaN power semiconductor devices such as HEMT FETs, and diodes. Low-cost manufacturing method combined with high-temperature operable devices seems to be the hallmark of their technologies among other claims.

Funding: Starting with a seed funding at the time of founding to three rounds of capital investment until 2013 are known but the amount of funding is not available to the public.

Investors: (i) MAZ level one GmbH, (ii) KfW Mittelstandsbank, and (iii) TechnoStart. The first one is standard venture capital. The second one is managed by ERP which is cofounded by this bank for innovative companies and for growth in Germany. The third is a venture capital company investing in mostly startups coming from university research.

Intellectual Property: Six US patents have been granted so far.

Products: Normally-On and -Off 600 V GaN HEMT (FETs), GaN Diodes 600 V, 4 A and 4–6 in wafer for customer integration and technical consulting. There are not very many products for different ratings available at this time. There is no indication of the sales outlet for their products and there are no extensive application notes and device information sheets comparable to their competitors' available at present.

Applications: Power electronics, sensors and actuators including an interest in MEMS.

Current situation as of 2020: It is cited as a government administration company and there is no information about them or their website.

Observations:

i. Startup from university-based research spinoff very similar to startups in England, France, Japan in this domain.

ii. Two of the founders developed their expertise and the basis of the startup as part of their research in the university (and not just a technology transfer from the university in the form of buying or licensing patents from the universities).

iii. Similar to other startups, they have both normally-off and -on GaN FETs in their product lines.

iv. They also claim that their GaN on Silicon gives them the comparable cost to silicon devices.

v. Funding seems to be adequate for the small number of products they have generated but may have to go for a bigger amount for expansion of their product lines.
vi. They have been funded from the seed stage to three rounds of investment capital with two of their investor entities specifically devoted to growing German startups.
vii. Investors like to be geographically nearer to their clients for closer interaction and advising and hence many VCs invariably invest in their geographical regions and similar is the case in almost all of the startups in this domain. This could be the reason that investors from other countries are absent in their backing.
viii. They seem to have developed a close relationship with potential customers seeking to find ways to work with GaN devices in their products which would go a long way in making them commercially viable in the future.
ix. Their patent position in terms of numbers is low compared to some of its competitors, for example, Efficient Power Conversion Corporation, GaN Systems. But the number of patents alone does not indicate the strength of the intellectual property nor its invincibility when challenged by others in a court of law. Higher numbers serve as a fence to block their competitors to certain innovations and for finding ways to spot infringement on their own property forcing other companies to be very careful in dealing with the startup's intellectual properties.
x. Knowing the power of the German industries in areas such as power management, power conversion, automotive and variable speed motor drives, this company is placed in an advantageous position should these industries go with GaN devices for some of their applications. MicroGaN will not have competition from startups as in USA, and the fact it is geographically nearer to their customers will be highly preferable to the customers as well.
xi. Such advantage cited in (xi) is applicable for startups in Japan (Flosfia) and France (Exagan) in their respective countries under those conditions.
xii. Very positive press and exposure in the media for this startup so far.
xiii. Important to follow the developments of MicroGaN to observe the impact of its nation (Germany) and its big industries, availability of further funds for its expansion, its international reach, emerging applications and revenue generation. They will provide worthy lessons for all future entrepreneurs and engineers.
xiv. It seems to be not functioning at this time of publication and it is difficult to know what happened to the startup.

4.4.9 Navitas Semiconductor

It is a GaN power device company with a market-oriented and needed product development. The WBG devices are becoming familiar but not as much to the product applications engineers in their specific gate driver requirements, their control and design of them. With the shrinking manpower at the engineering levels even in large user companies, it is quite attractive if the device along with their gate driver comes along in an IC package so that the effort in designing the gate drive and its integral packaging can be avoided resulting in both time and efficiency and hence cutting cost in developing their applications products in various areas of power electronics and motor drives. This company, among very few companies, fills this void in the market area.

It is founded in 2014 (whereas Bloomberg lists as 2013) and based in El Segundo, CA, USA. Its founders are Gene Sheridan, Dan Kinzer and Dr. Nick Fichtenbaum, and they all have degrees in electrical and/or engineering physics. Among them, they have 60 years of experience in top semiconductor device companies and high-level executive and management experience in them as well.

Technology: GaN power devices with fast and high-frequency operational gate driver integration and list their products as GaN power ICs.

Intellectual Property: One hundred plus patents issued as of June 2020.

Company Management: Gene Sheridan is the President and chief executive officer (CEO), Dan Kinzer is the chief technology officer and chief operating officer (CTO and COO), Nick Fichtenbaum is the engineering vice president (VP, eng) and all of them are cofounders. They have Stephen Oliver in corporate marketing and investor relations, who is leading also their media efforts and eye-catching presentations.

Funding: The startup has attracted $42.6M within three years of its founding and participating VCs in this are: (i) Atlantic Bridge Capital, (ii) Capricorn Investment Group, and (iii) MalibuIQ and possibly others. Such a fast and fairly high fundraising shows the technology's attraction but also the management's capability and visibility the founders bring to this startup.

Products: 650 V, 5 and 8 A IC GaN FETs, and additional but limited number of products and that is likely to change with their growth and time. Within 3 years, the company has managed to release products and that is a very short time in this line of business. Products are named such as NV6117 650 V Single GaNFast Power IC in market listing such as the electronic sales firm DigiKey. Capabilities claimed for their products are: 100 times increase in switching speeds with an attendant increase in energy savings by as much as 40% in the device operation. Because of the energy savings emphasis on their products, they named the company Navitas which in Latin means energy. Figure 4.22 shows the adoption of their product in a boost converter circuit to recognize the part reduction achieved in comparison with an external gate drive circuit and reduction of complexity in terms of its usage in the power circuit for power electronic designers. Note that GaN FET, its gate driver and gate logic are integrated into a single package in their product. Figure 4.23 summarizes the comparison of Navitas half-bridge product with other silicon and discrete GaN-based technologies in terms of size reduction in power devices and passive components, frequency of operation and overall efficiency. These advantages can be of immense use in major applications in power supplies, power conversion systems and electric motor drives.

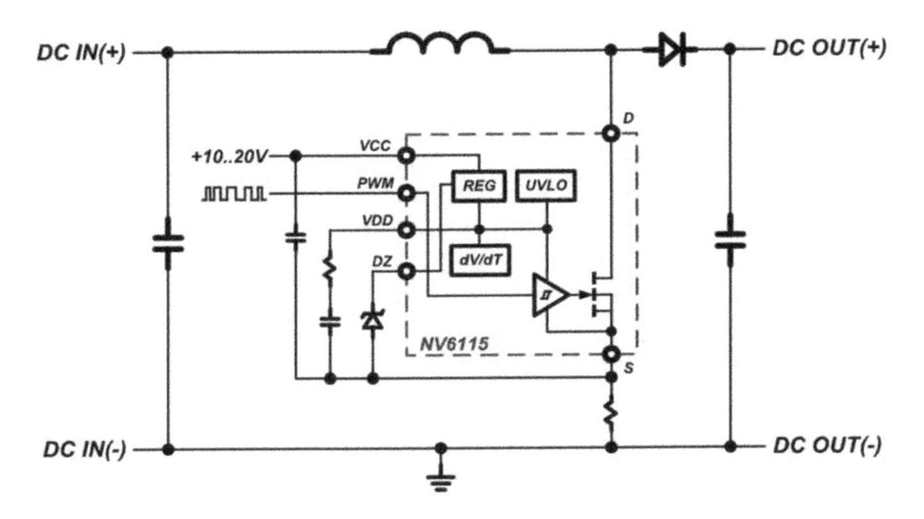

FIGURE 4.22 Navitas GaN product in a boost converter circuit. (Courtesy of Navitas Semiconductor.)

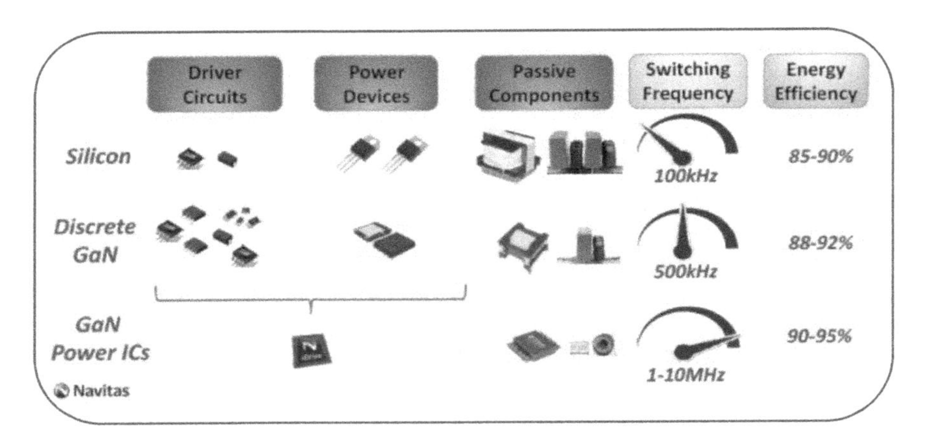

FIGURE 4.23 Capabilities of GaN Power ICs for half-bridge circuit in comparison with other technologies. (Courtesy of Navitas Semiconductor.)

Advertisement efforts of the startup are fairly standard in this sector of business by taking the information of their products to customer companies' engineering audience through trade shows and IEEE technical and trade shows such as IEEE Applied Power Electronic Conferences (IEEE APEC). This company has put in extraordinary efforts in these efforts apart from their website. There is no comparison with the info available on the website which is passive (like almost all of the companies in engineering do) as against the active meeting with the designers and architects of the engineering applications products from various companies. They stand out among the other two or three such startups in this domain.

Observations:

i. Founders strong in technology, innovation, product development and have a long and distinguished executive and management experience in semiconductor device related businesses, all pointing to clear advantages for Navitas.
ii. Raised large capital in a short time, whereas many companies with equally good vision are not in the same level of funding. That indicates the role of the management team, product development and customer acquisition.
iii. Released products in a short time of 2/3 years since founding, a great feat in this field.
iv. Very few products but focused on working for applications with industries and making life easier for the engineers working in these industries; for example, not having to design a gate driver for the GaN device!
v. Great exposure efforts in having a great presence in IEEE APEC and other trade conferences around the world. Their updating of the information is huge and voluminous that current and potential users have to be in a constant watch to learn more from them.
vi. New applications of their devices in high-power-density chargers, among others, make this company one of the leaders in the field to be working with many companies including well known Lenovo, Xiaomi and a host of others.
vii. Very high intellectual property for a company within its short time of existence.

This startup has all the attributes that are ideal such as a great speed of execution and greater exposure to user communities among other things. It has all the hallmarks of one of the most promising startups in this field and that too within such a short time from its launch.

4.4.10 NexGen Power Systems, Inc.

NexGen Power Systems, Inc. (hereafter referred to as NexGen) is a startup in the general area of GaN power devices and their applications in power conversion. It was founded in the year 2017 in Santa Clara, CA, USA.

Founders are claimed to be more than one and the only one disclosed in available information is Dr. Dinesh Ramanathan who also is its President and Chief Executive Officer. The background of this known founder is: He obtained his Ph.D. degree in information and computer science from the University of Irvine and held later senior management and executive positions such as director of business management in Raza Foundries and Raza Microelectronics early 2000, executive vice president of Cypress Semiconductor from 2005 to 2012 and as

President and CEO of Avogy, Inc., a startup in GaN power devices in the same field as NexGen.

NexGen has two directors on the board having a strong background in relevant technical fields, Mr. T. J. Rodgers, founder and ex CEO of Cypress Semiconductor corporation, which is a multibillion-dollar company, and Mr. Lou Tomasetta, the cofounder and CEO of Vitesse Semiconductor, Inc., which was acquired in 2015 by Microsemi. Their experience is of relevance to NexGen in its fundraising and growth.

Technology and products of the startup: Technology is in GaN vertical junction FET with low loss, high switching frequency capability and capable of higher breakdown voltages. Such claims are almost universal in some of the startup's competitors as well is to be noted. Intended product lines are to comprise of dc to dc and ac to dc power converter applications for data centers, EVs and energy management sectors. Currently, there are no products from the company available to the public. That may change in the future.

Intellectual Property: NexGen has two granted US patents and two applications pending.

Funding: Has raised $8M so far with investors undisclosed.

Manufacturing facility: Empire state department, NY has leased newly built 100,000 ft^2 (nearly) with cleanroom facility and has given $15M grant in addition to NexGen to promote economic development in Syracuse. NexGen expects to start production in 2019.

Observations:

i. GaN and semiconductor industry-based professionals founded this startup, and there is no university contacts/relationships as of present.
ii. Its current President and CEO was also in the same executive position for the failed Avogy, Inc. in the same technical domain and it consumed $73M before it ceased operation.
iii. The reason attributed to Avogy's demise is that in spite of its product generation, the market did not realize and more investment was required for continued operation and it did not turn up leading to its closure. It will be instructive to learn from this newer startup how they plan to avoid the same fate and navigate it to success. Hopefully, the next year will bring that to the public's notice through them or analysts.
iv. It has obtained a decent amount of funding to start its work but not sufficient to get into production and sales. Further rounds of investments have to be raised to meet the startup's ambitious plans. This will be one of its biggest challenges in the coming months.
v. As for the manufacturing facility, NY has offered it's specially built $90M cleanroom space and an additional grant of $15M to NexGen to start its manufacture. This is a very generous give away to NexGen from public money. Important to follow how this turns out depending on fundraising for operation.

vi. Their IP position is at a level for a 2-year-old startup. If they have purchased the assets of Avogy, Inc. (with 83+ patents) their IP position can be strong.
vii. The product emphasis is not just selling their GaN devices but integrating them in electronic power converters in emerging applications including EVs and data centers. It is ambitious but they may take on partners for converter assembly and integration of their devices to meet high-volume customers' needs should they arise. If it goes by itself, then this plan seems to be very hard to work and tough road ahead given the expertise needed and the amount of investment required to make it happen within the startup.
viii. NexGen is worth watching for future entrepreneurs to learn how the startup is navigating to various phases in its plans.

4.4.11 Nitronex Corporation

It is one of the earlier GaN power semiconductor devices company founded in the year 1999 in Raleigh, North Carolina, USA. Its founders and their backgrounds are:

i. Dr. Kevin Linthicum,
ii. Dr. Mark Johnson,
iii. Warren Weeks (master's degree holder), and
iv. Dr. Thomas Gehrke,

They all graduated in Materials Science from North Carolina State University (known in abbreviation as NCSU), Raleigh, NC. Dr. Johnson served as its founding CEO, Dr. Linthicum as CTO, Mr. Warren Weeks as its VP, and Dr. Gehrke in the product development. All of them were former research assistants in the labs of Prof. Bob Davis (Materials Science) and Prof. Jan Schetzina, Solid State Physics Lab of NCSU, Raleigh, NC where they developed novel methods of GaN realization on silicon. The technology that they contributed and built the startup was on Gallium Nitride-on-Silicon RF power transistors for communication industries, GaN HEMTs on 4-in Si wafers.

Funded: Around $85M (by one account) (other estimate around $100M+). Raised huge sums year after year after the start which was quite unusual for a startup being on the east coast of the USA while the big venture capital firms are in Silicon Valley in California. Some of the investors are Alliance Technology Ventures, VantagePoint Capital Partners, John Marren, David Norbury, Southeast Interactive Technology Funds, Contender Capital, Alloy Ventures, Diamondhead Ventures.

Exit: It came in 2012 to Gass Labs LLC., which is a private investment fund. The final exit was to MACOM in 2014 for a sale price of $26M.

Products: Gallium Nitride-on-Silicon transistors and an example of this product are given below:

Frequency: Dc to 6,000 MHz
Gain: 13–19.7 dB
Power Output (P3dB): 5–150 W
Supply Voltage 28 V dc
Power added efficiency: 55%–64% range in various devices

They were intended for applications such as cell phone base stations with a major emphasis on them.

Observations:

i. A university (NCSU) incubated startup and seed-funded by the university itself. NCSU graduates contributed to many semiconductor companies one of which is well known, i.e., Cree, Inc. which has four founders. Four of them came from NCSU. It may be termed as Wide Band Gap valley in terms of its concentration of the relevant startups and even comparable to Silicon Valley in California.
ii. It is one of the earliest GaN HEMT device startups.
iii. The startup had a highly technically trained team with its own breakthrough technologies in the process and the making of GaN on Si-based HEMT.
iv. Pipeline of patents from their alma mater in addition to opportunities to observe the ongoing research and to recruit their brilliant graduates to augment their team inside Nitronex provided this startup as well as others a great advantage over other startups far removed from research centers.
v. High levels of investment poured for the company in each succeeding years from 2000 starting series A, B and C and late in 2006 ($9.5M, $24.5M, and $11.5M and then $21.8M, respectively) and the total reached by 2010 to about $104M in total. This is unheard of at this early time for GaN investment and only one or two startups ever came to match this level of investment rush in this domain. Note that this startup is located far away from big venture capital companies in Silicon Valley.
vi. Products were not coming out into market place to gain revenue to match the huge investment and that was a serious concern for analysts.
vii. Their emphasis given their product capabilities of high-frequency operation, higher efficiency, compactness, perceived low cost or comparable cost to silicon device in RF and cell phone base station applications did not go well as this application was slow to emerge and communications companies did not rush to upgrade their base stations.
viii. The higher cost of the devices and very little and slow opening of the market additionally put difficulties on this startup and to their investors with high market expectations.

ix. Exit came in 2012 to Gass Labs LLC., a private investment fund and then on sale in 2014 to MACOM for $26M. Not a big payout for the investors as they may have lost money on it and other involved people who invested their time and efforts. No further information is available as to how the technology developed in Nitronex is being used by the acquirer.
x. There is no guarantee for the success of a startup given a technical base from university, the flow of graduates from the research group to the startup providing a very highly skilled workforce, and the highest levels of funding a startup can dream of, and that is something hard to grapple and understand, in general. But there are other factors to influence the success of the startup such as the speed at which the market opens to newer technologies and products, product delivery and the right target in the applications arena and some other factors known only to insiders.
xi. This is a startup that every entrepreneur should study in as much details that are available so that lessons can be learned to apply to their own cases.

4.4.12 Transphorm, Inc.

GaN devices and applications-based startup was founded in 2007 in Goleta, CA, USA. Its founders and their backgrounds are:

i. Prof. Umesh Mishra: He serves also as Chief Technology Officer in the company. Founder of Nitres in 1997 and acquired by Cree, Inc. in 2000. He is a professor in the University of California, Santa Barbara (UCSB) with active research in semiconductors and in particular on GaN. Highly recognized in this field and showered with top honors in professional societies.
ii. Dr. Primit Parikh: Served as Chief Operating Officer for a long time and now is Vice President. He obtained his doctorate from UCSB supervised by Prof. Mishra and worked as well in Nitres, Inc. and then with Cree, Inc. after the acquisition as Head of Advanced Technology. He has more than 30 patents and many papers.

Technology: Transphorm covers GaN-based semiconductor development for power devices, and their applications in the field of power electronics mainly in converters with power factor correction, inverters for ac motors, dc to dc converters, automotive applications and all applications that could afford the cost or technology changes being introduced by their product lines. It is one of the earliest GaN device startups that came out also with product lines since 2013 and later at 650 V rating thereby with a greater possibility to edge out SiC devices in the range but at lower current levels. With paralleling and futuristic higher current rated device development, the prospects of GaN devices being used for power levels of 50 kW arise in applications. All the devices that they build are all termed as normally-off which are made of depletion-mode GaN FET (which is normally on) with low-voltage silicon MOSFET in the same package so that the device is normally off.

Their technology has further expanded by technical collaboration and joint activities given in the following:

i. Fujitsu Limited (annual revenue of $47B plus) and Fujitsu Semiconductor companies putting their intellectual properties jointly in the future for product development and marketing as well at the international level. Further, they are a minority shareholder of Transphorm through additional investments and other arrangements.
ii. Furukawa Electric (annual revenue of $7.9B plus) has granted an exclusive license of its 150 patents on GaN power devices to Transphorm and carrying on joint R&D on their product lines.
iii. Yaskawa Electric Corporation (annual revenue of $4.6B plus) after introducing Transphorm's devices in their servo drives of 2 kW capability and after extensive testing entered into a collaboration with Trasnphorm and funded to the tune of $15M as well at that time. This also provides a pathway into Yaskawa's electric motor drives and solar product applications which are all high-volume applications beneficially positioned for Transphorm's product lines.
iv. Tata Power Solar is in joint partnership with Transphorm to develop power converters in this domain. Whether this is intended for the Indian or international market is unknown at this time but the strategic partnership provides an additional opportunity to move into another very high-volume application space for Transphorm.

Products: It has very limited products as of year 2019, seven of which are in 650 V with ratings of 10–47 A and, one at 900 V and 15 A rating. Note that very few of the other startups have products at these ratings. A few more are for release later. Their products meet the standards of JEDEC and others. An example of their inverter for motor drive application with their 600 V GaN is shown in Figure 4.24. It contrasts the current waveform of the pulse width modulated IGBT inverter with its GaN inverter with no freewheeling diodes and the efficiency and the superiority of performances is evident and in favor of their GaN inverter.

Intellectual Property: Transphorm claims to have more than 1,000 patents/applications and such a large number has become possible due to joint partnerships with Fujitsu and Furukawa Electric.

Funding: Investments of $245M plus are being claimed in various publications over its existence. The investors are highly known venture houses around the globe and they are Foundation Capital, Innovation Network Corp. of Japan, Kleiner Perkins Caufield&Byers, KKR, Quantum Strategic Partners which is an investment fund under Soros Fund Management, Google Ventures, Lux Capital, Yaskawa Electric Corporation, Japanese semiconductor manufacturer NIEC and others.

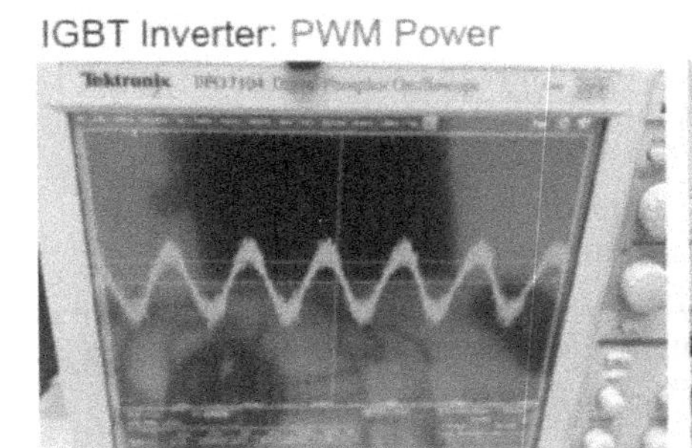

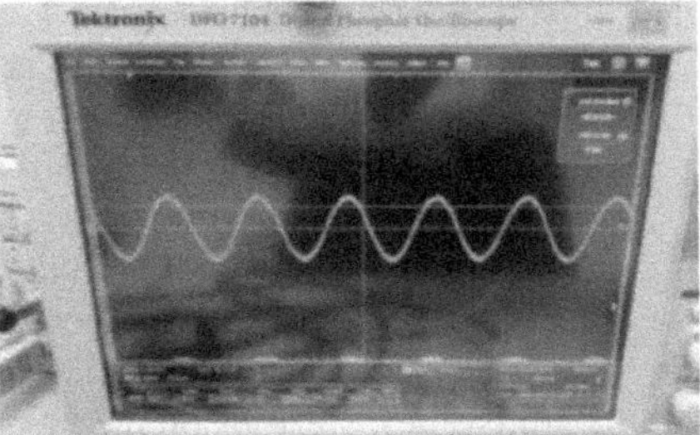

(a)

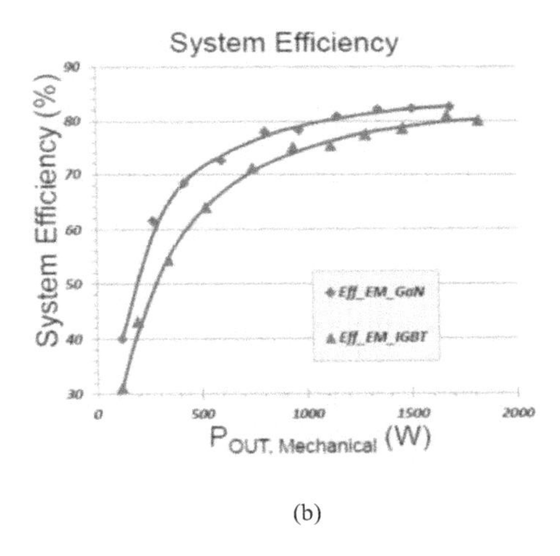

(b)

FIGURE 4.24 Comparison of IGBT and GaN (with no diodes) inverter: (a) Phase current waveform and (b) system efficiency vs. output power. (Courtesy of Transphorm, Inc.)

Observations:

i. Transphorm has very prominent researchers and one of whom also is a previous founder of a successful semiconductor devices company as their founders.
ii. University connected startup as prominent employees and officers came from Prof. Mishra's laboratory. Even though it is not officially a university spin-off, it benefits as though it is one in having a stream of graduates available for its staffing.
iii. This is the highest funded startup in GaN in the world netting around $245M plus. Well-known global venture houses and semiconductor companies have invested in this company. Compared to this level of

investment fund attraction, all other startups in this domain pale into insignificance. That does not surely mean and hence must not be interpreted that the competitors are not more or less equally as productive as this startup.

iv. It has the largest intellectual portfolio through its own work as well as through collaboration and joint partnership with multibillion-dollar established giants such as Fujitsu and Furukawa Electric. In addition, through them with the agreements, it has access to the foundries for mass production of their devices. That puts them at a big advantage compared to new startups on the block.

v. At some point in time, it is possible that with its gigantic IP portfolio lawsuits against potential competitors may emerge. It is something that established large companies, as well as startups, have to take precaution by building their own portfolio without any intersections with Transphorm's or any others for that matter. Note that this is going to create more expenditures for all other companies.

vi. Their product lines consist of a Cascode GaN structure to give it a normally-off state. It is quite likely that they may have the capability and the intellectual property to produce a normally-off p-type GaN HEMT that does not require a low-voltage silicon MOSFET in cascade.

vii. Their product line is claimed to meet JEDEC and other standards and it is the only one to have that at this time. That lead will sway potential customers to their side. Whether other competitors will follow through that kind of testing to make such claims is worth watching.

viii. The claim is that it shipped more than 250,000 devices in 2018 and likely to double it in 2019 and double the revenue thereby. They have the means to accelerate the mass production of their products as and when the market opens up.

ix. They, as is usual in this domain, provide excellent customer support in the form of advice, joint working, evaluation/demo boards, guides and application notes and share their in-house knowledge base.

x. High-volume applications are targeted for their product lines through strategic relationships with international companies in solar, motor drives and other domains. Opening of this market with such partners may push their revenue easily in hundreds of million dollars.

xi. Furthering their line at 900 V by increasing the current rating which is the direction they are pursuing from 15 A in 2018 to 34 A in 2019 will result in aggressively entering into the electric vehicle inverter market for main drive systems. That will open a market around the globe as the EVs are expected to have a phenomenal growth over the next 6–10 years. Transphorm is positioning itself very well for this very high-volume application of their product lines.

xii. If and when they or their competitors go to develop and make devices at 1,200 V and higher, a whole new set of applications will be open to them.

xiii. GaNPower International, Inc. (GPI), discussed in Section 4.4.6, has a 1,200V e-mode GaN HEMT in 2018 which directly competes with Transphorm's cascode. It will be interesting to see how that is going to play in the business strategy of Transphorm and also other companies.
xiv. VisIC, another startup, discussed in Section 4.4.13, also has a cascode 1,200V HEMT device with different control access to the GaN's gate itself to provide the normally-off operation also is in the market and it is possible that they may hike the ante by bringing in much higher voltage devices. How this will be tackled in the market place by Transphorm and GPI is to be seen in the future.
xv. Should items (x) and (xi) come to reality, Transphorm can be in an enviable position in terms of revenue and valuation.
xvi. What will the established multibillion-dollar semiconductor companies be doing when their silicon product lines get eroded in the market place? They will be arming themselves, as they have already, by entering this field through their manufacturing and R&D strengths to develop GaN product lines. They have options to buy out many of the startup competitors of Transphorm as well as Transphorm itself! Under this scenario, all the startups with good intellectual property and demonstrated products in this domain have a very good chance of having a successful exit.
xvii. Because of this scenario and the evolving market, it is crucial for budding entrepreneurs to monitor the development of various startups and learn valuable lessons for their own knowledge and hopefully to their benefit.

4.4.13 VisIC Technologies

VisIC Technologies (hereafter referred to as VisIC) is a GaN-based power device startup founded in the year 2010, in Ness Ziona, Israel. Its technology consists of normally-off GaN devices with lateral structures with GaN on SiC as well as GaN on Si capability for FETs at 650V and 1,200V levels.

Its founders and their backgrounds are:

i. Dr. Tamara Baksht (CEO): She obtained her Ph.D. from Tel Aviv University, Israel.
ii. Mr. Gregory Bunin (CTO): He worked in leading semiconductor centers in Russia and in research institutes in Israel. He has more than 30 years experience in compound semiconductor technology and has wide management experience in related field. He also had experience with GaN devices for RF application prior to his founding of VisIC.

Technology: D-mode GaN HEMT in series with a low-voltage Si MOSFET (which is usually named as Cascode HEMT) but drive control via the gate of the GaN transistor itself with the built-in manipulation of the MOSFET's conduction to get the normally-off operation.

Funded: \$21.6M in rounds C (in 2017) and D (in 2018) and previous funding amounts unknown. (Some sources indicate that the total funding has been \$23.2M so far).

Investors: Birch Investment, Federman Enterprises, Genesis Partners, Inversiones Plasma, Ministry of Industry, Trade and Labour of Government of Israel, and others may be undisclosed yet. The first four are venture capital companies.

Products: Currently there are very few (four) products at 650 V with maximum current 80 A and 1,200 V with maximum current 50 A on offer to the public. It is one of the few amongst the startups to have GaN FET at 1,200 V and being tested in a product at this time. They have half-bridge modules with temperature and current sensors in-built in their packaging making it easier for safe operation with control. This puts the company in an excellent position vis-à-vis other startups and even bigger established companies in this aspect. Their 6.7 kW onboard charger with a power density of 2.8 kW/L is shown in Figure 4.25. This onboard charger contains a dual boost and full-bridge LLC converter. Note the placement of their devices and cooling arrangement for the system giving a high power density. Additional details may be obtained from their website.

Intellectual Property: Six US Patents Applications and three granted US patents.

Partnership: VisIC and ZF Friedrichshafen AG, a global leading automotive supplier with revenues of \$40B, announced a partnership to develop inverters for EVs using devices from the former in 2020.

Observations:

i. VisIC is a startup in Israel and it is more or less modeled like US startups except that it had investment from the Ministry of Trade and Labour of the Government of Israel. Further, it may be not easy to find many power electronics companies to provide a strong basis of support to its efforts at the beginning.

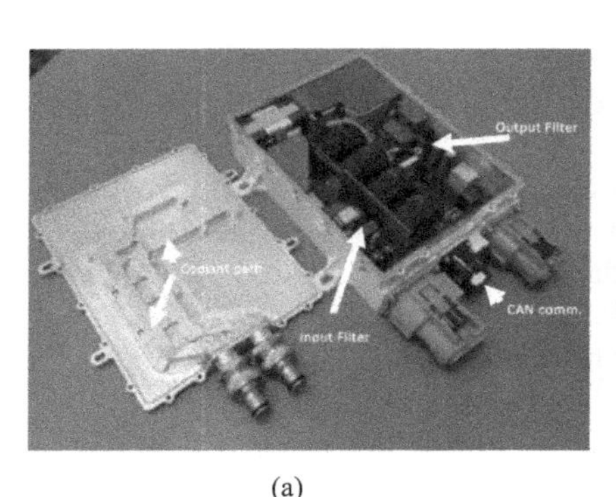

(a)

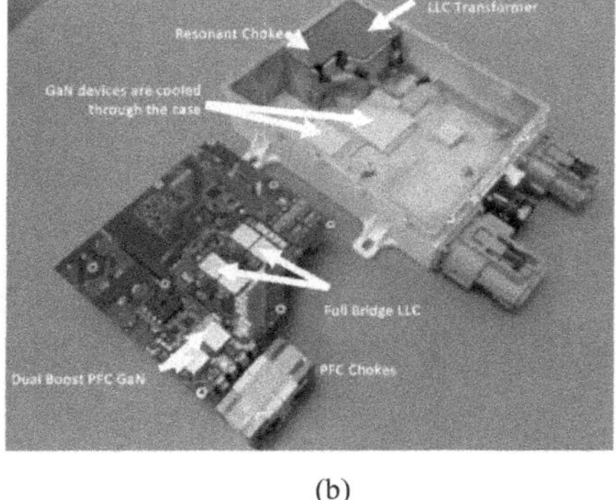

(b)

FIGURE 4.25 VisIC's GaNs in onboard charger. (Courtesy of VisIC Technologies.)

ii. Its existence was not well known until its products started to emerge.
iii. The product line is focused on normally-off GaN FET and that too for high power and at 650 and 1,200 V.
iv. The device is nothing but a cascode HEMT but has a different way of control that uses the gate of the HEMT itself and it is termed as direct control and that is also not known in the literature.
v. The difference between this control and cascode's control via the series MOSFET and how this plays out in the performance such as simplicity of circuit and any significant gain out of it is yet to be investigated. Unless this step is taken, it may be considered as a cascode GaN HEMT.
vi. It has very few products, four at this time, and some of them packaged in half-bridge configuration, i.e., with two devices in place. But it is open to custom design for customers, which is a much less risky path to bring in a product as there is already a customer willing to pay for that particular product.
vii. The packaging includes thermal and current sensors making it easy for adoption by customers.
viii. One of the founders is very highly experienced in this field and has substantial exposure to GaN related devices as it is usually common in this field of startups.
ix. This startup having 1,200V GaN FET makes it unique among its competitors. This allows them to compete with SiC devices at that voltage to garner substantial customers in the market place who are willing to pay for it to gain the advantages that come with it. But the differentiation with SiC at this voltage and higher power is yet to emerge to sway the customers and designers in favor of GaN immediately.
x. Partnership with ZF Friedrichshafen AG is a big boost for this startup and provides it with the greatest opportunity to enter a lucrative mass market.
xi. From its current path, it is less of a threat to companies like Efficient Power Conversion with their products at <400 V this time but may be a threat to other competitors who are all going up to 650, 900 and even higher such as Transphorm. But these competitors are well entrenched with collaboration and relationships with large companies and select customers in different applications in many countries outside of the USA. It will be a dynamic state of affairs in the coming years to see how this plays out.
xii. In terms of intellectual property, it has a low number of patents and applications and that may just serve their product line for the present.
xiii. The major customers for power devices are all over the globe but to a large degree concentrated in USA, Japan, China and Germany. Nearness to customers always helps a startup and from that point of view, the US startups who at least are very close to home customers as well as to customers in Japan and China through extensive tie-ups have an advantage in this regard compared to this startup. Unless such a tie-up comes along

either directly to itself or by aligning with US companies, this startup may have a big challenge to break into high-volume customers.

xiv. In terms of publicity to get their products known to the outside world, the startup is making some efforts to put their technical experts including their founder CEO in front of IEEE and trade conferences very recently. Their website gives adequate info to the general public and customers about their products, evaluation boards, technical notes, and presentations made at conferences.

xv. How far their focus in higher voltage devices will lead to market success and how soon and their impact on the market will be exciting to watch and learn from.

4.5 SiC DEVICES AND RELATED STARTUP COMPANIES

4.5.1 Anvil Semiconductors

This startup is aimed at commercializing its silicon carbide-on-silicon technology for power electronic devices. The long-term focus is both on the production of SiC materials and devices. Note the connection to anvil in its name which usually is associated with smithy and foundry and thus bringing back the memories of the old-time manufacture but indicating that they will also be associated with manufacture of the semiconductors. It is founded in 2010 by Dr. Peter Ward (founding CEO) and Professor Phil Mawby's team at the University of Warwick's School of Engineering, Warwick, UK, Mr. Kevin Marks and Mr. Stuart LeCornu.

Management: The chairman of the company is Mr. Martin Lamb and he has extensive experience in the semiconductor materials industry at high technical and managerial levels. Ms. Jill Shaw is currently the CEO and has more than 25 years of experience in a senior position from startups to corporates. One of the founders, Dr. Peter Ward, is also the managing director and director until March 2018 and the other co-founder Mr. Stuart LeCornu is a director from the investment firm of Minerva, UK. Dr. Ward's industrial experience is in the research, development and manufacturing of the silicon power and RF devices and he has worked with large companies in those domains. The other founders have no official management functions. Among them, Prof. Mawby is a power electronics and devices faculty member.

Technology and Intellectual Property: A process to grow device quality SiC epitaxy on silicon wafers to thicknesses that will enable fabrication of the vertical power devices and can be extended up to 150 mm diameter silicon wafers reducing the cost by a factor of 20 (in the words of Dr. Ward). This technology, its enabled device making and its compact packaging are ideal for targeted applications requiring low inductance modules for high-efficiency energy conversion. Applications, very similar to that of the other WBG devices, are equally open to Anvil Semiconductor's products also. Also its use of cubic form instead of the usual hexagonal form of material to process the device will give them cost savings and hence an advantage. Their technology application to GaN LEDs also constitutes additional property.

Products: 650 and 1,200 V Schottky Barrier Diodes and in 2016 GaN-based LEDs using the cubic form.

Funding and Revenue: Seed funding to the tune of $1.9M and then a stream of R&D contract funding from Innovate UK government agency around $3M over a number of years. The funding level is very low compared to other device startups at this time.

Spin-Off: Their collaborative work on Innovate UK government project with the University of Cambridge professor and his group and also Plessey Semiconductors resulted in cubic GaN grown out of cubic SiC and its use in LEDs for lighting with a claimed significant cost advantage. A spin-off of another startup to focus on LEDs was achieved in the year 2017 with these collaborators under the name of Kubos Semiconductors. The spin-off using Anvil Semiconductor's intellectual properties and its collaborative effort with other educational and industrial institutions is a big feather in its cap showing the underlying strength of its technology.

Observations:

i. Startup from a university laboratory and industrial R&D researcher. It has very credible technical and business teams.
ii. Technology is based on growing SiC from silicon wafer and using the cubic form to process the device fabrication both contributing to a very high-cost saving that could lead to significant opportunity in the market place.
iii. Funding seems to have remained at a very low level ($1.9M) for a device- and materials-based company which is a big surprise given the founders and their reputations.
iv. Operational funding (around $3M) mostly came from R&D contracts with Innovate UK agency, very crucial for its existence.
v. Because of (iv), collaborative efforts with the University of Cambridge professor and researchers and the company, Plessey Semiconductors, resulted in cubic GaN LED development and forming a spin-off with these collaborators for that product under the name of Kubos Semiconductors in 2017.
vi. Both Anvil Semiconductors and its spin-off Kubos Semiconductors are worth following closely to learn about this strategy of forming a spinoff based on a technology and its application to one product and whether there are any additional benefits to this mode of operation in company business formation.

4.5.2 Arkansas Power Electronics International (APEI)

One of the early startups with a background in device development and integration, it was founded in 1997 in Fayetteville, Arkansas, USA. The University of Arkansas-affiliated Arkansas Research & Technology Park in south Fayetteville incubated this startup as one of the principal co-founders was from the faculty of electrical engineering, University of Arkansas. The co-founders are Prof. Kraig Olejniczak and Mr. Bill Schirmer. Many of its employees came from the professor's laboratory as the focus of the company and their research and development (R&D) backgrounds coincided and thus was fortunate to identify and get the employees with the right qualifications for the company. Prof. Kraig Olejniczak's research was in the SiC-based devices and their applications to power electronics converters and systems contributed to the core strengths of the company and they formed the basis for the startup as well.

CEO: Dr. Alex Lostetter from the year 2003 and continuing as of now. He is also a University of Arkansas graduate and has identical technical and educational backgrounds required for the startup.

Funding: For a few initial years, the company grew and survived on the federal agency grants and private company R&D contracts. Subsequently, they opened their own manufacturing facility about 10,000 sq.ft to manufacture custom converters and power electronics systems.

Intellectual Property: One patent granted and six were pending as of year 2007, and many parts of their IP were presented in various conferences both engineering and trade types in about 150 presentations and papers. They have built a strong technical foundation and structure for the startup over a decade and more, and enhanced its image to potential customers as well as potential takeover companies.

Manufacturing and Products: Its specializations are in high-temperature, WBG (particularly in SiC), discrete device-based packages, power modules, and their use in power electronic converters and subsequently with embedded software for them in system-level applications including EVs.

Exit: APEI was acquired by Cree on July 9, 2015 and was made into a division of the acquirer under the name of Cree Fayetteville, Inc. Cree is a $ billion-plus company and noted for its strong position in making WBG devices. This acquisition gives them an added advantage in the application of these devices in systems requiring higher performance such as high-power-density realization and high-temperature operation.

Observations: Salient ones are:

i. The technology basis for the startup is very advanced and emerging with knowledge confined to very limited circles in the world and is provided by the founder with an excellent background and practice in this field which is essential for any startup to do well.
ii. The company is more based on applications of emerging technologies than technologies per se. That usually is an unusual basis for a university-based startup. It goes to prove that all university spin-offs need not only be based purely on technologies alone.
iii. The CEO also is technically trained in this field and contributed to its technology which gives a greater understanding to the steering of the company at the technical level.
iv. Because the company arose from the people with immense leading technical edge and familiarity and experience in grant writing with federal

agencies, the need to go to investors was circumvented and instead the federal agency R&D contracts paved the way for funding.

v. A strong basis in patents and technical applications enabled its success.

4.5.3 Ascatron

This startup was founded in 2011 in Kista-Stockholm, Sweden. It is a spinoff from research center Acreo which has a backdrop of 20 years of research in semiconductors and SiC in various aspects. Ascatron's technology is based on the SiC epitaxy (i.e., growth of crystal on a crystalline surface or base) using a proprietary technique known as the 3DSiC method and design of power devices with this material.

Founders, their current roles in the company and their and experiences are:

i. Christian Vieider, CEO: 25 years in management and semiconductor business development.
ii. Dr. Adolf Schöner, Director Technology and Sales: International expert with 30 years' in SiC technology.
iii. Dr. Wlodek Kaplan, Production Manager: 30 years in semiconductor and system quality.
iv. Dr. Sergey Reshanov, Development Manager: 20 years of R&D experience in SiC epi and devices.

Note that three of the founders have immense experience in semiconductors in various capacities and the CEO has relevant business experience in semiconductors.

Funding: 4M Euros including a grant. The investor, KIC InnoEnergy of Nordic Countries has put 75% of the funding and 25% of the funding is from the European Institute of Innovation and Technology (EIT) through KIC InnoEnergy. But they have also been supported by other research and development grants with industrial partners.

Products: Very small amount of products have been released very recently. They are:

i. Schottky diodes with ratings of 1,200 and 1,700 V, 20 A with high efficiency in operation among other standard claims to competitors' devices.
ii. PiN diodes with 10 kV (rms) blocking capability with low on-state voltage and fast switching speeds, mostly used in RF switching and high-frequency applications.
iii. SiC-based high-voltage devices such as MOSFETs for custom applications but that is yet to be released as of date.

Observations:

i. Strong research institution background to the company with the attendant benefits of flow of innovations in their path and hence making it easier for it to stay at the forefront of technology.

ii. An additional advantage of the item (i) is that it provides a good pipeline to graduates and researchers from it to work for the company.
iii. Product release has been recent and difficult to judge their use in customer applications (at least in USA).
iv. Very limited funds have been raised and early one-fourth of their total funding came from the European Institute of Innovation and Technology (EIT) through KIC InnoEnergy which is the only investor.
v. Makes use of grants and hence no necessity to give shares in the company and hence no dilution to be made in equity being held by the current investors. That serves well the current equity owners including the founders.
vi. It has excellent partners in research projects such as GE, ABB, Infineon, Norstel, and STM and LPE (from Italy) and more. By working with partners Ascatron wins not only the grants but also develops good access and working relationship with these partner company engineers and management which may serve its interests in the future more than the grant money.
vii. Joined forces with Norstel to make use of their high-volume fabrication of SiC materials, wafers and their epitaxy technologies and facilities and providing to the partnership with their customized SiC epitaxy and device designs. By this Ascatron has demonstrated that it is capable of attracting strategic collaboration to bring their products into the market place by joining with a well-known company. Point (vi) in the above is further reinforced by this collaboration for Ascatron.

4.5.4 GeneSiC Semiconductor, Inc.

GeneSiC Semiconductor, Inc. (hereafter referred to as GeneSic) is primarily a silicon carbide technology-based startup. It was founded in the year 2004, in Dulles, VA, USA. Its founder is Dr. Ranbir Singh who also serves as its President since its founding.

Background of the founder: He has a Ph.D. in ECE from North Carolina State University, published over 160 articles, authored a book and has 28 US patents. He served as a researcher on SiC devices with Cree, Inc., for eight years and about a year with the National Institute of Standards and Technology (NIST) prior to founding GeneSiC.

Technology and Products: SiC Schottky diodes, SiC FETs, SiC junction transistors and also silicon devices. Its contribution to SiC product areas are in materials, doping, reliability, sustained operation at rated currents at high temperatures without performance deterioration. Products are available to the public on sale from them but the magnitude of revenue is not published.

Demonstrated many unique features and advantages in their designs and products such as high-temperature operation at 300°C (and even 500°C), high and stable current gains under dc operation for a long time at rated currents, high speed switching characteristics (turn on and off transients of 15 ns or less), all attractive features for applications.

Funded: $13.4M plus (by various federal agencies and departments for over a period of 14 years) through research and development contracts and grants. Private investment details are not available in public.

Intellectual Property: Even though the founder is associated with 28 patents, they are assigned to various organizations. No information on patents of GeneSiC is available at present other than that they own leading patents on WBG power device technologies as per its website notification.

Observations:

i. Excellent technical background of the founder having relevant experience with a large semiconductor device company (Cree) previous to founding.
ii. One of the very few companies to raise close to $14M through federal agencies R&D contracts (and not grants!) within 14 years of its existence and it seems that it has built the startup with this as a major funding source.
iii. Item (ii) may be the reason that the company has no listed investor or other sources. This is very unique to this company compared to almost all of the competitors in this domain (and one may be found in the UK to a certain degree)
iv. It is considered to be one of the top 30 semiconductor companies around the world in the field of SiC devices even though its revenue may not be high since the market for SiC devices is still very low compared to the entire Si devices market.
v. Very well exposed to defense, aerospace, space and energy companies through its various contracts from respective and relevant federal agencies in these areas. This exposure is a big benefit to the company in that its products are known to them and likely to be favorably considered for applications.
vi. Its studies on reliability, high-temperature operation, high switching speeds, higher voltage operation, and the results being shared in public educate the potential users and thus favorably positions itself in their midst as a viable partner and/or supplier of their products.
vii. A large number of products in the categories listed is available through electronic part suppliers for some time, again a favorable factor for the company.
viii. Because of the high-voltage operation of SiC MOSFETs and SiC Junction transistors and diodes, high-power motor drives, and power conversion systems are the natural constituencies for these products and so far they use mostly silicon IGBTs and MOSFETs in their products. Therefore, the companies which are manufacturers of these devices, such as Infineon, ROHM, Mitsubishi et al., have also been working in the SiC field for more than 25 years and have an impressive array of SiC products as well. It is very difficult to sustain the competitive onslaught from them in time to come as the market expands for these products. This startup's, as well

as equally other startups' in this domain, strategies and tactics to overcome this critical challenge, if and when available at all, will provide an excellent insight for engineers.

4.5.5 SemiSouth Laboratories, Inc.

SemiSouth Laboratories, Inc. (hereafter referred to as SemiSouth) is a startup in SiC power devices domain. It was founded in 2001 in Starkville, Mississippi, USA. Its technology primarily consisted of realizing power devices from SiC and manufacturing methods for the same. This is one of the earlier startups in this domain. Its founders and their backgrounds are:

i. Dr. Jeff Casady: An assistant professor at the time of the start at Mississippi State University (MSU) in Starkville, Mississippi. He formerly worked for Cree, Inc. (a large power device company and also with interest and work in SiC devices). He published 70+ papers and was granted seven US patents (including co-inventions). He also served as SemiSouth's founding President and Chief Executive Officer.
ii. Dr. Michael Mazzalo: Also a faculty member in MSU. He has over 100 publications, and 14 granted patents and has varied expertise in semiconductor devices and their applications in power electronics. Currently, he is Duke Distinguished Chair in the University of North Carolina, Charlotte.

Funding: $70M+raised from private investors such as venture capital and large company (Power Integrations) and in addition to these obtained more than $25M in federal government agency research contracts and grants. All of them raised within about 10 years since its founding and that shows the strength of its intellectual property and business plans and their execution. Note that the federal grants are reviewed and the competition for those contracts and grants is fierce and only promising technologies usually win them. The amount of capital raised is phenomenal for this startup in this domain as compared to some of its competitors.

Intellectual Property: It was granted 30 patents by 2012 and this number is fairly high compared to other startups in this domain and to an extent, this puts at ease some concerns of the investors and potential partners.

Products: Developed products and released them to the public by 2008 which made it probably the first startup in the SiC domain to achieve this. The products are normally-on and normally-off JFETs with ratings of 1,200 and 1,700 V, respectively, both of which put this company into prominence. The other product lines are Schottky Barrier Diodes (SBD) of 60 A and 1,200 V rating and also gate drivers for their devices. Their plan was to increase the ratings of these devices so that they could be ideal for main electric vehicle drive systems.

Exit: It was closed down in October 2012 and its assets came up for sale. A number of factors could have made it come to this difficult exit. The primary factors are insufficient market thus giving them very little revenue whereas their operation came to almost 100 employees incurring a bigger recurring expense. The disparity between them, together with no further capital infusion probably drove the company to cease its operation. It is something engineering entrepreneurs have to come to accept that there are forces beyond their control such as the speed of market opening in a newer technology domain.

Observations:

i. A university spin-off with its attendant benefits.
ii. Mississippi State University spin-off and it is located in one of the least industrialized states in USA, Mississippi with very little semiconductor device-based industries.
iii. A great win and boost for Mississippi State University for public relations in terms of how the university incubates high tech startups leading to highly skilled workforce employment and tax base for the state.
iv. Its location may have helped in getting grants from the federal government for its high technology startups to promote disadvantaged regions.
v. Raised over $98M in private investments and federal contracts and grants within 11 years and that is a feat by itself in a location far away removed from the main centers of venture capital companies and mainstream high tech companies.
vi. All of these positive developments can be attributed to the attention that SiC technology for devices was getting and the timing was right.
vii. SemiSouth developed, ahead of almost all of its startup competitors, SiC devices for both normally-on and -off operation with 1,200 and 1,700 V rating, respectively and also Schottky barrier diodes of 60 A and 1,200 V rating and made them available to the public and customers.
viii. Power Integrations put in $30M into its development lending it high recognition both with investors and users.
ix. They built a big facility for production in Starkville, MS anticipating the market opening for SiC products and applications with high expenditure.
x. Its intellectual property position was excellent with 30 US patents granted by 2012 and this number is much better than many of the competitors in its domain.
xi. Keeping the startup alive with increased employees and bigger facility with cost consequences and a lagging market requires further investments which may have ended its journey. Power Integrations let its subsidiary SemiSouth go thereby choosing to accept the market dictate and reality.
xii. A comparison with GenSiC Semiconductor shows that it has survived without raising investment capital and mostly surviving on contract research and grants from federal agencies and bringing products much later than SemiSouth into market place and making a go of it now in the face of large companies being in the same domain. These two companies have two different strategies for their growth. It is important that future entrepreneurs look into them to make comparisons and draw inferences to further their own opportunities in the startup pursuits.

4.5.6 United Silicon Carbide, Inc.

United Silicon Carbide, Inc. (hereafter referred to as USCI) is a silicon carbide-based power semiconductor devices startup founded in 1998, NJ, USA. Its technology consists of SiC JFETS, SiC Diodes, methods of making the devices with novelties captured in many of their patents and use of their devices in power conversion and other applications. The founders' names are not available in public domain at this time. The technology on which it was founded came from Prof. Jian Hua Zhao of Rutgers University, NJ. USCI was acquired in 2009 by DOLCE Technologies which is a venture and advisory firm focused on technology start-ups and providing seed funding. The co-founder of DOLCE Technologies has been the President and CEO of USCI since the acquisition. There are two leading officers both of whom are investors in USCI and they are active in putting this startup on the map and they are:

i. Dr. Christopher Dries (President and CEO): He obtained his Ph.D. from Princeton University. Previously he managed Research & Product Development at Sensors Unlimited in photodiodes. Cofounded DOLCE Technologies which is a venture and advisory company for early stage tech startups. Their partners acquired USIC and invest in it.
ii. Dr. Anup Bhalla (Vice President Engineering): He obtained his Ph.D. from RPI. He is a co-founder of Alpha and Omega Semiconductor and held various management positions there. Sole inventor and/or co-inventor in 100+patents in his career at Harris, Vishay Siliconix, AOS, and currently in USCI.

Funded: $8M from some public documents and it could be more than this figure.

Investors: (i) Partners of DOLCE Technologies (includes Dr. Dries) and Dr. Anup Bhalla and some venture firms are undisclosed. (ii) They also obtained funding from the federal contract that helped to build relationships, products and the company.

Products: A significant amount of following devices and applications are their products at this time:

i. SiC JFETs (normally on) 650 V, 25–85 A and 18.4–65 A, 1,200 V devices
ii. SiC FETs (normally off because of cascade configuration): Devices with ratings of 4–85 A, 650 V; 4 and 8 A, 900 V; 20, 33.5 and 63 A, 1,200 V
iii. SiC Schottky diodes: 2–50 A, 1,200 V and 4–60 A, 650 V
iv. Applications ranging from ac to dc power factor correction system, wide range utility power supply, cascade resonant dc to dc converter and HV phase shift full-bridge converter.

Public and Targeting Customer Outreach: The company has strong interfaces with the engineering community via its website and presentations in both trade and IEEE and conferences, articles by their leading technical people on various items of interest to potential customers and users, and demonstration boards. They stand on equal footing with their competitors in these aspects.

Strategic Investor: Analog Devices, Inc. which has an annual revenue of $6.2B has made a strategic investment in USIC in March 2019 and entered into a long-term supply agreement. This is a big win and recognition for the USIC team, their work and products. Note that Analog Devices, Inc. is a very profitable company to the tune of 24% and this is after all expenses including spending about 16% on R&D.

Intellectual Property: 22 granted US patents.

Observations:

i. It is one of the early startups in SiC power devices.
ii. The current CEO and his partners in DOLCE after the acquisition of USIC in 2009 ramped up its development, product delivery and publicity.
iii. Originated from university-based research.
iv. The focus of the technology development is on cascade structured SiC FETs with a significant emphasis on them, SiC JFETS and SiC Schottky Diodes. Their devices are available in a wide range of current ratings and are offered for 650, 900, 1,200 V and a few at 1,700 V rating. The product range is very competitive with others in its domain.
v. A big boost to their business has landed in the form of strategic investment from Analog Devices, Inc. and with a long-term supply agreement.
vi. Their presence in the public by various media means is at a high level and much more attractive to the readers than some others in its domain.
vii. Their technology presentation to the user audience is good but they may benefit by sharply delineating the differences between their products and others, if possible (which many companies do not for reasons only known to them!).
viii. Their IP position is comfortable and comparable to other startups.
ix. They are at a point when the market opens up (i.e., a lot more users mostly power electronics companies open up), they are ready to be a strong contender for their orders.
x. The startups in this domain are not too many and hence they do not have much competition as in the GaN domain but their biggest threat is the large companies who are also deeply involved in this technology and product development for more than 25 years. How that is going to play out is worth following closely so that technology entrepreneurs can learn how to cope with that dynamics and still can emerge successfully.
xi. Compared to typical semiconductor companies, USIC does not seem to have taken a lot of investment at this stage and in some ways, it has some resemblance to GeneSiC Semiconductors, Inc. There is much to learn from these startups.

4.6 DEVICE CONTROL ORIENTED ENTREPRENEURSHIP

4.6.1 AMANTYS

This is a startup in the power electronics area based to change current design, implementation and control of power electronic systems practice and to a larger extent focused on medium- and high-voltage applications. Located in Cambridge, UK, the company was founded by Pete Magowan, Bryn Parry and Mark Snook in 2010 and they are executives from ARM Limited. ARM Limited is a large British company worth more than $30B (in 2015) providing the ARM processors for various industries including the phones. Naturally the executives from it who are the founders can be safely assumed to be well versed both in hardware and software and their potentials in any emerging application in electronics but not necessarily in power electronics.

Intellectual Property: Licensed from Warwick Ventures and the technology was developed by Dr. Angus Bryant from University of Warwick, in partnership with Professor Patrick Palmer of Cambridge University. The technology of interest here is the prediction of failures in the electric power conversion systems. It will impact the motor drive systems and their power conversion systems in naval and air force defense systems, merchant navy, power grids, to name a few.

Company and Management Team: The founders served in the following capacities:

i. Chairman: Pete Magowan
ii. CEO: Mr. Bryn Parry until 2012, from January 2013 Ms. Karen Oddey, and from Oct 2013 Mr. Erwin Wolf
iii. Technical Director: Mark Snook

Mr. Erwin Wolf's CEO appointment brings a huge experience in the semiconductor industry through his Managing Director and vice presidency employments in Infineon, a well-known company in power semiconductors, and also from previous Siemens employment as well. His appointment is an ideal one for a startup to open many doors for it with the investors, customers and collaborators. Note it took about 3 years to get here.

Funding: Investors are MoonRay Investors, ARM Holdings, plc., and later Avago, Inc. for a total of $20M all raised in <5 years. The connection and familiarity of the founders to their previous employer ARM Limited and its investment arm Holdings, plc., is very beneficial to the startup. Further their previous employer earned access to the evolving technology which may be of technical and/or business interest in the future. From that point of view, their investment is worth in this start-up or any startup of their interest and it is only a very small amount compared to its huge size.

Products: Gate driver ICs for IGBT modules from 1.2 to 4.5 kV levels and products to evolve to develop high-reliability medium and high-voltage power conversion systems.

Exit: The company did not raise funds to support its operation around 2015 and ceased operation. The assets of the company were bought by Maschinenfabrik Reinhausen GmbH (Regensburg, Germany), a large conglomerate. It is interested in keeping the division at Cambridge with all its possible connections to Cambridge University and develop products aggressively.

A number of observations can be made and they are:

i. There is no guarantee for success of a startup even with an ideal management team with excellent experience, connections to various other companies and investment houses is amply demonstrated by this startup amongst many others seen in this chapter.
ii. The startup had the right academic connection of a very high caliber through a faculty member and a researcher. Many startups do not have such a fortune in their favor.
iii. Huge investment funding is primarily due to factor (i) and partially to factor (ii) for this startup. Many startups would love to be in such an enviable position.
iv. The product focus seems to have been not received well in the market place. Gives the entrepreneurs a pause to think about this more and brings to focus the necessity to develop ways to strive for convergence between their starting perception and the market. How soon that can be achieved spells the success of the startup.
v. The speed of execution is very critical to inspire continued confidence of the investment community and that usually means bringing the product to the market at a faster pace with wider recognition among many customers. That goes hand in hand with the technical team and marketing team working together. Not knowing exactly what happened within a startup, it is very difficult to attribute very specific reasons for its coming to an end.
vi. It is important for aspiring entrepreneurs to read and research further about this company to learn more valuable lessons.

4.6.2 General Observations on Startups in Power Devices

Startups mostly in USA, two each from United Kingdom and Canada, and one each from France, Germany, Israel and Japan have been covered in the WBG device sectors of silicon carbide and gallium nitride. Common and some unique features of this sector, its nature and environment may be summed up as follows:

i. The immense activity in WBG devices field is a natural consequence of two driving forces. They are:
   a. Silicon-based power semiconductors are not the end though they have been dominating the market for a very long time, i.e., until now. Science, engineering technology and engineering change, and

they play a role in trying to find the next generation not only in the devices but in every field.

   b. The current market size is estimated to be around $15 to $20B per annum for silicon devices. The future given green revolution and use of solar to electric vehicles etc. for high-efficiency operation, may make the market size even bigger and estimates are thrown around for $30B. That leaves the scope for many startups to assume the transient nature of silicon devices dominance and better than those devices finding an immense role in the future and to realize a market share for themselves now and in the future.

ii. The knowledge base for moving into WBG devices rests with:
   a. Some (note that all universities may teach these subjects but not necessarily conduct research) research-based universities around the world and they happen to be high in the rank as well. They are, to name a few from the startups discussed earlier:
      - North Carolina State University, Raleigh, NC, USA
      - Massachusetts Institute of Technology (MIT), Cambridge, MA, USA
      - University of California, Santa Barbara, CA, USA
      - Stanford University, CA, USA
      - Rutgers University, NJ, USA
      - Cambridge University, Cambridge, UK
      - Warwick University, Warwick, UK
      - Kyoto University, Kyoto, Japan
      - University of Ulm, Germany
   b. Major and well-known companies, to name a few, such as General Electric, Rohm, Cree, Infineon, Mitsubishi, Panasonic, Samsung, International Rectifier, Fujitsu, Furukawa Electric, Toshiba, ABB, Fairchild Semiconductor, etc.

iii. Knowledge transfer from universities comes in the form of patents held by the university, their faculty serving as chief technology officers, graduate students and faculty going to found or become key employees.

iv. Expertise from industries comes as the individual researchers leave these companies since they do not see an opportunity to pursue their research and business direction within the constraints of the large companies' management and their strategic goals.

v. Founders' educational and professional backgrounds: In this field, most of the founders have a Ph.D. degree background, and very few only with bachelor degrees. But almost all of them have experience in research, and some with senior business and technology management experience at visible levels. The knowledge base of this field is not extensively taught in-depth at the undergraduate level and therefore, the opportunities for bachelor degree holders to be at the cutting edge and to be recognized are highly remote. In spite of this disadvantage, some have learned and advanced their knowledge in the companies that they worked for and emerged to launch startups, such as Navitas Semiconductor. It is more of an exception than the rule in this field. Note that such depth and breadth of research knowledge requirement is demanded in this sector is unique. Such is not the case in many fields of startup, say, many aspects of information technology.

vi. The startups have staffed their upper management from professionals hired from large companies and with immense field and/or corporate experience and extensive contacts but with engineering backgrounds and experience (but not master's in business degree!).

vii. Most of the startups have raised a phenomenal amount of funding showing the prospects in this field. They are all located in places where the venture capital companies are or very close by, i.e., in major cities and in and around Silicon Valley in USA except Raleigh, NC, USA which is a medium-sized city. The foreign companies have a similar pattern in terms of their location. As the real estate people say, it is location, location and location that matters the most not only in their business but in the startups as well.

viii. Strategies on seeking investment capital: Not all startups go for large investments as seen from UK startups and GeneSic Semiconductor, USA. Some startups win government research contracts and grants and steadily develop their product lines through that means. It may be due to the founders' desire for independence from investors and they want to charter their own path at their own pace. Note that these companies could raise substantial investment as they have strong intellectual property and product array either emerging or already in the market. Strategy on investment capital is individual startup founder(s) based and in many countries there are options like the one commented upon.

ix. The startups have all a fair amount of patents (Transphorm as much as 1,000 patents/applications) covering their technologies and this is an important requirement in this field since they all have to stand against (at least until they are acquired by) the giant companies in this field with thousands of patents, trade secrets. The creation of innovations and acquiring the rights to them via patents is a long, and expensive process with the result that huge investments are required ($245M in the case of Transphorm). There is another technique to broaden the base and increase the intellectual property portfolio by entering into a strategic partnership with big companies in the field.

x. Strategies for products: Many companies pursue different strategies. For example, Efficient Power Conversion (EPC) has focused on GaN devices for <350 V rating even though they have the expertise to go to higher voltages and with their normally-off enhancement-mode GaN devices. They did that because they see immense opportunities in many high-volume applications. Contrast to that is Transphorm efforts in higher voltages of 650 and 900 and probably even more in their plans and mainly focused on cascode structures for their GaN devices. They seem to target applications at this voltage level because of opportunities in the grid, solar, electric vehicle drives related applications, again all of them high volume, partly may be influenced by their strategic partnerships with giant companies. VisIC Technologies is focused on 650 and 1,200 V devices and not at all in the lower voltages even though may have the capabilities to do so. This is an area where each company has its own vision and the guiding reasons for that may only be inferred and guessed at the best from outside but they remain exclusively with the upper management.

xi. There is an outstanding outreach to potential customer base through demonstration/evaluation boards, application notes, technical papers and presentations in various trade and IEEE conferences and one on one working with customers for high-volume applications. The amount that is spent on creating such tools/entities and supporting them with their employees is very high and in many cases, if all else fails, they can start manufacturing their application products to the customers directly and has the potential to bypass many power electronic companies in the market for selling power converters for various applications. That is made feasible since many of the startups have patents on applications also apart from the devices, their composition, designs and manufacture. The fact that they are not resorting to these application sales is that they need these customer companies to buy their primary product of devices.

The take away for future entrepreneurs from these observations are:

i. Equip oneself with the cutting edge research and development experience by pursuing an advanced degree in an institution where well-recognized faculty and facilities exist to learn from. This is almost a must for startups in electrical engineering.

ii. Find employment with large companies, preferably in their R&D groups (which are all shrinking nowadays) to learn more about the modus operandi and research pursuits.

iii. Cultivate strong connections through research during the study and during working in companies with people of similar capabilities and as well in capabilities that the founders lack.

iv. Strategies on intellectual property development, product development, customer outreach and raising investment capital need intensive work. It is better to have partners to start with so that they all could be handled with competence from day 1 of planning the startup.

iv. The choice of startup location is very important for success and that should not be under-estimated.

4.7 EXERCISE PROBLEMS

1. Research and estimate the combined IGBT and MOSFET devices market size in your country.
2. Estimate the market size of power conversion systems, particularly in dc to dc converters, ac to dc converters including power factor correction systems and dc to ac inverters in your country of study.
3. Estimate for China alone on the topic of exercise problem 1.
4. Estimate for China alone on the topic of exercise problem 2.
5. Compare the market size of China, USA and Europe in categories 1 and 2.
6. Based on the result of problem 5, discuss techniques for market success when a startup is founded in the categories 1 and 2 from the understanding gained from the startups described in this chapter and from observations on them.
7. The market in India is not as big as these countries discussed but estimate the size now and in the future, say by 2030.
8. Discuss the startups in China in the power devices market. Note that it is difficult to obtain such information in the English media. It is important for any future entrepreneur to have this information and may serve one well.
9. Why is that there are many GaN device startups in North America compared to SiC device startups?
10. Why there are a very small number of startups, in the rest of the world, in GaN and SiC devices?
11. If there are no startups in WBG devices in your country, is it attractive for the teams to start one there? What will be the advantages and disadvantages that the teams will face in starting, making it successful and exiting?
12. Even advanced countries such as Japan, Germany, France and United Kingdom have a few (one or two) startups in WBG devices. Will they have a better chance of success compared to a large number of startups in the USA and give your reasons and back them with any data and rationale.
13. Analyze the varieties of GaN FETs, and their realization, cost implications and market scope.
14. Silicon substrate is much preferred for their low cost and ability to use current fabrication methods again at a lower cost. Are there any barriers to this for going to higher voltages, say, 3,000 V in GaN devices?
15. SiC devices are mostly aimed for high voltages >600 V to start with and desired for making them to ratings of few kV. How will this impact power electronic converters and their control?
16. What are the chances of SiC devices ever to come to use in low power electronic converters?
17. What are the chances of GaN power devices ever to come to use in high-power electronic converters, say in 50 kW and higher levels, and if so, give an estimate of time.

18. Because of lower input and output capacitances in the WBG devices, it becomes necessary to be very conscious of leakage inductances in the circuit layouts as they will affect the losses as well as limit the switching frequencies. Will that affect their adoption in newer products? If not, why?
19. Enhancement and depletion-mode GaN FETs are normally-off and normally-on devices. Which is preferable in applications and why?
20. Depletion mode FETs are also made to be normally off when they are in a cascade structure with low-voltage Si MOSFET in series. Will that add to the cost and if so, in practice will it affect their adoption in high-volume applications that are known to be very cost-sensitive?
21. Very few products are available in GaN related devices from a small number of startups and one or two large semiconductor device companies. Is such a situation conducive for power electronic companies to invest in new product development? If yes and if no, enumerate your reasons.
22. Gate drive control incorporation within the device packaging will be of great help to the designers of application products. Discuss how many startups do that and some have walked away from that option, at least for now. Discuss the merits and demerits of such approaches and how they are likely to impact on the success/failure of the startups. What approach will you advocate if you are assigned the responsibility to make a decision in this aspect.
23. Device startups spend significant resources to popularize their products through evaluation boards, design guides and application notes for many high-volume applications in power electronics and other fields. Will they help to launch new startups in the power electronics market around the world that use only WBG devices in the applications?
24. With all the expertise the device startups have which they freely give away, why can't they commercialize them, say, by launching subsidiaries? Will they cut into their main objective of getting into the devices market? What will you or your team do in such a circumstance as this?

4.8 CLASS PROJECTS

1. Select any two startup companies in each sector of SiC and GaN in the USA and compare them on the basis of the following:
   i. Technology
   ii. Products
   iii. Initial market entry issues
   iv. Their outreach to customers and means and how well they do
   v. Funding and investors
   vi. Any additional information you can gather from the media

Give your comments indicating which is better and how these factors will impact their success and which one is more likely to succeed in the market and what likely is their exit scenario and give your rationale.

2. The class may divide into many groups and each group can tackle the Project in 1 and then discuss it in the class. Based on the discussion, each group with the winning startup will be taken to the next round by combining two groups that will have two startups for comparison study and likewise it can proceed until the entire class gets to estimate which one startup is the most likely to be the biggest winner. Discussions will increase our understanding of startups. It is hoped that every startup will win, but the class has to rank them based on the progressing stages as the number of members in the group goes from two to four, four to eight, etc. and each stage winner goes up in the rank with the final winner being the most winning startup.
3. Suppose the team wants to start a company based on SiC power devices. How will you go about it and develop an elementary business plan to accomplish it?
4. Complete a project similar to Project 3 with the modification that the startup is to be in GaN power devices instead of SiC devices.
5. Develop a marketing plan for WBG power devices and how that will be different from what the startups are currently doing. You may choose one startup for your basis of comparison. Choose the best marketing plans from the class and take the initiative to communicate with the startups and see the response. Note that it may lead to the team/individuals getting an invitation to present to the company and may end in internship and/or employment.
6. Each group comes up with a startup in the technology domain and discuss various ideas for your technologies. Then build on to the feasibility study and let two groups merge into one after discussion and choosing only one out of the two. Likewise, keep moving the process of elimination until the class gets to choose the final startup idea from the groups.
7. Conceive a new application for GaN devices (excluding all the applications cited in the startup's websites), simulate to assess its feasibility and bring out the positives compared to existing silicon device-based solutions for the application.
8. Examine Project 7 whether it could stand out as a startup by doing market research, writing a business plan and sharing it with seed investment fund companies in your area for their feedback and possible funding. Most of the investment groups may not fund until experimental verification is made available but it is worth listening to their feedbacks and learn from them.
9. If the study results from Project 8 is not favorable for a possible startup, then try to contact the GaN startups and see whether anyone will be

interested to work with the team and, if so, whether they could give grants and/or assistance to the university and team to continue this work to experimental setup and verification of the conceived application and assist the funding startup to benefit from the team's contribution and efforts. This is likely to lead to internships for the team members and possible job offers from the funding/supporting startup.

10. Let teams pick whether Project 9 should precede Project 8 or another way about depending on which option will best serve their interests. Discuss the pros and cons of the order in which the project will be chosen.
11. Apply Projects 7, 8 and 9 for SiC-based devices.
12. Intellectual property (IP) is one of the key factors that will determine the success of the startup and its viability in the market place vis-à-vis its competitors. Each team can pick on one topic of IP such as device structure, manufacturing and packaging, and power device application for a problem of interest in the market domain. For the chosen topic compare any two startups' granted patents and pending patent applications and discuss the merits and demerits of them. Each team can choose a different topic and can make individual presentations to the class. That will get a good picture of the strengths and weaknesses of the chosen startups.
13. The outreach and publicity approach seems to be very standard with all the WBG devices startups. Are there ways that could be formulated and worked on so that they can be individualistic and distinct for each startup? Could teams or groups take this assignment and improve them very specific to some startups and get in touch with the marketing teams of the startups to share the results of their studies? By doing so, the team members may find internships and future employment opportunities with the selected startups. An implementation of the studies for integration into startups' websites will pave the way to attract the attention of their management.
14. Research on fabrication costs for new devices. Explore how the decision to work with an existing company with foundry, thereby going fabless, is made. Identify the foundries that are open to startups and their modes of operation in working with potential clients. If possible, to get an insight into this very difficult part of the business, approach experts in the area, both professors and practicing electrical engineers, to mentor and guide the teams on this topic. This is one of the ways future entrepreneurs can learn the intricate practical aspects of manufacturing in the power semiconductor device.
15. Consider an existing application of SiC or GaN devices such as a buck synchronous converter or of the team's choice and assume the team wants to get into that business by having a startup. Assume that there are customers willing to buy form your startup and therefore there is no need to worry about demand. To come with a business plan, the team may estimate the required financial and manpower resources to

manufacture it in various quantities of (i) 1,000/year, (ii) 50,000/year, and (iii) 1,000,000/year. Include all choices: (i) including location that can be in your own country and abroad, (ii) part or full outsourcing such as the boards, assembly of the circuit on the board, packing for shipping, invoicing and accounting, marketing and sales, and their coordination from a central location, and (iii) fully under one roof by the team's startup. Include all expenses that may possibly be incurred such as legal services, rent, communication services, office equipment, etc. Multiple teams can work on this project sharing various tasks and a final report is assembled from their individual reports giving a full appreciation of this immense and important aspect of entrepreneurship to all the teams.

REFERENCES

GaN Devices and Applications

1. A. Lidow, J. Strydom, M. De Rooij and D. Reusch, *GaN Transistors for Efficient Power Conversion*, John Wiley & Sons, Hoboken, NJ, 2015.
2. I.C. Kizilyalli, "Vertical power electronic devices based on bulk GaN substrates", Avogy, Inc, http://cusp.umn.edu/assets/DC_Workshop_2015/Kizilyalli_Avogy_Day2.pdf.
3. Z. Liu, "Characterization and failure mode analysis of Cascode GaN HEMT, M.S. thesis, Virginia Tech, 2014.
4. H. Peng, "Gallium nitride power devices: Switching characteristics, gate drivers and applications", *Seminar, IEEE Power Electronic Society*, March 28, 2019.
5. E.A. Jones, "Review and characterization of gallium nitride power devices", M.S. thesis, University of Tennessee, August 2016.
6. U. Mishra, "Compound semiconductors; GaN and SiC, separating fact from fiction in both research and business", *IEEE-APEC*, June 2013.
7. R. Mitova, R. Ghosh, U. Mhaskar, D. Klikic, M.-X. Wang and A. Dentella, "Investigations of 600V GaN HEMT and GaN diode for power converter applications", *IEEE Transactions on Power Electronics*, vol. 29, no. 5, pp. 2441–2452, 2014.
8. D.B. Reusch, J. Strydom and J.S. Glase, "Improving high frequency DC-DC converter performance with monolithic half bridge GaN ICs", *IEEE Energy Conversion Congress and Exposition*, Montreal, pp. 381–387, 2015.
9. A. Lidow, D. Reusch and J. Glaser, "Getting from 48 V to load voltage: Improving low voltage DC-DC converter performance with GaN transistors", *EPC Presentation in IEEE-APEC*, March 2016.
10. J. Glaser, "Kilowatt laser driver with 120 A, sub-10 nanosecond pulses in <3 cm^2 using a GaN FET", EPC applications note related to LIDAR.
11. L. Xue, "Design considerations of highly efficient active clamp Flyback converter using GaNFastTM Power ICS", *Navitas Presentation in APEC*, 2018.
12. H.C. Wilhelm and D. Kranzer, "Application of a new 600 V GaN transistor in power electronics for PV systems," *In Proceedings of 15th IEEE EPE-PEMC*, Novi Sad, Serbia, pp. 1–5, 2012.
13. X. Zhao, "High-efficiency and high-power density DC-DC power conversion using wide bandgap devices for modular photovoltaic applications", 2018.

14. B. Hughes, J. Lazar, S. Hulsey, D. Zehnder, D. Matic and K. Boutros, "GaN HFET switching characteristics at 350V/20A and synchronous Boost converter performance at 1MHz," *In Proceedings of IEEE 27th APEC*, pp. 2506–2508, February 2012.
15. Y.-F. Wu, D. Kebort, J. Guerrero, S. Yea, J. Honea, K. Shirabe and J. Kang, "High-frequency, GaN diode-free motor drive inverter with pure Sine wave output", *Power Transmission Engineering*, vol. 6, no. 5, pp. 40–43, 2012.
16. E. Personn, "Practical application of 600 V GaN HEMTs in power electronics", *Professional Education Seminar S17, IEEE- APEC*, March 2015.
17. IEEE Power Electronics Society, "International technology roadmap for wide bandgap power semiconductors", September 2019.

SiC Devices and Applications

1. G. Vacca, "Benefits and advantages of silicon carbide power devices over their silicon counterparts", http://www.semiconductor-today.com/features/PDF/semiconductor-today-apr-may-2017-Benefits-and-advantages.pdf.
2. B. Sahan, et al., "Enhancing power density and efficiency of variable speed drives with 1200V SiC T-MOSFET", *PCIM Europe 2017,* Nuremberg, Germany, pp. 196–203, 16–18 May 2017.
3. M. Schulz, L. de Lillo, L. Empringham and P. Wheeler, "Pushing power density limits using SiC-JFET based matrix converter", https://www.infineon.com/dgdl/Infineon-PCIM_2011_Power_density_with_SiC_JFET-ED-v1.0- en.pdf?fileId=db3a3043399628450139b550fd257554.
4. B. Passmore and C. O'Neal, "High-voltage SiC power modules for 10–25 kV applications", *In SiC Power Modules Section*, www.wolfspeed.com.
5. R. Singh, "Ultra high voltage SiC bipolar devices for reduced power electronics complexity", https://www.arpa-e.energy.gov/sites/default/files/documents/files/SolarADEPT_Workshop_NxtGenPwr_Singh.pdf.
6. J.L. Hostetler, X. Li, P. Alexandrov, X. Huang, A. Bhalla, M. Becker, J. Colombo, D. Dieso, and J. Sherbondy, "6.5 kV enhancement mode SiC JFET based power module," *In Proceeding of WiPDA*, pp. 300–305, 2015.
7. S. Ji, Z. Zhang and F. Wang, "Overview of high voltage SiC power semiconductor devices: Development and application", *CES Transactions on Electrical Machines and Systems*, vol. 1, no. 3, pp. 254–264, 2017.
8. H. Alan Mantooth, M.D. Glover and P. Shepherd, "Wide bandgap technologies and their implications on miniaturizing power electronic systems", *IEEE Journal of Emerging and Selected Topics in Power Electronics*, vol. 2, no. 3, pp. 374–385, 2014.

Power Electronics Devices Market

1. Power electronics market size worth $39.22 billion by 2025, April 2017, https://www.grandviewresearch.com/press-release/global-power-electronics-market.
2. GaN semiconductor market to reach $4.3B by 2025, December 15, 2017, https://epsnews.com/2017/12/15/gan-semiconductor-market-reach-4-3b-2025/.
3. Gallium Nitride (GaN) Semiconductor Devices Market Analysis By Product (GaN Radio Frequency Devices, Opto-semiconductors, Power Semiconductors), By Wafer Size, By Application, By Region, And Segment Forecasts, 2018–2025, November 2017, https://www.grandviewresearch.com/industry-analysis/gan-gallium-nitride-semiconductor-devices-market.

4. Annual Report of Power America, 2017, https://www.poweramericainstitute.org/wp-content/uploads/2017/10/Power America_2017_Annual_ Report_ONLINE.pdf.
5. Si C & GaN Power Semiconductor Market Forecast, Trend Analysis & Competition Tracking – Global Market insights 2017 to 2026, January 2018, https://www.factmr.com/report/381/sic-gan-power-semiconductor-market.
6. Global Silicon Carbide Power Semiconductors Market is Expected to Reach US$ 1,164 Mn and growing at a value CAGR of 17.68% throughout the period of forecast 2017 to 2025, February 7, 2019, http://www.amecoresearch.com/press-release/global-silicon-carbide-power-semiconductors-market-296.
7. A. Lidow, J. Strydom, M. de Rooij and D. Reusch, *GaN Transistors for Efficient Power Conversion*, Wiley, Hoboken, NJ, 2014.

Arkansas Power Electronics International (APEI)

1. https://www.arkansasbusiness.com/people/aboy/478/arkansas-power-electronnics-international-inc.
2. https://www.cree.com/news-media/news/article/cree-acquires-apei-arkansas-power-electronics-international-inc.
3. https://talkbusiness.net/2015/07/cree-inc-acquires-fayettevilles-apei/.
4. https://talkbusiness.net/2015/07/apei-acquired-by-cree-will-remain-at-tech-park/.

Amantys

1. www.amantys.com.
2. https://www.crunchbase.com/organization/amantys#section-overview.
3. https://www.bloomberg.com/research/stocks/private/snapshot.asp?privcapId=111949062.
4. https://www.businessweekly.co.uk/news/hi-tech/german-owner-plans-new-amantys-powerhouse-cambridge-base.
5. https://www.transformers-magazine.com/issue1/2018-mr-acquires-power-electronics-business-amantys-transformer.html.
6. https://warwick.ac.uk/services/ventures/news/older_warwick_ventures_news/2011/amantys3_11_11/.
7. https://arm.com.
8. https://www.electronicsweekly.com/news/business/finance/amantys-appoints-new-ceo-2013-01/.
9. http://www.amantys.com/news/pr/release/25/en.

Anvil Semiconductors

1. https://warwick.ac.uk/services/ventures/news/older_warwick_ventures_news/2011/anvil/.
2. https://www.gov.uk/government/case-studies/anvil-semiconductors-lighting-the-way-on-led-technology.
3. www.anvil-semi.co.uk.
4. https://www.crunchbase.com/organization/anvil-semiconductors#section-overview.
5. https://www.bloomberg.com/research/stocks/private/snapshot.asp?privcapId=137242283.
6. https://gtr.ukri.org/organisation/71A79D5E-4B54-41DF-83B3-FD2A36E73F69.

7. https://www.cbinsights.com/company/anvil-semiconductors-patents.
8. https://www.gov.uk/government/case-studies/anvil-semiconductors-lighting-the-way-on-led-technology.
9. https://www.eenewsanalog.com/news/startup-formed-develop-cubic-gan-leds.

Ascatron

1. http://ascatron.com/.
2. https://www.bloomberg.com/research/stocks/private/snapshot.asp?privcapId=340548004.
3. https://www.acreo.se/media/news/ascatron-introduces-its-first-silicon-carbide-power-device-products.
4. https://www.crunchbase.com/organization/ascatron#section-overview.
5. https://www.owler.com/company/ascatron.
6. http://www.semiconductor-today.com/news_items/2017/sep/ascatron_180917.shtml.
7. https://arcticstartup.com/swedens-ascatron-raises-e4-mln-wrapping-a-round/.
8. https://pitchbook.com/profiles/company/155685-88.
9. https://www.viktoria.se/media/news/acreo-spins-off-the-new-company-ascatron-for-fabrication-of-silicon-carbide.
10. http://www.norstel.com/norstel-ascatron-join-forces-provide-complete-offering-sic-epitaxy/.

Avogy, Inc.

1. https://www.bloomberg.com/research/stocks/private/snapshot.asp?privcapid=269870795.
2. http://www.intelcapital.com/portfolio/company.html?id=19075.
3. https://www.crunchbase.com/organization/avogy#section-overview.
4. https://www.axios.com/mueller-investigation-donald-trump-reaction-38b664a0-a7cf-4ed3-9cce-2b4de08cef8c.html.
5. http://www.marketwired.com/press-release/avogy-inc-raises-40-million-series-b-round-led-by-intel-capital-1945095.htm.
6. https://www.democratandchronicle.com/story/money/2018/02/02/avogy-photonics-rochester-syracuse-canal-ponds/301744002/.
7. https://www.pntpower.com/avogy-inc-dead/.
8. http://www.chipworks.com/about-chipworks/overview/blog/a-peek-inside-the-avogy-zolt-laptop-charger-is-gan-really-there-yet.
9. https://wikidiff.com/heteroepitaxy/homoepitaxy.
10. https://en.wikipedia.org/wiki/Epitaxy.

Cambridge Electronics, Inc.

1. https://pitchbook.com/profiles/company/162201-88.
2. https://www.zoominfo.com/c/cambridge-electronics-inc/369723822.
3. http://www.semiconductor-today.com/news_items/2015/aug/cei_040815.shtml.
4. https://www.engine.xyz/founders/cambridge-electronics/.

Efficient Power Conversion

1. https://www.bloomberg.com/research/stocks/private/snapshot.asp?privcapId=99399790.
2. https://epc-co.com/epc.
3. https://www.owler.com/company/epc-co.
4. https://www.marketscreener.com/news/Efficient-Power-Conversion-EPC-Announces-Spirit-Electronics-as-Distribution-Partner-for-Defense-an--26649325/.
5. https://www.crunchbase.com/organization/efficient-power-conversion#section-overview.
6. https://pitchbook.com/profiles/company/83983-69.
7. https://markets.businessinsider.com/news/stocks/efficient-power-conversion-corporation-epc-receives-2018-top-10-power-products-award-from-electronic-products-china-magazine-21ic-media-1027534149.
8. https://www.barrons.com/articles/a-silicon-pioneer-plays-taps-silicon-and-power-cords-1467239722.

Exagan

1. http://www.exagan.com.
2. http://www.exagan.com/en/news/releases/release-20/.
3. http://www.exagan.com/en/company/management-team/.
4. https://powerelectronicsworld.net/article/104624/Exagan_targets_GaN_charger_markets.
5. https://www.eenewspower.com/news/control-material-key-gan-success-says-exagan-boss/page/0/1.
6. https://www.electronicsweekly.com/news/products/power-supplies/electronica-exagan-demos-multi-kw-gan-psus-2018-11/.
7. https://www.xfab.com/about-x-fab/news/newsdetail/article/x-fab-and-exagan-successfully-produce-first-gan-on-silicon-devices-on-200-mm-wafers/.
8. https://automotive.electronicspecifier.com/power/electronica-2018-high-power-conversion-solutions-for-evs.
9. http://www.exagan.com/en/news/releases/stmicroelectronics-to-acquire-majority-stake-in-gallium-nitride-innovator-exagan-35/.

Flosfia

1. https://www.bloomberg.com/research/stocks/private/snapshot.asp?privcapId=270357221.
2. http://flosfia.com/english/.
3. http://rocakk.sakura.ne.jp/index/struct/wp-content/uploads/Company-Profile_20170728.pdf.
4. https://www.ut-ec.co.jp/english/portfolio/flosfia.
5. https://www.denso.com/global/en/news/news-releases/ 2017/20180104–g01/.
6. http://japan-product.com/denso-flosfia/.

7. http://www.semiconductor-today.com/news_items/2018/jul/flosfia_170718.shtml.
8. https://patents.justia.com/assignee/flosfia-inc?page=2.
9. https://www.eenewsanalog.com/news/japan-startup-reports-first-normally-gallium-oxide-mosfet/page/0/1#.
10. http://flosfia.com/struct/wp-content/uploads/Press-Releae_FLOSFIA_20170302.pdf.

GaNPower International, Inc.

1. http://iganpower.com/.
2. http://www.ganpowersemi.com/gsyg.
3. https://en.wikipedia.org/wiki/Midea_Group.
4. https://www.ic.gc.ca/app/scr/cc/CorporationsCanada/fdrlCrpDtls.html?corpId=9325891.
5. http://www.ganpowersemi.com/cpqj.
6. https://patents.justia.com/assignee/ganpower-international-inc.
7. https://www.knowmade.com/downloads/gan-devices-for-power-electronics-patent-investigation/.
8. https://compoundsemiconductor.net/article/103968/Cree_Licenses_GaN_Power_Patents_To_Nexperia%7BfeatureExtra%7D.

GaN Systems

1. http://www.gansystems.com.
2. https://www.teledyne-e2v.com/news/e2v-and-gan-systems-global-alliance-extends-the-power-of-gan-to-aerospace-and-defense-industries/.
3. https://www.bloomberg.com/research/stocks/private/snapshot.asp?privcapId=142687796.
4. https://www.crunchbase.com/organization/gan-systems#section-overview.
5. https://markets.businessinsider.com/news/stocks/powersphyr-and-gan-systems-lead-the-wireless-charging-revolution-1027460178.
6. https://www.businesswire.com/news/home/ 20150507005393/en/GaN-Systems-raises-USD20-Million-Series-Financing.
7. https://obj.ca/index.php/article/ottawa-tech-firm-gan-systems-revved-about-bmw-investment.
8. https://www.mouser.com/pdfDocs/343654_GaNSystems__GN001_How_To_drive_GaN_EHEMT_Rev_20160426.pdf.
9. https://www.sensorsmag.com/components/gan-systems-demonstrates-wireless-power-transfer-and-commercial-power-systems.
10. https://www.nucurrent.com/ 2018/04/12/nucurrent-partners-with-gan-systems-to-expand-wireless-charging-product-offerings-with-150-watt-solution/.
11. http://www.semiconductor-today.com/news_items/2019/mar/gansystems_200319.shtml.
12. https://www.globenewswire.com/news-release/2019/03/05/1748178/0/en/GaN-Systems-Debuts-Suite-of-Low-Cost-High-Performance-GaN-Power-Transistors.html.
13. https://patents.justia.com/assignee/gan-systems-inc?page=2.
14. https://stks.freshpatents.com/Gan-Systems-Inc-nm1.php?archive=2015.

GeneSiC Semiconductor, Inc.

1. http://www.genesicsemi.com.
2. https://www.bloomberg.com/research/stocks/private/snapshot.asp?privcapId=204538365.
3. https://entrepreneurialenergy.wordpress.com/category/power-electronics/.
4. https://www.sbir.gov/sbirsearch/detail/833093.
5. http://www.semiconductor-today.com/news_items/2014/OCT/GENESIC_281014.shtml.
6. http://www.power-mag.com/news.detail.php?STARTR=0&NID=50.
7. https://www.powerelectronics.com/discrete-power-semis/sic-rugged-power-semiconductor-compound-be-reckoned.

MicroGaN GMBH

1. https://www.bloomberg.com/research/stocks/private/snapshot.asp?privcapId=29535111.
2. https://pitchbook.com/profiles/company/57184–12.
3. https://ieeexplore.ieee.org/document/6776783.
4. http://venture-q.com/pdf/Venture-Q%20Article-Web.03.05.13.pdf.
5. http://www.microgan.com/includes/about/iis.pdf.
6. http://www.microgan.com/index.php?site=company§ion=team.
7. https://www.electronicsweekly.com/market-sectors/power/gan-is-coming-silicon-beware-2012-05/.
8. https://compoundsemiconductor.net/article/88690/MicroGaN_takes_nitride_transistors_into_the_third_dimension/feature.
9. http://www.thebusinessinvestment.com/2019/03/20/wide-bandgap-semiconductors-market-demand-is-increasing-rapidly-in-recent-years-2017-2027/.
10. https://patents.justia.com/assignee/microgan-gmbh.
11. https://www.apollo.io/companies/microGaN-GmbH/556cf0d4736964l1debd5500.

Navitas Semiconductor

1. https://www.navitassemi.com/.
2. https://www.ganfast.com/in-the-media/.
3. https://www.cbinsights.com/company/navitas-semiconductor-funding.
4. https://pitchbook.com/profiles/company/119523-61.
5. https://www.crunchbase.com/organization/navitas-2#section-overview.
6. https://www.businesswire.com/news/home/20160316005634/en/Navitas-World%E2%80%99s-GaN-Power-ICs-Leave-Silicon.
7. https://www.electronicspecifier.com/companies/navitas-semiconductor.
8. https://www.owler.com/company/navitassemi.
9. https://www.bloomberg.com/research/stocks/private/snapshot.asp?privcapId=306401012.
10. Products list from Digikey, https://www.digikey.com/en/supplier-centers/n/navitas-semiconductor?utm_adgroup=Integrated%20Circuits&slid=&gclid=EAIaIQobChMIjZ_9l_-L4QIVAR-GCh1zwgZyEAAYAiAAEgL98vD_BwE.
11. On China office opening, https://www.eenewseurope.com/news/power-gan-pioneer-opens-china-0.

NexGen Power Systems

1. http:// nexgenpowersystems.com.
2. https://www.crunchbase.com/organization/nexgen-power-systems#section-overview.
3. https://www.bloomberg.com/research/stocks/private/snapshot.asp?privcapId=556109329.
4. https://evertiq.com/design/44544.
5. https://patents.justia.com/assignee/nexgen-power-systems-inc.
6. https://i3connect.com/company/nexgen-power-systems.
7. https://www.electropages.com/ 2018/08/gan-technology-aiding-power-supply-miniaturisation/.
8. https://policybynumbers.com/nexgen-in-syracuse-throwing-good-money-after-bad.

Nitronex Corp.

1. https://electroiq.com/2001/01/nitronex-producing-gan-based-hemts-on-4-in-si-wafers/.
2. https://projects.ncsu.edu/research/results/vol4/10.html.
3. https://www.edn.com/electronics-news/4061660/Nitronex-raises-21-8-million-for-GaN-RF-chips.
4. https://compoundsemiconductor.net/article/80342/Nitronex_Lands_Additional_Venture_Capital.
5. https://www.bizjournals.com/triangle/news/2012/06/26/durhams-nitronex-sold-to-gaas-labs.html.
6. https://www.crunchbase.com/organization/nitronex#section-overview.
7. https://www.bizjournals.com/wichita/stories/2002/11/25/focus2.html.
8. https://electroiq.com/2001/01/nitronex-producing-gan-based-hemts-on-4-in-si-wafers/.
9. https://projects.ncsu.edu/research/results/vol4/10.html.
10. https://www.edn.com/electronics-news/4061660/Nitronex-raises-21-8-million-for-GaN-RF-chips.
11. https://compoundsemiconductor.net/article/80342/Nitronex_Lands_Additional_Venture_Capital.
12. https://www.bizjournals.com/triangle/news/2012/06/26/durhams-nitronex-sold-to-gaas-labs.html.
13. https://www.crunchbase.com/organization/nitronex#section-overview.
14. https://www.bizjournals.com/wichita/stories/2002/11/25/focus2.html.

SemiSouth

1. https://www.crunchbase.com/organization/semisouth#section-overview.
2. https://www.sandia.gov/ess-ssl/docs/pr_conferences/ 2008/ritenour_semisouth.pdf.
3. https://www.businesswire.com/news/home/ 20080410006219/en/SemiSouth-Laboratories-Announces-Efficiency-Improvement-Solar-Inverter.
4. http://www.woodsidecap.com/woodside-capital-partners-serves-financial-advisor-semisouth-laboratories-inc//.
5. http://www.epdtonthenet.net/article/46368/Highest-output-power-density-inverter-uses-SiC-JFETs-from-SemiSouth.aspx.
6. http://www.epdtonthenet.net/company/21721/SemiSouth-Laboratories--Inc-.aspx.

7. https://www.businesswire.com/news/home/ 20060824005611/en/SemiSouth-Celebrates-Grand-Opening-New-Silicon-Carbide.
8. https://www.businesswire.com/news/home/ 20080129006181/en/SemiSouth-Announces-New-SiC-Power-JFET-Patents.
9. https://www.businesswire.com/news/home/ 20110729005302/en/SiC-JFETs-SemiSouth-target-high-end-audio.
10. https://msbusiness.com/ 2013/03/high-tech-manufacturer-semisouth-laboratories-closes-after-decade-plus-run/.
11. http://www.semiconductor-today.com/news_items/2012/OCT/SEMISOUTH_241012.html.

Transphorm, Inc.

1. https://www.transphormusa.com.
2. https://www.crunchbase.com/organization/transphorm#section-overview.
3. https://www.bloomberg.com/research/stocks/private/snapshot.asp?privcapId=104930879.
4. https://www.greentechmedia.com/articles/read/transphorm-raises-70m-to-scale-deployment-of-its-gan-based-semiconductors.
5. https://www.businesswire.com/news/home/ 20181213005155/en/Transphorm-Ships-Quarter-Million-GaN-Power-Devices.
6. https://www.transphormusa.com/en/news/transphorm-secures-investment-yaskawa-electric/.
7. https://www.zoominfo.com/c/transphorm-inc/345606240.
8. https://www.businesswire.com/news/home/20180604006523/en/Seasonic%E2%80%99s-New-High-Efficiency-1.6-kW-Platform.
9. http://www.finsmes.com/2017/11/transphorm-secures-15m-investment-from-yaskawa-electric.html.
10. https://www.fujitsu.com/jp/group/fsl/en/resources/news/press-releases/ 2015/0127.html.
11. https://www.furukawa.co.jp/en/release/ 2014/kei_140514.html.
12. https://www.fujitsu.com/global/about/resources/news/press-releases/2013/1128-04.html.
13. http://chipdesignmag.com/lpd/blog/2013/12/22/transphorm-and-fujitsu-integrate-their-gan-businesses/.
14. https://patents.justia.com/assignee/transphorm-inc.
15. https://www.transphormusa.com/en/gan-training/.
16. https://www.transphormusa.com/en/document/transphorm-pr-one-sheet/.

United Silicon Carbide, Inc.

1. https://pitchbook.com/profiles/company/99262-36.
2. https://www.bloomberg.com/research/stocks/private/snapshot.asp?privcapId=260665982.
3. https://patents.justia.com/assignee/united-silicon-carbide-inc.
4. https://www.ecomal.com/fileadmin/Datenblaetter/USCi/App_Notes/Overview-and-User-Guide-to-United-Silicon-Carbide_s-_Normally-On_-xJ-Series-JFET.pdf.
5. https://electronics360.globalspec.com/article/13616/apec-2019-united-sic-announces-partnership-with-adi.

6. https://electronics360.globalspec.com/article/13603/apec-2019-unitedsic-sees-greener-possibilities-with-silicon-carbide.
7. http://unitedsic.com/wp-content/uploads/2016/02/SiC-Research-and-Development-at-United-Silicon-Carbide-Inc.-Looking-Beyond-650-1200V-Diodes-and-Transistors-March-2015.pdf.

VisIC Technologies

1. https://visic-tech.com/.
2. https://visic-tech.com/products/.
3. https://patents.justia.com/assignee/visic-technologies-ltd.
4. https://www.zoominfo.com/c/visic-technologies-ltd/357329156.
5. https://pitchbook.com/profiles/company/58752-46.
6. https://www.crunchbase.com/organization/visic-technologies#section-funding-rounds.
7. https://visic-tech.com/about/investors/.
8. https://www.prnewswire.com/news-releases/visic-technologies-ltd---new-product-generation-industrys-record---gallium-nitride-device-with-highest-efficiency-at-highest-frequency-300279825.html.
9. https://en.globes.co.il/en/article-power-electronics-co-visic-technologies-raises-116m-1001172050.

5 Entrepreneurship in Power Electronics

5.1 INTRODUCTION

Applications continuously emerge in every field, and power electronics is not an exception. They are ushered in by technological demands of, for example,

i. Renewable energy both stand-alone and grid-connected types,
ii. Electric vehicles for their charging from the grid and feeding the grid as well,
iii. High efficiency and performance enhancements with wide-bandgap power semiconductor devices in newer converter applications as well as in older equipment,
iv. Power system grid performance when distributed energy sources are connected and to handle rapid load variations with no significant variations in their voltages, and controlled spikes in them, and frequency,
v. Bidirectional power conversion and large energy storage systems,
vi. Microgrids and home energy management systems,
vii. Modernization of circuit breakers from electro-mechanical to solid-state base, and
viii. Wireless charging for mobile phones, lights, laptop computers, electric vehicles, and other applications.

Power electronics entrepreneurship is starting to flourish in the above applications and this chapter presents some of them, mostly from USA and some from abroad for which a reasonable amount of information is available. The market potential for each application is researched and included in the chapter providing the reason for springing startup efforts and possibly more for them in these applications. Evolving a new technology base for most of the application sectors is presented in a way that their connection to their early power electronics knowledge at undergraduate level is made obvious. This removes the mystery surrounding most of the evolving technologies and opens their eyes to their own effort to tackle these application challenges. Even though very few startups from outside of USA are presented, it certainly does not mean that there are no activities in this domain anywhere. Readers are encouraged to dig into them on their own to learn as well. The methodology and presentation follow that of Chapter 1 and objectives similar to that given in Chapter 4, to the extent possible and they could form the basis for their own studies.

5.2 CONTROL CIRCUITS

Most of the control circuit development, production and market are with giant multibillion-dollar companies. The high cost of development requiring fabrication facilities discourages startup in this domain with small funding. The environment for such activity is tough and it is reflected in the lack of startups. The ones that come along do not survive on their own for a long time and get absorbed by the big companies. Partly power electronic and control circuit-based company, Powervation Ltd., is presented in the following.

5.2.1 Powervation Ltd.

This is in the area of high-frequency, high-efficiency dc to dc power conversion. It is founded in 2006 in Limerick, Ireland. It is a technology spin-off from the University of Limerick, Ireland. The founders and their backgrounds are:

i. Dr. Karl Rinne: He also served as its Chief Technology Officer. He is a lecturer in the University of Limerick in its Electronics and Computer Engineering Department. He is also the director of the Circuits and Systems Research Centre in the department where he supervises doctoral researchers.
ii. Dr. Eamon O'Malley: He was a doctoral student of Dr. Karl Rinne. He is currently associated with Dr. Karl Rinne's research group at the University of Limerick after having served in Powervation Ltd., for a long time.
iii. Mr. Alan Dunne: He is in the business area and served as General Manager of Powervation Ltd.
iv. Mr. Antoin Russell: He also is in the business area and served as Chief Strategy Officer and Director.

Technology: Digital dc/dc controllers, IC solutions for power point-of-load systems and controller modules.

Products: Broadly can be classified into three subareas as:

i. Digital DC/DC controllers with Auto-Control technology
ii. Digital IC solutions for communications, computing, and power point-of-load systems
iii. DC/DC controller modules used in FPGAs, ASICs for networking, telecom and computers.

Funding: Raised $68.4M in total during its operation. Some of the investors are big companies (and not all) or their venture arms as can be seen from the following:

i. Scottish Equity Partners
ii. Intel Capital
iii. VentureTech
iv. Alliance, Braemar Energy Ventures
v. 4th Level Ventures, and
vi. Semtech

Note that this may not contain the full list of investors.

Exit: Acquired by Rohm Co. Ltd, in July 2015 for a sale price of $70M.

Observations:

i. This is a university spin-off with all the associated advantages of a pipeline for advanced researchers' availability for the company workforce and seed funding in some cases as well.
ii. Its computer chips are used in the data centers of Google and Amazon lending them great credibility.
iii. They raised large investment capital for a fabless company. Intel Capital and Semtech, both big companies were investors giving the startup an aura that any startup can dream of bestowing instant technical credibility.
iv. One of the investors, 4th Level Ventures, is chartered in its creation to supporting Irish university spin-offs and that is a big boon for this startup as for any other too.
v. Raised $68.4M (in <8 years) as per available information and established product lines and a customer base.
vi. Rohm, the multibillion-dollar Japanese company well known and reputed internationally, has been interested in having Powervation as its property is a feather in the caps of the founders and a recognition of their technology and product lines. Note that this also bestows a tremendous recognition for the Powervation's technical founders that may come handy one day, possibly in the near future.
vii. The sale price ($70M) indicates that the investors could not have made a profit if the investment figure ($68.4M) is right. But the consolation may be that partial investment has been regained and not everything has to be written off. The investors before the stage of IPO are not ordinary people and therefore the present investors at this private stage of the company may not have much regrets. Note that the founders assuming that they had shares in the company would have benefited and they can be winners provided there is no clause with the investors that their investments have to be paid first and then only the shareholders get their share!
viii. A number of reasons must be there for the sale and difficult to know them from outside of the startup.
ix. The technical founders of Powervation Ltd. (Drs. Rinne and O'Malley) have founded again a startup, PEK Semiconductor Ltd., based on dc-dc conversion technologies in 2018 and that indicates a belief that there

is a scope yet for this product area with technical advancements in the market place.

x. Tracking this new startup (PEK Semicondcutor Ltd.) to see how serial entrepreneurship in similar areas fan out for the founders and how easy or difficult will be the path to final exit is likely to yield valuable lessons to future entrepreneurs.

5.3 SOLID-STATE CIRCUIT BREAKER

Circuit breakers isolate the power circuits from the grid supply to various individual loads, say under short circuit or over the current limit set by the users, without creating destructive arcs, thus giving a safe working environment around electric circuits. The circuit breakers have been mechanical so far. The opening and closure of the circuit breaker are manual at low currents and voltages, and for higher ratings, it can be actuated electromechanically. It takes about 4 ms for these circuit breakers to open, and its speed of response is limited by it. When it opens, the inductance in the line and the load releases its stored energy to the metal contacts of the circuit breaker. A huge current results dissipating a sizeable amount of energy in the contacts and in possible partial or full melting of breaker contacts, subsequent deterioration and invariably in a short circuit of the supply itself. Much faster action has eluded the circuit breakers so far because of the limitation of the technology. The availability of wideband gap semiconductor materials and devices, particularly that of SiC opened up an opportunity into a solid-state circuit breaker (SSCB) application. Such an SSCB guarantees programmability, remote control, and faster action; all features not available with conventional circuit breakers. SSCB has been under development by a startup, Atom Power and is described in the following section.

5.3.1 ATOM POWER

Atom Power is founded in Charlotte, North Carolina, USA in 2014. Its cofounders with brief information about their current roles and backgrounds are:

i. Ryan Kennedy: Founding CEO, has a bachelor's degree in electrical engineering from UNC, Charlotte. He served as a project manager working on electrical projects in data centers, trading floor spaces, commercial and industrial buildings for about 10 years and has hands-on experience.

ii. Denis Kouroussis: He is the founding CTO of Atom Power. He obtained his Ph.D. in computer engineering and an M.A.Sc in electrical engineering specializing in power electronics and control circuits from the University of Toronto, Canada. He worked as a VP of engineering in Virtual Power Systems, in its early stage, with a focus on software-defined infrastructure for data center power delivery.

Technology and Products: They consist of fast-acting semiconductor-based circuit breakers with programmability, remote control and fast action with the use of the latest power electronic technologies. It is all based on their patent (pending) approach monitoring the current and detecting a potential fault and short circuit and then opening the circuit by means of semiconductor-based devices in a few microseconds thus protecting the system and the load. It has various products and they are SSCB, distribution panel and operating software named Atom Switch, Atom Panel and Atom OS, respectively. Atom Switch's earlier development version and current product in the market place are shown in Figure 5.1a and b, respectively. Additional applications of these products to cater to the soft start of motors, etc., with the programmability both at the start and at the end of the operations are in place. They have also developed a custom SiC-based module giving their SSCBs the capability to interrupt 100,000 A within a short time of few microseconds. Their product has acquired UL certification (as seen on their product in Figure 5.1b) which is an important and final step to the market place for electrical products.

Patents: Seven patent applications pending from 2017 onwards and some of them describing their SSCB subsystem using semiconductor devices with one or two in parallel for their operation.

Funding: Raised $6.7M and investors are Siemens, ABB, Eaton and Next47. The former three investors are the largest manufacturers of electrical equipment including circuit breakers and the fourth investor is a venture capital company backed by Siemens.

Observations:

i. SSCB is a challenge waiting to be tackled for the last 50 years and Atom Power's team came out with an innovative concept and implementation. In terms of innovation, it seems to stand out vis-à-vis the largest power sector mammoths such as Siemens, ABB, Eaton and others.

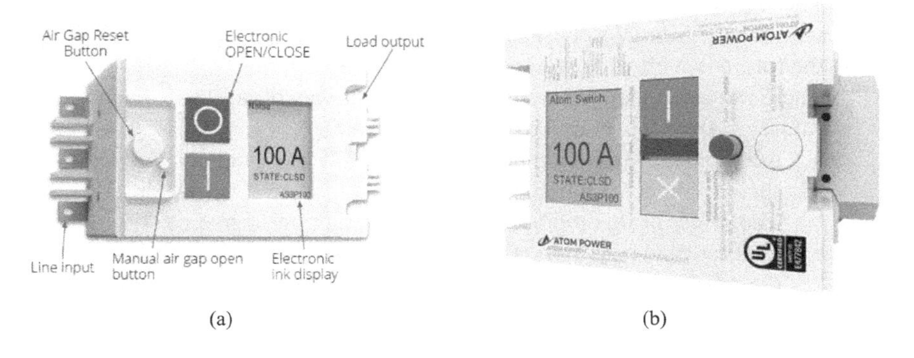

FIGURE 5.1 Atom power's three-phase SSCB. (a) Early version and (b) current product in the market place. (Courtesy of Atom Power.)

ii. The innovation has taken a very bold approach in incorporating the advantageous silicon carbide semiconductor in its product implementation thus gaining the ability to scale up to larger SSCB ratings, and faster response.
iii. Given the inherent capabilities of the SiC material and devices, the product designs need not be updated for a longer time to come with the result that the product development expense for the future is highly contained resulting in higher profits and greater flexibility to reduce the prices as against any current potential competitor's products.
iv. Atom Power's approach and innovation earned its recognition with funding from large companies. Their funding does not mean that they are not pursuing their own efforts in-house.
v. The product has been through rigorous testing and earned the UL certification which will accelerate its market introduction and hopefully its adoption.
vi. The cost of the SSCBs as against the conventional ones is not available for analysis and comparison. That is an important factor determining product adoption by customers. Particularly, this will be the case for big building electrical contractors. Note that Atom Power's products for the present are aimed at homes and buildings but can be extended to factories in which case the market size not only becomes larger but also the cost factor may not be as restrictive as the case of adoption in residential buildings.
vii. The hands-on experience of the CEO in this sector of the business has served the company in its product development, inventions, patents and packaging to address the needs of the targeted clients.
viii. CTO's background in power electronics has benefited in innovations, patents and implementations.
ix. The patent activity is reasonable for the size of its workforce and given its priority to get the products through testing and certification.
x. The aspiring entrepreneurs with the background in power electronics will benefit from closely reading the patent applications and see where the future is with wide bandgap devices and not silicon.
xi. There are a number of interesting factors to watch and learn from the future of Atom Power and they are as follows:
    a. Relationship evolution between itself and the funding companies such as Siemens, Eaton and ABB.
    b. The effect of this relationship on its market performance. Will that facilitate and help market entry and penetration and if so what forms the relationship may evolve?
    c. The product development within these large companies in SSCBs and how they will impact Atom Power.
    d. Will Atom Power sustain its independent status for time to come or will it be a potential candidate for acquisition by these companies or entirely another entity?

The investors and Atom Power's management may know some or all the answers and budding entrepreneurs will get to know in the future by tracking the developments. There may be interesting lessons to learn from them and could become useful in structuring a deal with giant companies in their own future entrepreneurial activities.

5.4 MICROGRIDS, HOME ENERGY MANAGEMENT SYSTEMS (HEMS)

Market for renewable solar power systems for homes is a growing one. The technology has been in place for solar power either tied to the power grid or off-grid for remote places where the utility grid is inaccessible or not available. The later version of off-grid solar system and its integration with the power grid, and battery energy storage and to provide two-way energy transfer from the battery and/or solar panel to the power grid and vice versa has been in development over the last two decades for homes and offices. It is a growing sector of business and application driven by: (i) awareness of clean power generation opportunities and needs for a clean environment for the future of mankind, (ii) partial or full independence from the power grid, (iii) opportunity to make revenue by drawing power when the utility rate is low and storing it in the battery and sell to the utility at its peak load/demand times when it pays money to the homeowner at a higher rate than it sells the energy, and (iv) opportunity to save money as the cost of the energy from home solar battery storage system is expected to be lower than that of the cost of the utility-supplied electric energy.

5.4.1 TECHNOLOGY

The system technology is described in the following. There are many different types of them:

i. The well-known and least expensive system is that of the photovoltaic (PV) panel connected in the back of the panel with an inverter (known as microinverter in the literature) whose output is tied to the grid and let it be termed as Scheme 1 and shown in Figure 5.2. The main advantage of the system is that the grid buys the power and pays the customer for it and the customer does not do anything beyond buying and installing the system. The disadvantages of the system are: (i) power for the home is

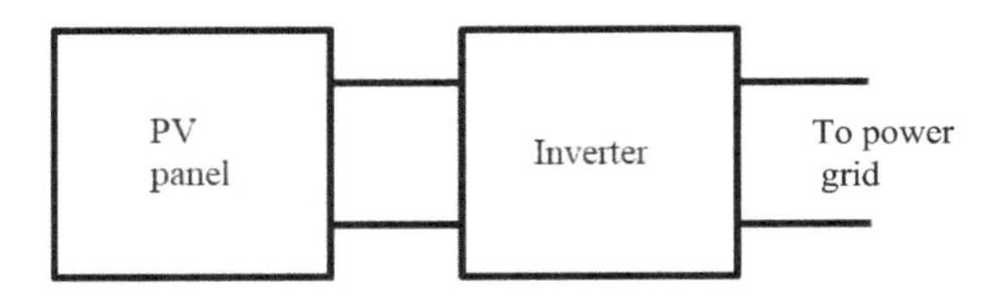

FIGURE 5.2 Scheme 1.

purchased through the grid only and hence may have to accept different rates for different times in some places, and (ii) in case of grid outage such as storms and hurricanes, solar power even if it is available may not be tapped for use by the customer unless a circuit isolator is in the system to disconnect utility supply during these times and another switch connector is available to the loads from the microinverter's output.

ii. Solar power with battery storage and off-grid system is for remote communities or where the grid is inaccessible. This version has the following system components: PV panel, dc to dc converter to regulate the output voltage of the solar panel, the output of dc to dc converter is to charge the battery as well to supply the home lighting and appliances including TV, fans, etc., via a dc bus (say from 40 to 48 V dc) and shown in Figure 5.3 and is termed as Scheme 2. The advantage of the system is that it is fairly low cost with a power capability of less than 500 W in general and targeted for rural households in developing countries, mainly in the far east and some parts of Africa. Its disadvantage is that the appliances have to be dc-based only and many such are available in the market, and in the case of ac motor-driven appliances, they have to come with an inverter for fans and refrigerators that are available with dc inputs also.

iii. Consider Scheme 2 but connected to the grid for charging the battery as and when necessary. Then the system with the considered modification will require an uncontrolled ac to dc, i.e., a rectifier, and a dc to dc conversion system to control the voltage for charging the battery. This allows the home solar system to have power at all times, day and night, which is its big advantage. Note that the system equipment grows to make it possible and hence the cost also increases. This scheme may be termed as Scheme 3 and shown in Figure 5.4, and note that the arrows in the figure indicate the direction of power flow between the subsystems.

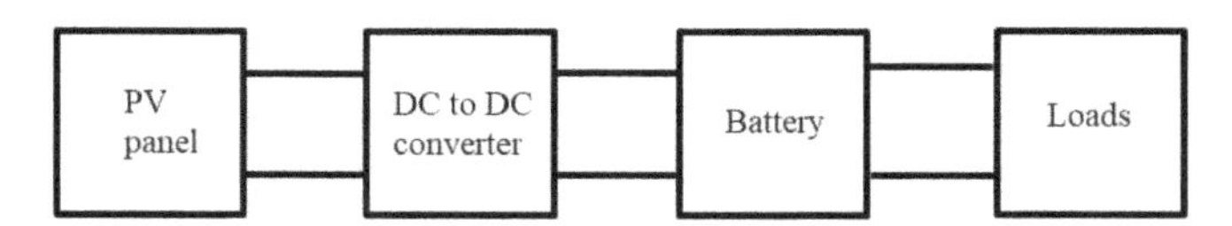

FIGURE 5.3 Scheme 2.

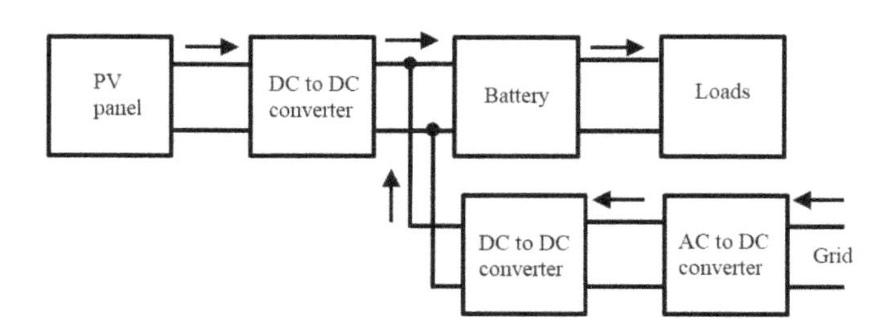

FIGURE 5.4 Scheme 3.

The main advantage of this scheme is that when the battery fails or being replaced, power is still available to the load through the grid.

iv. Another scheme where the output of the PV panel is given to both an inverter and a bidirectional (usually a buck-boost) dc to dc converter. The inverter output feeds the home load. Solar electricity when available charges the battery and when it is not available, the battery voltage is boosted and fed to the inverter so that its output continues to supply the loads. The inverter output can be connected to the grid through an isolator switch so that power can be transferred to the grid from the solar and battery system also. This is known as dc-coupled system and let it be referred to as Scheme 4 here and shown in Figure 5.5.

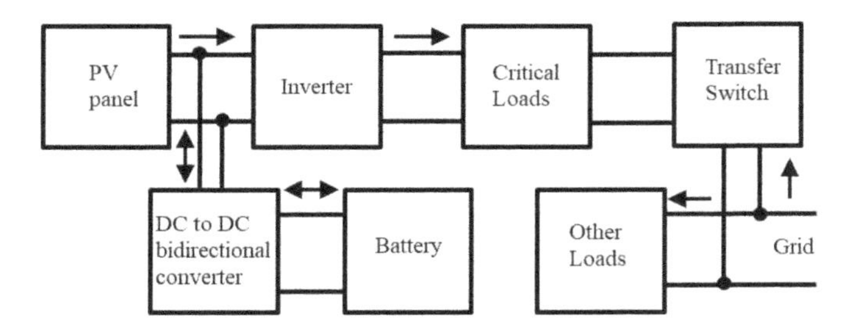

FIGURE 5.5 Scheme 4 (dc-coupled system).

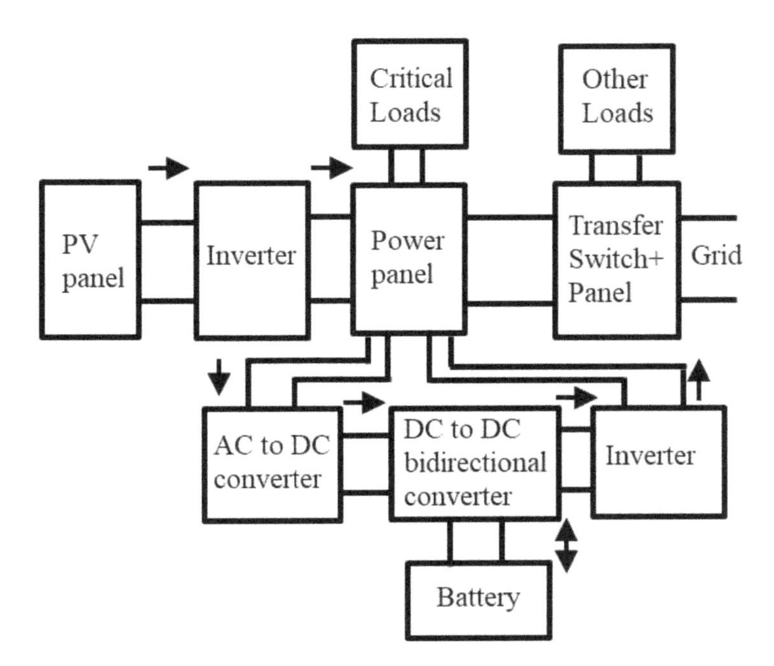

FIGURE 5.6 Scheme 5 (ac-coupled system).

v. A variant of the dc-coupled system is known as an ac-coupled system (Scheme 5) shown in Figure 5.6. Here the battery system can draw power from the grid regardless of whether solar electricity is available or not thus making it independent of each other. This then requires an inverter and a dc to dc converter so that grid side ac is converted and then controlled for charging the battery and when power needs to be drawn from the battery to the grid or to the home loads, the dc to dc converter boosts the voltage of the battery and feeds the inverter which then feeds the loads and/or the grid. Many isolator switches may be introduced at appropriate points depending on what types of options are required. The disadvantage of the system is higher cost due to additional inverter in the system and lower efficiency compared to the dc-coupled system.

5.4.2 Market for Home Solar Power System

Global market for solar, both utility-scale and rooftop categories is of interest to the businesses and more so to the startups. In the renewable arena, only the solar industry is relevant to the startups under discussion in this chapter. It is considered here for its market size and scope in the future. Other renewable sectors such as wind also require power electronics, machines and control, but that is not treated in this chapter as there is not much startup activity in that arena at least in USA. The total global installed capacity of the solar is around 509 GW, China leading with 34% and USA with 12% of that. The majority of that installation is at the utility scale and for commercial operation. A minority of that installation is in the rooftop category and it is a growing sector of business. Utility-scale solar installations are projected to grow annually from 70 GW in 2019 to 187 GW by 2023 (the low estimates are: 55–81 GW during the same period) around the world. Global annual growth in rooftop solar is 25–44 GW (low estimate) and 40+ to 76 GW (optimistic) from 2019 to 2023 (estimates).

All solar power installations require power electronic converters and control regardless of whether they are utility or rooftop solar categories. Utility-scale ones require larger power electronic converter units and hence only established large companies can meet their demands and it is very difficult and almost impossible for startups to penetrate that sector of the market.

Rooftop installations (both off-grid and on-grid types) require generally more of the power processing as they are installed in homes and offices. It is because of the interface with the utility for taking advantage of differential rates tied to the time of use. That invariably requires energy storage with battery banks and for such grid-edge systems, the power processing needs more than one power converter as seen in the home energy management system. Since the rooftop installations for homes and offices are of small in power, usually <10 kW, it is expected to be of lower revenue per customer, unlike the utility-scale category. Also, the rooftop categories cater to various segments of the population with different affordability and therefore require many models both in the system consisting of software, hardware, etc., and in finance. In many instances, such offerings are not available to meet all the customer base from established companies and that opens up ideally an area for startups to focus, innovate and benefit.

With the projected price of lithium-ion battery to go from $150/kWh to $96/kWh by 2025 and $76/kWh by 2030, rooftop units will grow immensely as their price level will come down and hence the market share will grow. Then their global growth if it is optimistic will be to the tune of 75 GW by 2023. US market is bound to add 2 GW capacity in rooftop solar capacity by 2023.

Conversion of this growth in kWh to sales figures in dollars is a different thing for different countries: In USA, the approximate cost is $1,200/kW for a HEMS with energy storage and ability to work with the grid, that will result in $2.4B for 2 GW capacity over the rooftop market category. That is a substantial market size in 2023.

For countries such as India (higher potential for rooftops), China, Europe and Australia, this market is bound to be much bigger and a multiple of US market potential. The potential in African countries is a different matter and sale prices to be much less per household can be expected and given the population size, that market will also be large. The potential for Middle East countries is immense for a number of reasons: governments can afford to subsidize much more than any other country in the world because of oil revenue and it is already taking off significantly in higher installation capacity.

Given such a scenario, the market size is in tens of $Bs around the globe for the rooftop category alone while that of the utility-scale type it will be at least five to ten times that of the rooftop market. The readers may get to work out market figures for any country from respective government reports which are much more rigorous and usually available for free.

5.4.3 Startups

Startups considered for this section target from remote villages in India and Africa to USA with their different requirements and markets.

5.4.3.1 Mera Gao Power

Mera Gao in the Hindi language means my village and the startup's electric power is meant for villages untouched by power grids. Mera Gao Power (MGP) is founded in 2010 by three co-founders (and some sources cite only the first two given below) and they are:

i. Nikhil Jaisinghani, Co-founder and past Executive Director: He is US born and worked in Africa and Asia and did consulting on energy in Nigeria. He worked with Brian Shaad in energy-related business ideas and moved to India.
ii. Brian Shaad, Co-founder and Director of Business Development: He is US born and worked with first co-founder from 2004 since their Nigeria days.

iii. Sandeep Pandey, CEO: He worked for 7 years in the micro-financing industry in Uttar Pradesh, India and he developed the payment system for MGP and very knowledgeable about the rural community dynamics of the Indian state in which MGP is focused on.

Technology: It is an off-grid company with a microgrid formed with solar power. Energy is obtained from a solar panel with a peak power rating of 1.2 kW, harvested and stored in batteries at 24 V, carried over aluminum wires (instead of copper wires which are more expensive) to supply about 50 homes or even more in small villages and each home being given 2 LED lights of 1 W capacity and a phone charger connection of 2 W capacity with a total of 4 W power rating. Note that LED lights made it possible for this scheme to be in a realistic application more than anything else. There is a power electronic connection to this technology for extracting maximum power at any given time of the day from the solar panels and in charging the batteries.

Products: Instead of kerosene lamps with very little light output and costing sizeable money for poor families, lighting homes and small shops with solar-generated power in a village, and providing charging energy for cell phones which are essential but not yet available in small villages in India thus serving millions at affordable prices. Currently, each home/shop owner pays about $0.42 per week with no other charges to MGP. Usually, the electric supply's maximum coverage distance from the solar energy station is not beyond 600 m but usually falls within 200 m so as to keep the line losses lower. For a village, it only costs about approximately $1,000 for the grid. MGP's service has a great impact on the villagers who are not very well off and how it is positively impacting them can be seen from Figure 5.7.

FIGURE 5.7 Impact of MGP's microgrid on villagers in India. (Courtesy of Mera Gao Power.)

Intellectual Property: There is none in this technology.

Funding: For the first 6 years, the capital was raised as part of grants and initial investment came from the first two co-founders. The investors are ElectriFi, Engie, Insitor Impact Asia Fund, National Geographic Corporation, The Energy and Resource Institute Corporation. Recently, it raised \$2.5M and had a grant from USAID.

Observations:

i. MGP is one of the finest examples of semi-social entrepreneurship where the products and services are directed to the most under-served population with elementary lighting and phone charging facilities. It seems that it is guided more by social service than making money.
ii. The founders are driven by entrepreneurship but are more inclined to serve the poorer population in power grid deprived villages. This is an inspiring story for many graduates to find a course of life that can bring joy and happiness to a lot of homes.
iii. This facility brought in by MGP has allowed a lot of children to study at night and prepare for the school, mothers to cook and shops to have extended hours in the evening to cater to customers without having to worry about kerosene lighting accidents.
iv. The founders' ability to put together a system of the solar-based grid to supply reliably to the population and control the cost is phenomenal.
v. MGP has currently 400+ employees and plans to serve nearly 50,000 households and 300,000 people with the last round of raised funds.
vi. The company has streamlined the installation (in 1 day), troubleshooting and training of the workers, weekly bill collection from the households and shops, and keeping a solid workforce in the villages. Note that MGP has become a substantial employer in villages where it is not easy to find jobs as the agricultural jobs are on the wane with significant automation and note they are seasonal too.
vii. This is not a traditional type of startup that engineers are surrounded by but there is something that they can contribute by improving the technology in some parts of it in the future.
viii. The expansion of the business can be along this line: To make each home independently have a mini microgrid sufficient to supply power for its use and may have more than the current allocation of 4 W but in the range of, say 50 W, to cover an electric fan and more lights. Such a mini microgrid may have its own small solar panel and battery system. It will lead to a compactly packaged product and result in manufacturing operations and enhanced revenue. There could be other opportunities that can be explored. It is interesting to see where MGP goes in the future.

5.4.3.2 Me SOLshare Ltd.

Me SOLshare (hereafter referred to as MSS) is a solar energy sharing facilitator company founded in Dhaka, Bangladesh in 2014.

Founder: Dr. Sebastian Groh who serves as its chief executive officer. He obtained his Ph.D. degree in economics from Aalborg University, Denmark and the Postgraduate School Microenergy Systems at the TU Berlin studying the role of energy in development processes, energy poverty and technical innovations, with a special focus on Bangladesh.

Technology: About 4.2 million solar homes are in Bangladesh. They are not interconnected among themselves and some of them are off-grid, and they do not utilize about 30% of their solar energy. MSS came out with technology to link up these solar homes and facilitate energy sale and purchase among themselves so that self-sufficiency and independence from the power grid can be achieved or energy could be sold to power utility also. Sometimes the integration among solar households is termed peer to peer (P2P) utility. This technology illustration by them is shown in Figure 5.8 with their product SOL box which is described later in the product section. It has the following features:

- Direct-Current (DC) bi-directional power meter,
- Solar charge controller, and
- Machine-to-machine (M2M) communications-enabled end-user device (that functions as an individual node of the electricity trading network).
- Connects both homes with and without solar energy.
- Off-grid network.

Another part of technology, shown in Figure 5.9 is to include small businesses with the following features:

- Powers solar and non-solar homes
- Powers small industries: Tailoring (sowing), rice mill, water pumps, fridges and eateries

Their technology encompasses tie-up with grid also as shown in Figure 5.10.

Products: Main product is an SOL box that connects the individual solar home grids among themselves in a neighborhood with the facility to meter the

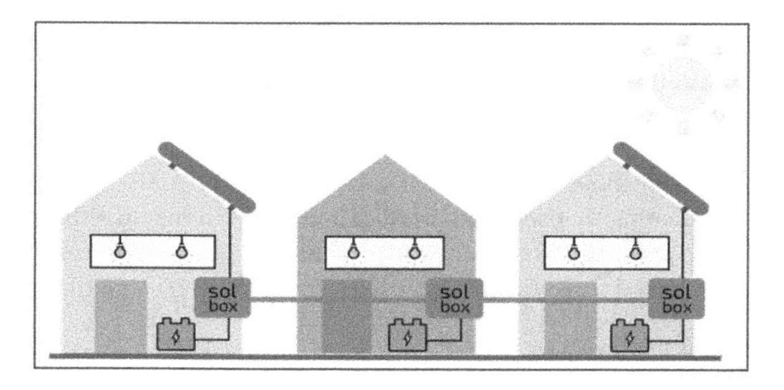

FIGURE 5.8 Basic technology. (Courtesy of MeSOLshare.)

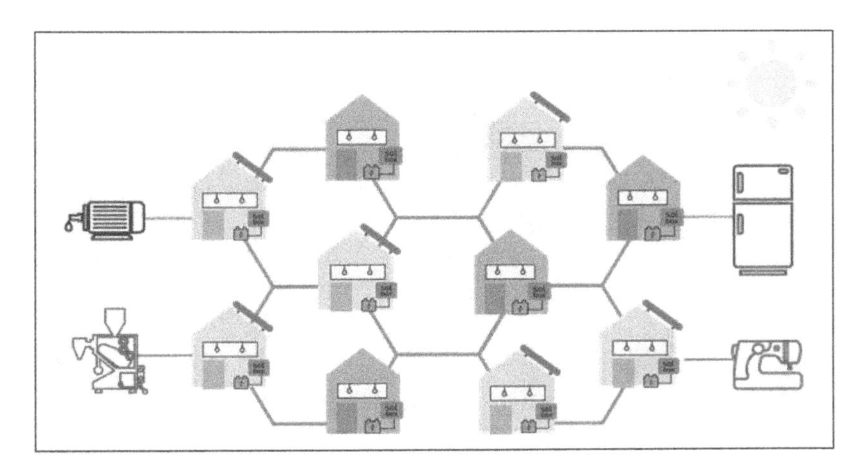

FIGURE 5.9 Solar microgrid including small business loads. (Courtesy of MeSOLshare.)

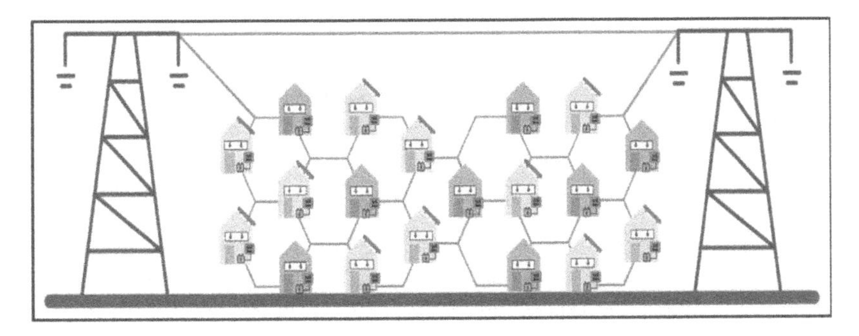

FIGURE 5.10 MSS technology deployment visualization also with power grid. (Courtesy of MeSOLshare.)

power flow among themselves so that buying and selling among themselves as well as to the utility can be monitored, logged and billed to every homeowner's advantage. It has both hardware contents to connect two solar grids and that is where power electronics plays a part and also in tie-up to the utility grid and a strong software part not only for control but also business and financial side of the P2P utility including billing. Their SOLbox is shown in Figure 5.11 and that is self-explanatory.

Their product line may incorporate blockchain technology to register energy transfer between the solar households to make all the transactions secure and transparent.

Intellectual Property: No patents can be found at this time but it is made clear that the products are based on their proprietary technology.

Funding: Raised \$1.66M in 2018 and the investors are: (i) Singapore-based IIX Growth Fund, (ii) Silicon Valley-based innogy New Ventures LLC, which

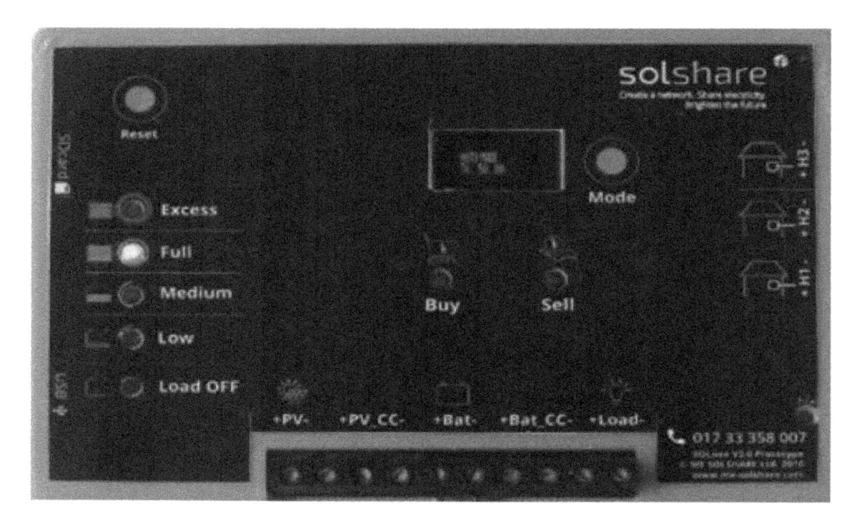

FIGURE 5.11 SOL box. (Courtesy of MeSOLshare.)

is the venture capital investment arm of the German utility firm innogy SE, and (iii) EDP, a Portuguese utility. They may not require much more capital except for bigger expansion as they seem to have revenue.

Observations:

i. The study focused on Bangladesh by the founder and CEO in his thesis has given him insight into the market place and role of solar households.
ii. Nearly 15% of the households have solar power in Bangladesh mostly and they are also off-grid, and their interconnection provides one of the biggest market places awaiting the arrival of startups. MSS came up with products that seemed to have attracted the solar households for their own advantage and benefit.
iii. There is not only a revenue potential for solar households but also in the off-grid areas, there is a high probability for power self-sufficiency among the users from excess capacity or vacationing households.
iv. A platform for such a product which they produced in the form of SOL box is very viable not only for that market in Bangladesh but ideally in Southeast Asia including Japan. Note that Panasonic and others have been developing home energy management systems (HEMS) which have a lot of similarities to the MSS technology and its deployment in Bangladesh.
v. The scope for expansion to include as many of the solar households is a long way to go and will result in scaling up the company with its value rising.
vi. What would be the next product within this technology sector of their focus is something that budding entrepreneurs can take it as a productive challenge and study.

vii. How MSS will keep its competitors off their market place in Bangladesh is a development to watch and learn from.
viii. The scope for their products in other countries may be there and how MSS would tackle it by licensing off their technologies and products to offshore companies or start their own divisions in those countries by themselves or go into collaboration with established power electronic companies in these countries – all of them open up opportunities for budding entrepreneurs to get to explore with MSS.

5.4.3.3 Cygni Energy

Cygni Energy is founded in 2014, Chennai, Tamil Nadu, India and headquartered in Hyderabad, India. It is a collaborative effort with the Indian Institute of Technology (IIT), Madras (which now is known as Chennai). The founders of Cygni Energy are:

i. Venkat Rajaraman, Founder and Chief Executive Officer: Venkat Rajaraman has an MS in Electrical Engineering from Stanford University, USA. He worked for Sun Microsystems, Portal Player and NVIDIA in USA and as CEO in Su-Kam Power Systems, India. He has 16 issued patents, 4 pending and 4 filed in USA and Japan.
ii. Paramjit Singh, Founder and Chief Operating Officer: He has a B.Tech. in instrumentation and control and an MBA from MDI, Gurgaon, India. His 23 years experience, prior to founding Cygni Energy, is in the power and telecommunications sectors having worked for the leading companies such as NTPC, BSNL and Airtel.

Technology: Technology is for solar power for homes and businesses with only a dc grid and no ac in the distribution inside the home or on the solar power system. Solar power output is in dc which is then connected to a battery with a dc to dc power converter to handle the difference in voltages, and the battery output directly feeds the dc grid to the home, its loads consisting of lights, fans, mixer, TVs, computers, phone charger, etc. with no ac system at all in this network. If utility grid power is available through the utility system, then it will be connected to the dc grid system after the ac input is rectified and conditioned to match the grid voltage which is established as 48 V in this technology. As the solar power is not intended for feeding the utility grid, there is no need for an inverter at all and there is no need for another inverter to provide for an ac distribution system to cater to the loads from the battery when the utility power is off. Therefore, the company's technology does away with two inverters and can save considerable cost and they call it Solar Inverterless. Further, they claim that their dc distribution is much more efficient compared to the ac in the homes, but yet to be given much data from engineering studies, which may be proprietary and not made available to the public at this time. Such a technology-based home solar system is shown in Figure 5.12. The problem with this dc grid approach inside the home is that starting from power switches, sockets, dc appliances instead of prevalent ac appliances, have to be available to the customers to buy in to this scheme.

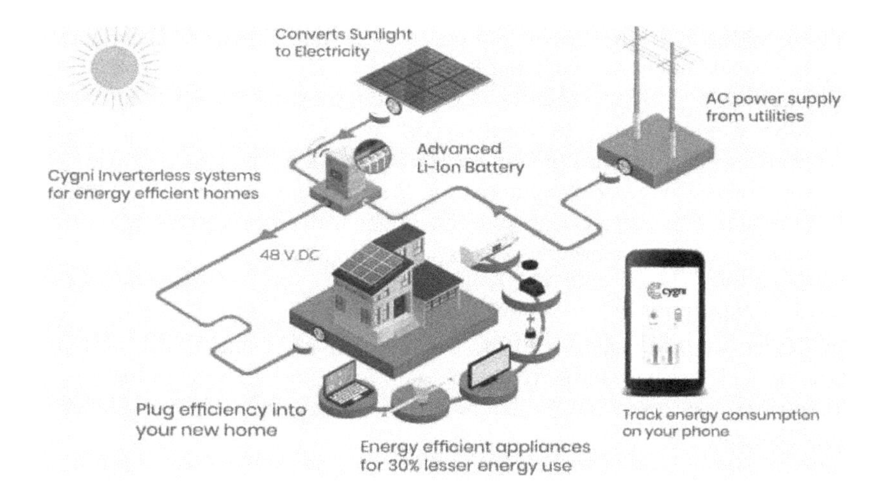

FIGURE 5.12 Cygni energy solar home system. (Courtesy of Cygni Energy.)

Products: They make the switches and sockets for dc and many of the dc appliances are from established manufacturers and they are all made available through their e-shop itself so that customers are not left hanging high and dry. They make available batteries and solar panels to their customers. Their bread and butter are the products that deliver solar power at 48 V and one such is Solar Inverterless 500 intended for homes, shown in Figure 5.13, and it has the following features:

FIGURE 5.13 Solar inverterless 500 (Cygni IBIS). (Courtesy of Cygni Energy.)

i. 48 V dc system,
ii. Solar charge control is through maximum power point tracking (MPPT) based which is from their patent,
iii. Dedicated battery management system,
iv. Data can be sent to a server or can be read by Android app,
v. Load control SMS based,
vi. Normal load cut off when the battery state of charge <25%, and
vii. Li-Ion Battery for 1.25 kWh energy storage.

The price is $602.00.

Their product line includes similar units but with slick packaging and with higher energy levels for shops and businesses.

Intellectual Property: They claim patents for their technology and no further information is accessible at this time.

Funding:

i. 2016: 100 Million Indian Rupees (≅$1.53M) invested by Atyant Capital as pre-series A investment.
ii. 2018: 300 Million Indian Rupees ($4.3M) as debt from IndusInd Bank, and 150 Million Indian Rupees ($2.1M) as Series A from Endiya Partners, a venture capital company.

Observations:

i. The company's founders have an excellent engineering background and management experience with various companies. This is a boon for any startup and Cygni Energy is no exception to this rule.
ii. They have from the very beginning aligned themselves with an institution well known for incubating startups and with an engineering group within that capable of realizable and usable technology generation. Cygni Energy has benefited immensely from this association to generate product lines.
iii. Their product and technology have found traction with state and central (federal) governments in India leading to very many opportunities for deployment in rural and hilly areas that are all off-grid at this time.
iv. A steady revenue stream has been established since 2016/2017, and they have a revenue of around $7.8 M for the financial year 2018.
v. They may not need any more fundraising for survival but may do so for expansion.
vi. They are not limiting themselves in the future to India alone for their products and looking at markets in North Africa and South East Asia where their products are likely to have good reception.
vii. They have deployed in 40,000+homes their solar power system within such a short time since their founding. It is an enviable achievement for any management team.

viii. Scaling up of their products with high power consuming appliances with the dc bus may pose problems of line losses, and switch designs for interrupting larger dc currents. Those technical challenges have to be overcome before high-end installations are contemplated. These customers, though smaller in number compared to middle-class customers will come from upper-income groups. The fact that they can afford to spend a lot makes them an attractive market segment. The market to cater to their taste and desire opens up further opportunities for startups in this line leading to augmentation of their technologies. It may further lead them to western markets with the resulting high financial rewards.
ix. A combination of MeSOLshare of Bangladesh and Cygni Energy technologies may lead to a combination of individual homes with the latter's technology into a peer-to-peer (p2p) trading of excess power within a community of solar households making them almost independent of fossil fuel-based power. Given possible opportunities for growth, service and profit with p2p trading and linking in a solar home community, this field of solar energy for homes and small businesses is full of intellectual challenges and satisfaction apart from financial rewards. It is important to note that some Japanese companies such as Panasonic are trying to put this into practice.

5.4.3.4 Zola Electric

Zola Electric (formerly known as Off Grid Electric) is a US company founded in November 2011 in San Francisco Bay Area, CA, USA with its primary office in Arusha, Tanzania, and its founders are as follows:

i. Xavier Helgesen, current Chief Technology Officer: He obtained his BA in Management Information Systems from the University of Notre Dame, USA, and MBA from the University of Oxford, England. He is also the co-founder and director of Better World Books in South Bend, Indiana and has been entrepreneurial in effort since his graduation.
ii. Erica Mackey: She has a BS in evolutionary biology from the University of California, Los Angeles and an MBA in Business from Oxford. She co-founded also My Village, in the area of home-based child care and related tools, which has attracted successful funding. She is a social entrepreneur.
iii. Joshua Pierce: He has a BS in construction management and building science from California State University, Chico and studied in the Executive Education MDE program at UCLA Anderson School of Management. He co-founded Zola Electric and served as its CTO from 2011 to 2018 and its Chief Product Officer from 2018 to 2019. He also serves as the president of Pierce Consulting. Prior to that, he was the director of technical services at Richard Heath and Associates.

Technology: Individual home-based off-grid dc grid with a solar panel, LED lights, a meter for measuring power, and a USB charger without connection to the utility grid forms their initial technology. Technical content wise it is very simple that electrical engineering students should be able to do that in a very short time. Recently, solar power grid for homes with utility tie is introduced to have uninterrupted power and at higher power than their initial technology that was capable of very few watts. The technology incorporates inverter to take its solar and battery power into the home ac network.

Products: (i) Zola Infinity: It integrates solar power, energy storage (i.e., battery) and utility grid power and scalable to higher levels of power guaranteeing pure sinewave power to the user. The inverters in the system, it is claimed, have 300 years mean time between failure. Therefore, high reliability built into the system is the hallmark of this product. (ii) Zola Flex: Designed to power the basic needs of homes and businesses and works on- or off-grid. They offer customers easy to buy in schemes spread over a period of time. Their systems serve 1M people and the total installed power capacity is 1.17 GW.

Intellectual Property: No patents can be traced at this time.

Funding: Raised phenomenal funding of $271.6 M with 32 investors including equity-based investments, debts, grants, etc. from its inception to May of 2019 (over a period of nearly 8 years) and this is the most funded solar home systems (SHS) company that caters to homes and small businesses in the world. Some of its investors are well known and they are GE Ventures, Tesla, EDF, DBL Partners, Helios Investment Partners, TOTAL, SunFunder and Omidyar Network.

Observations:

i. The founders are all new to this technology area but with good fundraising and/or entrepreneurial backgrounds. These skills have come quite handy to launch this startup.
ii. Their motivation to take this technology and light up people's lives literally in places that are off-grid and for people not having access to great financial resources is inspiring. Some of the solar startups to serve villages in developing countries also fall under this category.
iii. The scaling up their technology from a few watts of power to a few kilowatts and assuring almost failure-proof operation show that they aim to stay in this business for a long time.
iv. Investors driven either by profit or by public good have thronged to support them and that is something that the founders and management team can be proud of and feel fortunate.
v. Armed with a vision, immense funds and execution of plans, Zola Electric has started to expand from its base of Tanzania to Rwanda, Ivory Coast, and Nigeria. With this continued expansion, it is quite likely that it will have the major share of the market in the African continent.
vi. Aspiring entrepreneurial electrical engineers have to draw a number of lessons from this startup and some of them are: (i) basic knowledge of startup is very important but not necessarily true in the case of previous experience in founding startups; (ii) knowledge of emerging needs of segments of population is a must and being sensitive to that can generate

ideas for entrepreneurship; (iii) startups do not have to serve the needs of the place in which they are founded but can be conceived to serve populations and markets in different and faraway places, as Zola Electric was conceived by three Americans but serves mostly populations in faraway countries; (iv) founders must move to stay and work for the startup in places where their company's products are targeted for to get to know the nitty-gritty of the environment, market and people to serve and finding potential candidates from the locals to lead and work for the startup as well; and (v) raise capital at the earliest and required amount to grow and keep growing the startup.

vii. Zola Electric is the largest solar home systems company and it attained that position within 8 years of its founding, not a small feat.

viii. Given the variety of products it has introduced, it may be able to license or export its products to southeast Asian countries where solar home systems have a greater market potential and where there are not many big companies/startups to keep Zola Electric away from their markets.

ix. Marketing-oriented entrepreneurs can explore to work with companies including Zola Electric in solar home systems.

x. It will be interesting to watch how the technology developed by Zola Electric or for that matter many in its category protect their technologies from copying and then selling the products in various markets including in the markets they are already rooted.

xi. The founders and the team members working in this business sector have much more soul satisfaction than in some other sectors as they help to make a cleaner environment and also assist people having no access to a utility grid or enough up-front money to afford it.

5.4.3.5 AlphaESS

AlphaESS Co. Ltd. (hereafter referred to as AlphaESS) is founded in 2012 and headquartered in Nantong, Suzhou, China. Its founder is not mentioned in available open documents and the CEO from inception is Thomas Yuan. He studied in Friedrich Alexander Universitat Erlangn, Nurnburg, Germany and for his bachelor's degree in Nankai University. He has limited experience of 3 years before joining AlphaESS. There is not much to learn about the team that could be of help for analysis as to the founding.

Technology: It is based on power electronics-controlled energy storage systems up in the 32–50 kW range working with solar and utility power. It is way above the lower middle-class affordability and is in the middle- and upper-class homes, and businesses having microgrid system supply that come with the startup's own lithium-ion batteries. It is above the power ranges, missions and markets targeted by Mera Gao, Cygni Energy and Zola Electric and gets in the 10 kW power range in solar renewable microgrids with on- and off-grid connections. Their technology is offering a system with all the communication capability to customers via phone, tracking energy sales to the utility system and data analysis for viewing as well as for maintenance of the system. The technology caters to:

i. solar panel power extraction with and without maximum power point tracking,
ii. battery charging from it,
iii. conversion of that power to ac to feed the ac loads in the home and option to pump excess solar power to the utility and tracking of the energy flow,
iv. operation from a battery without the utility power to meet the home/business power needs, and
v. products that can store energy from solar panels in battery banks for supplying the utility during peak load hours for commercial purposes.

Note that, all these functions require specific power electronic converters and they can be identified for each function identified and are given below:

i. Dc converter (buck or boost or any of that family) to reduce or increase the dc output, V_1, from the solar panel output voltage, V_p, so that the output voltage is compatible for direct ac conversion.
ii. Battery voltage for home use is lower than V_1 and to charge it requires a dc buck converter. When the battery energy has to supply to the ac system at home or to utility, that requires to be boosted to V_1 necessitating a dc converter to boost the voltage. Overall, therefore, to cover both the functions, this stage requires a bidirectional dc to dc converter for this stage.
iii. Ac conversion from voltage V_p through a filter and an inverter and then an ac filter for this stage.
iv. The converters are covered in (ii) with boost capability and inverter covered in (iii).
v. Charging of the battery banks from utility supply via the inverter and a dc to dc converter is fairly standard and this stage may not require any more converters than mentioned in stages (ii) and (iii).

Invariably circuit breakers to isolate the PV output feeding the dc converter, battery bank, and the utility supply are required and placed at the appropriate points. Overall, the system may be represented as shown in Figure 5.14. Such schematic is common in this business sector and the differences come in certain specifics including control objectives and software implementation with features built specifically for the applications. Note that there are many products that have been in place for a long time in the market within this schematic. Taking only the upper part of the diagram gives a solar renewable system exporting energy to the utility system. If the solar part is ignored overall in this schematic, then an uninterruptible power supply (UPS) system is identified. Hence, the technology that evolved for this emerging sector in renewable energy for homes is a combination of various products that have been in existence for more than 30 plus years in the market.

Products: Many products in the renewable energy sector are available and they are: 3 and 10 kW residential PV energy storage system with on-/off-grid

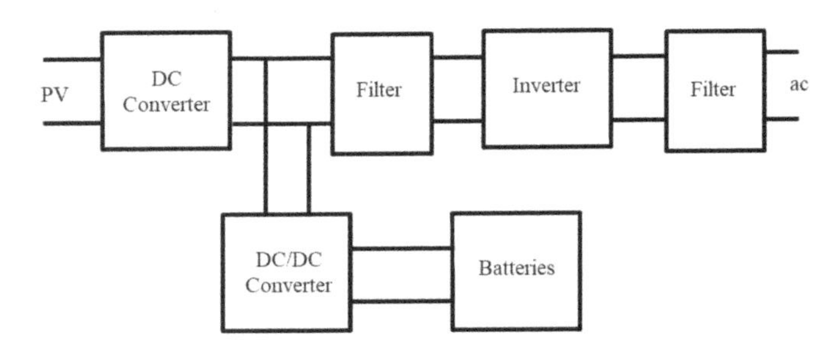

FIGURE 5.14 Residential solar power and energy storage system building blocks.

integrated; Small and medium commercial energy storage products of the startup are from 30 to 100 kW with on-grid for low power and for high power with ac/dc grid adaptation. Some of these products up to 5 kW have intelligent control meaning that maximizes PV self-consumption, export power, programmable charge and discharge of the batteries, and remote control with apps, upgrade energy and battery management systems, etc. Another product, Storion Hybrid Inverter (a storage lithium-ion 5 kW inverter), contains all the subsystems shown in the previous schematic except batteries in its package, as shown in Figure 5.15.

Growth: They have subsidiaries in Australia, Germany, and collaboration in Taiwan, and New Zealand. For a company of this recent region, the company has entered into significant renewable energy markets and countries.

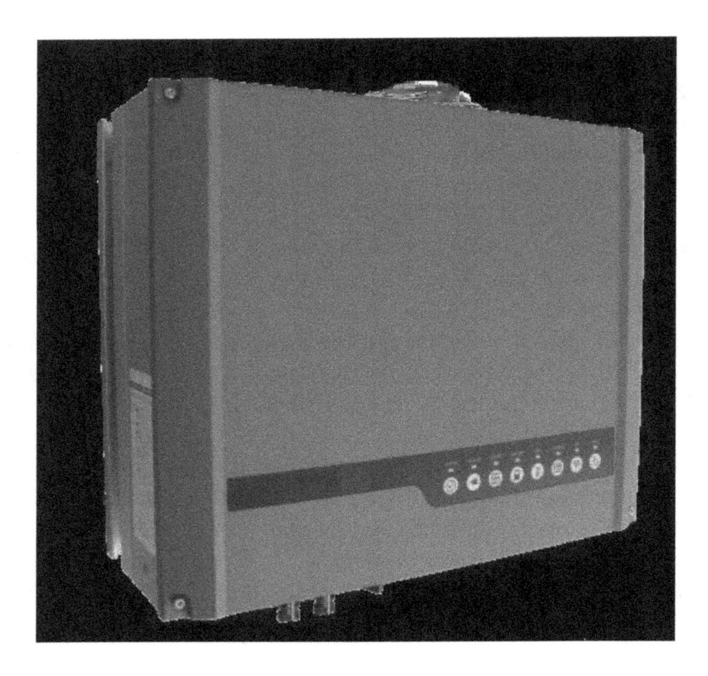

FIGURE 5.15 Five kilowatts hybrid inverter. (Courtesy of AlphaESS.)

Intellectual Property: Listed to have 100+ patents (but not cited as to where they are granted from and they are not easily accessible in the open). Only five are identified in European patents office filings. There is no patent traceable in USA.

Funding: Not known.

Observations:

i. AlphaESS is a classic case where a company comes into fill a void in the power range of 5–200kW solar and energy storage types of business, the lower power levels catering to homes and higher power to businesses.
ii. Note that its technology is based on feeding mostly an ac load for which the technologies are abundant and companies in that domain are also very many across the globe.
iii. Comparing to Cygni Energy which is a dc-based home system with loads all being dc, the amount of technology development required for standard ac-based energy management system with renewables is not very significant as many of the subsystems and appliances are in place for a very long time.
iv. Zola Electric has similar technology for homes but limited to a few kilowatts power is to be pointed out. If and when companies such as Zola Electric and others decide to make a presence in the market of AlphaESS, it may be within their reach.
v. While technology base in the form of various subsystems and some integrated as well for products of AlphaESS and other related companies may be available and known in the literature, their implementation and bringing in the software access to the system are domains in which intellectual property may be embedded. That task is not too demanding when teams are in place. This may apply to this startup as it brought out a fair amount of products in such a short time of 5–6 years.
vi. Their expansion with subsidiaries in Australia, Germany to cover Europe, and joint venture within USA demonstrates their strategic vision in marketing and a strong base of investment funding to move toward these objectives.
vii. Also, their patent position is very high given the time frame of their existence. To be noted is that their number of patents in the European Union is only 5 out of this 100+patents claimed with none in the USA, and therefore, the rest must be in other countries.
viii. This company does not give enough information about its founders and funding and therefore, it is an enigma and difficult to learn from their background or to know about the investors with the result that budding entrepreneurs are deprived of possible sources of contact when they embark on their funding quest.
ix. In turn, a lack of information may also discourage the budding and future entrepreneurs to seek their fortune in such an environment and hopefully, engineers do not fall into that mode of thinking!

5.4.3.6 Electriq Power

Electriq Power, Inc. (hereafter referred to as Electriq Power or shortly as EP also) was founded in late 2014 in San Leandro, CA, USA. Its co-founders are:

i. Chadwick Manning, Co-founder and CEO: He obtained his Bachelor of Arts/Science from the University of San Diego, CA and worked for Pricewater Coopers as an associate over 2 years.
ii. Jim Lovewell, Co-founder and Chief technology Officer: He obtained his B.S. degree from Iowa State University and M.S. degree from Stanford University, CA, both in Electrical and Electronics Engineering. He has varied experience in engineering and management with a number of companies for nearly 35+ years before cofounding ElectrIQ Power.

Technology: The technology resembles that of AlphaESS but only directed toward home and small commercial installations at present. It is based on a battery energy storage system having an inverter that links with the utility. A part of the technology is claimed to be in their monitoring, analysis, user customization and real-time optimization of the energy management system. It may not seem like it is very different from a few of the companies competing in this market place but it has its distinct features.

Products: Their products consist of: (i) Electriq Power energy management platform named the Dashboard. It is capable of coordinating the solar generation, utility rates for the time and self- consumption of solar power, backup power and uninterruptible power operation when the utility is down. (ii) PowerPod that has the batteries, charging converter and an inverter to the utility which can be connected to the solar renewable energy source that can be dc- or ac-coupled and the control system for the entire system using a micro-controller and link to the energy management system via Dashboard. The features of the PowerPod are 5.5 kW inverter, battery capacity of 13.4 kWh, with usable capacity of 11.4 kWh, power out of 4.5 kW, an inverter efficiency of 96%, and the entire package weight of 375 lbs. This costs about $8,999 at this time. This is their baseline product at this time, and higher battery capacities can be installed up to 33.5 kWh. Two wall-mounted PowerPods with the energy management system accessible over the cell phone is shown in Figure 5.16.

Intellectual Property: One patent application pending and it is related to the control of an uninterruptible power source keeping high efficiency at its standby operation and correcting for voltage variations of the grid voltage. The inventors of this application are the two co-founders listed above.

Funding: A total of $9M and the major investor is GreenSoil Building Innovation Fund.

Observations:

i. One of the founders (Jim Lovewell) is a long-time practicing engineer with hands-on experience and that factor seems to have played a key role in product development.

FIGURE 5.16 Two wall mounted PODs with cell phone monitoring. (Courtesy of Electriq Power.)

ii. Their entry time in the home energy management market space could not have been better as it was opening up. The earlier ones by a decade were not as successful due to a number of factors such as: (i) the battery was very expensive (about two to three times more than now), (ii) lack of opportunity to make some revenue by selling power to the utility at the peak times, and (iii) cutting home consumption at certain times when there was a surcharge for using utility power thus reducing the electric bill.

iii. While Zola Electric can size up and compete with this startup and the fact that they did not do played well for it. There are many companies with similar technologies and capabilities such as Tesla with its Powerwall product of battery energy storage with an inverter system at very low prices, and they can ramp up other features in a short time should they choose to do so. This factor must be in the backdrop of ElectrIQ Power and its peers.

iv. The major competition may come, price-wise, from companies such as AlphaESS with its own batteries and lower cost of manufacturing from China. The business and engineering strategies have to be addressed by this as well as others in the domain in USA. These issues provide an opportunity for budding entrepreneurs to come up with solutions and to make their marks.

v. It is important to observe that the founders and investors do not have a background in power electronics which constitutes a major subsystem of the products in their portfolio and it will play a key role in future development as well. There is a great opportunity to reduce the size of the power electronic converters and inverters in these systems using wide-bandgap devices and improving on the efficiency that may introduce a disruption in the market and if not at least a clear differentiation among product lines of the companies.

vi. The intellectual property portfolio under its own name seems to be slim which is not very different from its competitors. It is quite likely that it is away from the public eye under a different company or a subsidiary which is a practice with some companies including large ones. Given the very low number of intellectual property for many companies in this sphere, investor confidence may be proportional too. It may be good to watch and learn how they build up their successes without having to go through intellectual property disputes with other companies in the future.

vii. Their strength in software control of the energy management system in many of the features described is a major weapon in their hands. But it is a product that is embedded only in their hardware such as PowerPod now. Will there be a market opportunity for the software alone and if so, will it threaten their own market in the hardware part of their major product lines? But there are startup companies coming in to provide many features of this software but without the hardware products and it is a threat to be contended within the present and future for this startup and its peers.

viii. The company has brought out products in a record time with very limited resources (around $3M), and then twice that resources came in as an investment. If they continue to do this at the rate that they have done in the past, there is every chance that they can be a major player in this market in the future.

ix. Their display of the system as shown above is very attractive that is something budding entrepreneurs could learn from them.

x. Engineers interested in entrepreneurship have a lot to learn from this startup such as identifying the importance of the software management solution with the hardware part of the product, the rising opportunities provided in the market place by the utility and its willingness to go distributed power operation in buying power at certain times, etc. Having stated that it is important to realize that the value of the product line of Electriq Power is not solely dependent on utility revenues to the homeowners but more on the energy storage, its utilization in the home efficiently and autonomy from the utility with solar renewables.

5.4.3.7 Geli

Growing Energy Labs, Inc. (hereafter referred to as Geli) is founded in 2010 in San Francisco Bay Area, CA, USA. Its founders are:

i. Dr. Ryan Wartena, Founding Chief Executive Officer: He obtained a doctoral degree in chemical engineering from Georgia Tech in 2001. He served as a post-doctoral researcher and scientist in Naval Research Lab for about 2 years, and also in MIT for 3 years related to microbattery and in battery assembly technologies. He served as CEO for 8 years and is currently President and Director of Product with Geli.
ii. Crispell Wagner, Chief Technology Officer: Varied experience in industries in software, hardware and design among other areas including 2 years in Harvard University as a software engineer. Served as CTO for 8 years and currently as a senior staff engineer in Geli.

Technology: It is to address the planning, design and deployment of energy storage systems for home, office and virtual power plants through software simulation, analysis, system monitoring and integrating tariffs from 1,300 US utility territories. Further, it integrates solar power source to the energy storage system and the utility. The ultimate interest to the customer is the financial data such as utility rate, at a given time both for buying energy from and selling to utility and the revenue from their energy storage system. The technology includes the aggregation of many energy storage systems scattered around homes, offices with and without solar panels around customers' neighborhoods and then using that to leverage with the utility for maximum return for their systems. The claim is that their software is based on machine-learning algorithms.

Products: Geli ESyst is their system planning and analysis software. It was made available to customers for free. It also provides monitoring and control from anywhere at any time and with all the user-friendly features that are required in the market place at present. To serve their customers who use their software for planning and execution, they bring to them products from various proven companies. For example, their home energy storage system promotion is from Tabuchi Electric known as Eco Intelligent Storage System. For large power installations, Geli recommends using suppliers such as Delta Electronics, Dynapower, Princeton Power. Lockheed Martin and LG Chem, etc. for hardware solutions and for implementation. By taking advantage of such suppliers, Geli stays away from the hardware part of the energy storage solutions and concentrates only on software aspects.

Intellectual Property: One granted patent by the co-founders and one patent application in USA.

Funding: $16.5M in seven rounds. Their major investors are Southern Cross Venture Partners, and Shell Ventures.

Observations:

i. Founders bring a strong background in the battery, its assembly and chemistry, and in software to Geli to start with. They do not have the background in power converters and stayed away from it and hence from the hardware to focus on the software part of the subsystem of the energy storage management system.

ii. Almost all energy storage system startups and established companies for smaller installations such as homes and offices provide the whole system including the software but they do not usually provide the planning and evaluation in many geographic areas and this is a big gap in the business.
iii. With peak load reduction using an energy storage system becoming financially attractive, many consider going in and it is difficult to find resources to plan for them and to evaluate their financial returns in their geographical area. Geli addresses such a vital need of the customers and it is one of many evolving ones in this sector of the business.
iv. Their intellectual property is limited, but they seem to have built a wall with it to protect their business and products.
v. The competition is likely to be more severe as more and more established companies focus on the software part of the business. Even big companies, such as Advanced Microgrid Solutions (AMS), have shifted their focus to software in order to address the needs of mega energy storage systems. But there is no guarantee that they will refrain from invading the smaller installations including home and commercial market by tuning their software to address their needs.
vi. The independent operation status of Geli and its peers for a long time may be at stake as many system (hardware and their control software) manufacturers could take them over thus providing a successful exit.
vii. A number of conclusions can be reached from the point of electrical engineers based on the success of this company and its peers in the market place. Some of them are:
    a. Opportunities are immense in the software part of the general management system than only in hardware and its control.
    b. Software knowledge not only embedded kind is essential but more importantly, software supported online with a reach from anywhere and everywhere to the system is required and it is in the best interests of engineers and students to be prepared in it as part of their education also.
    c. Funding raised for software-based business is much lower, and turnaround time to put products is much shorter than for hardware electrical products because of their rigorous testing requirements and getting approval from many agencies for them. An additional advantage is that the talent pool for coding of the software around the world is available much more than in the hardware and embedded skill sets.

5.4.3.8 Sunverge Energy

Sunverge Energy (hereafter referred to as Sunverge) is founded in 2010 (some documents cite it as 2009) and headquartered in San Francisco, CA, USA. Its founders are:

i. Kenneth Munson, Chief Executive Officer and President: He worked as VP of business and marketing for Kinetek, Inc. also served as adjunct lecturer in St. Mary's College in its graduate school of economics and business administration. He has an MBA from St. Mary's College of California and a Bachelor's degree in marketing communications from California State University, Sacramento. He served with Sunverge until 2017.

ii. Dean Sanders, Chief Technology Officer and Chairman of the Board of Directors: He also founded and serves as President in Inertia Engineering, Inc. for the last 24 years whose main product is medium voltage switchgears, automation products and special equipment for smart grid applications. He is an inventor and developed most of the patents for Sunverge. He has a BS degree in mechanical engineering from California State University, – Sacramento.

Technology: The technology is in home energy storage system with features such as accommodating solar energy, on- and off-grid operations, battery charging and self-consumption as well as selling power to the grid. Not much different from many other companies except that it has its own power electronics hardware with batteries (not its own) and software for integrated management in the house/office but interacting with the grid (utility) buying and selling energy depending on the time of use with financial return in the forefront while providing and making sure energy is available to home/office users. Software for grid integration and for working with utilities for their use as well as to homeowners is novel, thereby securing strong support for its system of hardware and software with them. The company like many others is aggressively positioning itself as more of a major software company also in this domain.

Products:

i. Hardware Products: Two types and they are: (i) Ac-coupled solar energy storage integration system and (ii) Dc-coupled solar energy storage integration system, but both having grid ties. The continuous power ratings for 4.5 and 6 kW with battery capacity options from 7.7 to 19.4 kWh and each of them comes in one integral package. In an ac-coupled system, power is obtained from solar panels and converted to ac with an inverter and then it is connected to the backup battery via inverter/charger for charging and then inverted from the battery with that charger/inverter tied to the grid with isolating switch panels. On the solar end, the ac output from it feeds the critical loads and the remaining energy is routed to the battery system and to the grid. When the grid is lost, the battery inverter feeds the critical loads after opening the grid side breakers. The energy conversion involves, solar dc to ac, and then to dc via battery charger and then inverter to feed it to dedicated loads in case of grid failure and three power stages of conversion are involved when solar is generating energy. It is very slightly inefficient compared to the dc-coupled system but has the advantage of easier retrofits and upgrades.

Dc-coupled system involves no inverter on the solar side at all except only the dc to dc converter remains, thus saving one stage of energy conversion in the system. It is marginally more efficient because of it and is very popular.

ii. Software Product: Addressing the energy management systems and providing user-friendly interface to customers and control similar to some of the startups such as Geli. Focus is on: (a) grid services by providing a platform to address peak demand, congestion and power quality issues, and (b) energy management of homes and businesses and positioning the EV battery system as a virtual power plant and its control to buy and sell power to grid. Note that their software product is not made available free of cost and comes with their services and purchase of their products.

Intellectual Property: Three granted patents and four patent applications in the USA.

Funding: Raised $54M total. Investors include Siemens, Mitsui Co., Southern Cross Venture Partners, Softbank China Venture Capital, Total, Kokam, AGL, Australian Renewable Energy Agency (Govt. of Australia), Total Energy Ventures International, and VisIR.

Observations:

i. One of the co-founders (Dean Sanders) is well experienced in starting a company and running it also. He has developed intellectual property for Sunverge and played a major part in its portfolio.
ii. The other co-founder (Kenneth Munson) is well experienced in managing and expanding the operation and his leadership resulted in positioning Sunverge in a very good market position.
iii. The founders have brought professional experiences that are very advantageous to any startup and Sunverge has benefited from them.
iv. Its intellectual property position, in terms of its numbers, is good compared to its peers in USA.
v. A very high amount of capital has been raised for this startup, and in this, it stands head and shoulders above all its peers.
vi. Most importantly the investment came from very big companies who have a clear interest in this business and one of these electrical partnering/funding firms may go for acquisition in the future. That is a very good position to be in for the company, its founders and other investors.
vii. Their positioning in the market place not only with hardware products but also software with much more than integration tasks but to be part of the grid analysis puts them at a great advantage in the market place.
viii. They may want to expand their market to high energy storage systems beyond their traditional strength of homes and businesses with the help of their software integrating other companies' hardware as it is the only option at this time as they do not have products at that level. This may

lead to many advantageous tie-ups with mega electrical firms when it happens.

ix. Utilities such as Con Edison and Puget Sound Energy have selected to work with Sunverge for its capability and energy platform monitors, to work with distributed energy resources. This is a big vindication of their software products and gives them a solid footing in the evolving market place.

x. Its work with Alectra Utilities is in the use of Sunverge's dynamic virtual power plant platform to assist the utility to optimize the use of electric and gas resources to minimize Green House Gas emissions and such applications are likely to increase in the future due to concern for the environment and growing movements in many and advanced countries. It is a good omen for all startups in the domain and more for the ones that could find traction with utilities.

5.4.3.9 Bboxx

BBOXX (hereafter referred to as Bboxx) is founded in 2010 in London, United Kingdom. Its three founders are all fresh Master's in Engineering (M. Eng.) degree holders in Electronic and Electrical Engineering from Imperial College, London and they are as follows:

i. Mansoor Hamayun, Chief Executive Officer,
ii. Laurent Van Houcke, Chief Operations Officer, and
iii. Christopher Baker-Brian, Chief Technology Innovation Officer.

A student charity in Imperial College called equinox, started by Mansoor Hamayun brought electricity access to 400 households in Rwanda and through this project, he gained knowledge about the requirements of poor and off-grid communities, NGOs and working experience with them in African countries. Realizing that there is an immense need and opportunity for low-cost, clean off-grid power led the way to the for-profit startup, Bboxx.

Technology: Solar PV panel power with battery storage is the basis for the technology but takes a dc bus-based approach for their major product with dc appliances made available to the users. This is somewhat unique in that period and only one another company seemed to have this model and that is Cygni Energy in India. The technology behind is very elementary in that the solar power is extracted from its dc output and regulated to charge a battery which then is connected to the household appliances such as lights, charger, fan, radio, and refrigerator. The number of appliances and power requirements for each household may vary and accordingly many products with different levels of power are available from them. Another scheme where ac power is made available to drive ac appliances at higher powers than normal dc appliances is also provided by Bboxx. There is a very intelligent user-friendly interface to the equipment via phone as well as a central control through online monitoring and control. Its technology and product lines are shown in Figure 5.17.

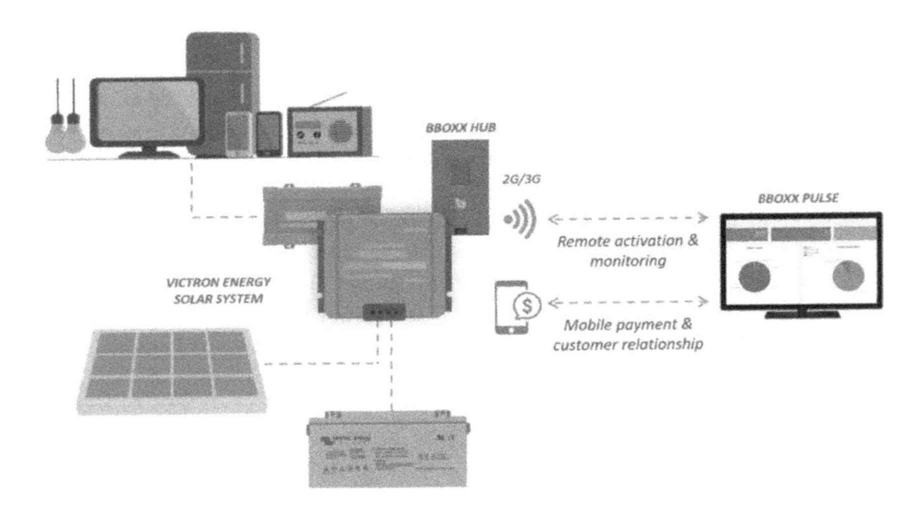

FIGURE 5.17 Company technology and products in brief. (Courtesy of Bboxx.)

Products: There are three types of products as follows:

i. Bboxx Dc solar power system is standardized for home and micro-business use and in different levels of power ratings.
ii. Bboxx Hub for monitoring and control of larger systems.
iii. Bboxx Pulse to assist next-generation utility business in terms of pay as you go, collection of bills, means to remotely disable and enable power, real-time display of operations, etc.

Intellectual Property: Not readily available in the public domain at this time.

Funding: $165 M raised so far in 13 rounds with the latest round D in 2019 for $50M. Its major investors, to name a few are: MacKinnon, Bennett & Company, Mitsubishi Corporation, Engie, KawiSafi Ventures, Khosla Impact fund, Bamboo Finance, and DOEN Foundation.

Observations:

i. Founders, all electrical engineers, with very limited internship and a brief exposure to industrial experience set a very good example to other electrical engineering graduates in entrepreneurship.
ii. Their market focus mostly in the African continent has added a new dimension to their challenge in terms of understanding the economy and people far away from their home, organizations that could help them such as NGOs, banking practice and operation, consumers' capability to pay and how to make it happen with pay as you go like financial schemes and most of all to convince investors to put funds to support

their activities and expansion. They have risen to the challenge and did extremely well.

iii. Zola Electric, which was founded by all business majors that had to overcome the lack of engineering knowledge and technology in this arena in the founders, is the opposite of Bboxx in terms founders' talent pool.

iv. Many of the successful companies in the domain of Bboxx started around 2012–2014, and this company had a lead of at least 2 years in that aspect and hence acquired experience that has been put to its current success.

v. Their expansion has been in 12 countries so far and some of them are Togo, Democratic Republic of Congo, Guinea, Ivory Coast, Kenya, Mali, Rwanda, Senegal, Pakistan and Nigeria. Notice the concentration in the African continent where the market is large that it may be difficult for regular utilities to capture that market and hence the interest of the utilities in Bboxx. Their interest has a positive benefit to Bboxx in that the utilities are also their leading investors with probably a business motive to learn from the startup and use it to their advantage in the future.

vi. The ability to raise such a large amount of funds for a sector with slowly rising market potential at least in the first 10 years but having a dramatic potential increase in the face of environmental concern and pressing need for clean power is very unique to this engineering sector. Note that Zola Electric has also raised an even larger amount of investment funds.

vii. There is a certain uniqueness about Bboxx in terms of its dc grid-based products, thus having to provide dc appliances for using this dc bus-based power that they supply. This enabled faster development to get solar energy with the only dc to dc converter in the middle between battery and PV panel output but required sourcing and/or making on their own dc appliances to exploit such a dc grid. Note that Cygni Energy in India around the same time or a little later took a similar approach to the home energy system with solar energy.

viii. They have also introduced an inverter at the output only to supply high-power ac appliances and apart from that which may be used also to give an ac grid access, if necessary to the customers. That strengthens their position in the market place so that customers do not have to go and look for other providers for that service.

ix. Their introduction of software-based control to financial methods to pay as you go and electronic payment from customers, etc. demonstrates that the startup has been closely following developments in other business sectors and takes advantage of them. That demonstrates a far-sighted upper management and teamwork.

x. Students of entrepreneurship will benefit following this company and others in this domain to see their evolution, market expansion and their nimble steering required over the coming decade.

5.5 MOBILE POWER PLATFORMS

Electric power is required in some instances and places where the utility power is absent. For example, power is required on a temporary basis for certain events such as in entertainment, disaster relief areas where the utility power might have failed, and temporary camps for housing migrants in remote border areas. It could be addressed by mobile power platforms with solar panels and energy storage batteries. There are also applications for mobile power within the living area and/or construction sites exist such as for charging vehicles in situ. Two startups that address the needs of mobile power platforms with a completely different approach are presented in this section. The market for each of the categories is not available in the open at this time but from the needs of the community, they stand a chance to grow and prosper.

5.5.1 Multicon Solar AG

Multicon Solar AG (hereafter referred to as MS) is selected here to illustrate that solar power does not need to be confined to home, business/office and industrial applications but can be for mobile applications. The Multicon Solar's focus on mobile applications came around 2010 when other applications such as home were taking off in the market place. MS is located in Duisburg, Germany.

Technology: The technology is based on energy extraction from solar PV panels and using a battery energy storage system and its output is to support a stand-alone ac microgrid supplying loads. The innovation comes in assembling them on a mobile platform in a container, very similar to the shipping container type, for deployment at any desired location within a short time with the installation time to be as short as an hour and not more than 3 hours. Further in the case when the energy requirement at a given location is finished, the disassembly and packing back into the mobile platform are accomplished very fast and then put on the road for travel. Further, they are in the business of development, production and assembly of solar power plants that are mobile and stationary as well.

Products: The product lines are as follows:

i. Solar Container: Container-based packaged solar power plant arranged in a way that enables the PV panels with their supports dragged out of the container in a systematic way resulting in deployment in the field in a short time. Multiple products up to 50 kWp PV output and up to 100 kWh storage with battery are available in this category.
ii. Solar Power Trailer: Mounted on a trailer and can be pulled by vans and other vehicles and intended for smaller power levels such as for camping and the installation time is only 3 minutes.
iii. Energy Power Rack: Each rack has 6 kWp PV panels which can be opened from the top and therefore easy for installation. Multiple energy racks can be interconnected to augment energy storage. The power controllers and batteries are inside the racks.

iv. Solar Power Plants: They also offer planning, design and construction of large solar power plants with their own products.

Observations:

i. The company has been around since 1993, but its focus on solar power plants and mobile platform-based approach came about in 2010. There were not many solar power plants during this period and that might have contributed to the company's strategy to seek alternative products and market with the good end result.
ii. Niche market using the mobile platform as well as with the ease of installation developed by the company placed them in a unique position.
iii. Note that the technology in power electronic converters, battery energy storage and maximum power extraction from PV panels all existed with many companies but not the positioning of the system for a niche market that this company targeted. It is important to distinguish this aspect, i.e., the market demand and coming up with innovations such as in mobile platforms, reducing the time for deployment from minutes and an hour to 3 hours for larger solar power systems and that are completely off-grid. That is a lesson to take home for all future entrepreneurs.
iv. Their products are claimed by the company to be patent-based and -protected. This again emphasizes that even simple idea centered enterprise has to be protected to discourage newcomers to invade their market place. Even though the business idea seems to be simple, note its implementation with the container, packing the panels, drawing them out in the installation site, locating the controllers and batteries in the space-optimized container, and developing methods for finally making them work in the shortest time some in less than 3 minutes, all of which requires innovations every step of the way and they are the backbone of their property.
v. They have installed a very high number of systems and that bodes well for them. Note that they have achieved definite uniqueness in their products, and attention in the market place. They are all very valuable as much as their product sales, particularly when and if the time comes for an exit.
vi. All the above points are valuable lessons to be learned that apply to their lines of entrepreneurial activities. That is the main reason that this company has been included in the lineup on energy storage system startups.

5.5.2 FreeWire

FreeWire was founded in 2014 in San Leandro, CA, USA. Its founders are:

i. Arcady Sosinov, Chief Executive Officer: His prior experience is in the investment firm GMO for 7 years in financial database analytics for trading. He has a B.S. in business administration and management of information systems, a B.A. in economics from Boston University and an MBA from the University of California, Berkeley.

ii. Jawann Swislow, Chief Information Officer: His prior experience is in investment accounting software in Eagle Investment Systems, a subsidiary of BNY Mellon. He has a B.S. in business administration from the University of Vermont and an MBA from the University of Colorado-Boulder.

iii. Richard Steele, Chief Engineer: He has worked in ithium-ion battery industry in managing pack technology at Leyden Energy, in testing of missile and satellite components, in design and build of general payload system for U.S. space shuttle mission and in medical devices redesign, testing and manufacturing. He has a B.S. degree in electrical engineering from DeVry University.

iv. Luv Kothari: He worked for Intel, Broadcom and now at Google where he is a manager. He has a B.Tech. degree from the Indian Institute of Technology, Varanasi, India in electrical engineering, an MS degree in electrical and computer engineering from the University of Illinois at Urbana-Champaign and an MBA from the University of California, Berkeley.

v. Sanat Kamal Bahl: Served as chief technical product officer during 2014–2017 in FreeWire. He has working experience in internet of things and worked for PwC and Jasper IoT Cloud and as a senior product manager in Intel for 8 years. He has a B.S. degree from electronics engineering from VJTI, University of Mumbai, India, an MSEE degree from the University of Southern California, MS in digital VLSI design from the University of Maryland, Baltimore County and an MBA degree from the University of California, Berkeley.

vi. Sameer Mehdiratta: His expertise is in IT and is currently senior director, global business applications in Renesas Electronics and has around 18 years of experience including stints in Hewlett-Packard, Applied Materials, Flex, and for 6 years in the electronics and automotive industry. He has a B.E. in production engineering from Punjab Engineering College and postgraduate manufacturing management from S.P. Jain Institute of Management & Research and an MBA from the University of California, Berkeley.

Founders (i), (iv), (v) and (vi) met while they were doing an MBA at the University of California, Berkeley where they started the work on FreeWire.

Technology: FreeWire started out to transform second-life electric vehicle batteries into intelligent and mobile energy storage systems. Their chargers use battery energy storage systems with the advantage that they can be charged when the electricity rates are lower and with the additional advantage that they need lower power level for the infrastructure grid since they draw a fraction of power at any given time to charge the battery but the energy required for charging other vehicles and loads comes from the battery only. For a given infrastructure of grid power-based charger, six boost chargers of FreeWire can be accommodated. The technology is standard as for power conversion from ac to dc to charge the battery and/or from battery dc to ac output in the case of their generators. The technology is integrated with software to find low-cost charging time from the grid and its control, coming up with a mobile platform so that charger with the battery goes to the automobiles instead of automobiles coming to a permanently fixed stationary charger, and the control of the charging from the mobile platform to the automobile and for receiving a charge from the grid to replenish the battery charge. Various aspects of user-friendly features are introduced, as usual in almost all of the customer interfacing products, and they find their place into their intellectual property domain.

Products: Three products of this company are:

i. Boost Charger: Stationary platform-based for charging simultaneously two automobiles having battery storage system with connection to the grid for charging the battery, shown in Figure 5.18a. Provides two dc outputs of 60 kW each dc for charging the automobiles with an output voltage range of 200–500 V having a total energy capacity of 160 kWh. Its ac input consists of a three-phase 208 V and a single phase 240 V system with a power rating of 27 kW. This product application is in EVs.
ii. Mobile Charger: It is a cart packaged with batteries and an inverter for charging the automobile with all the accessories such as the cable, handle, etc. and also a charger to connect to the grid supply so that the battery can be recharged. It is shown in Figure 5.18b. It has a continuous power output of 11 kW and provides 120/240 V single phase with a power factor of 0.98 and an energy capacity of 80 kWh. Its charger for batteries receives an input of 90–265 V ac single phase over a frequency of 45–65 Hz. This product application is in EVs.
iii. Mobile Generator: It is on a mobile platform with batteries and inverter to provide a single-phase ac output for customer use, as shown in Figure 5.18c. It also has a charger for batteries to recharge them from

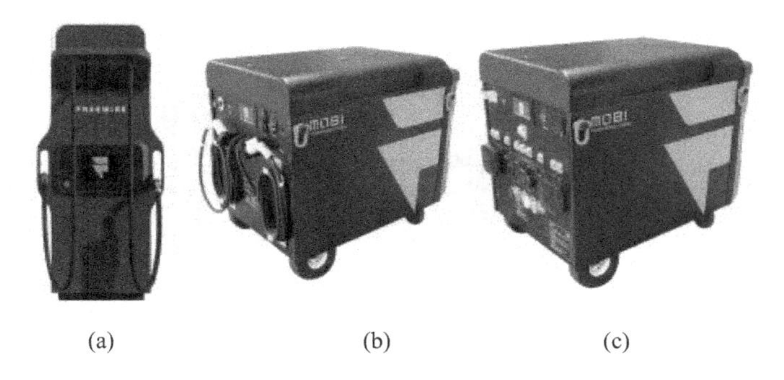

(a) (b) (c)

FIGURE 5.18 (a) Boost charger, (b) mobile charger, and (c) mobile generator. (Courtesy of FreeWire.)

grid supply. Its ratings, inputs and outputs are very much similar to the mobile charger. This product has applications in Disaster relief centers, construction sites, entertainment platform and equipment needs, utilities, aviation and telecom.

Their products in EV charging using their mobile and fast station charger are shown in Figure 5.19. Note that their mobile charging technology is the basis for extending to charging station applications and given the market opportunities, their foray into this market is natural.

Intellectual Property: Three US granted patents and the inventors are: Arcady Sosinov, Sanat Kamal Bahl, Luv Kothari, and Sameer Mehdiratta, all of them co-founders of FreeWire. Note that the last three of the patent co-inventors are not with FreeWire anymore.

Funding: $30.2 M from investors and they are BP Ventures, Volvo Cars Tech Fund, Stanley Ventures (an arm of Stanley Black & Decker company), Blue Bear Capital, Oski Clean Energy Partners, Strawberry Creek Ventures and others.

Observations:

i. The idea and concept of FreeWire technology and products came out of the four patent co-inventors (also co-founders) during their MBA study together. It was made possible that the majority team members had engineering degrees and experience and one in business so that they can carry out the concept to a practical end. A business background of one may have also helped in tuning the product to arising customer needs.
ii. The original team brought two additional co-founders with a background in product development, innovation to fit the customer base and testing which resulted in faster rollout of products, i.e., in less than 3 years from the concept.
iii. The major difference to this startup came not from technology alone but in identifying and moving in the direction to catch up with the customers' needs and interests.
iv. The concept of charging associated with the car coming and standing in front of the stationary platform is reinforced into one's consciousness

(a) (b)

FIGURE 5.19 EV charging with mobile charger (a) and fast charging station (b). (Courtesy of FreeWire.)

by everyday viewing of gas/petrol station with fixed hoses for pumping the gas/petrol into the car's fuel tank. The founders identified that the charging station does not have to be stationary but can be on a mobile platform and that idea constitutes the underlying intellectual property for this company. This is a classic example of thinking "out of the box", as they say in common parlance. Once this concept has been arrived at, the evolution of products seems to be automatic and free-flowing.

v. The fact that a sizeable amount of funding has been raised for this company in such a short time shows the traction with its striking concept of providing a mobile platform. It has found interest from other industries including a gas company such as BP through its venture arm BP Ventures, automobile maker such as Volvo, and tools and accessory maker Black & Decker Stanley.

vi. Intellectual property, so far, published is from the co-founders, and the quantum is normal for a company that has been in existence for about 5 years. Whether their intellectual property is sufficient to block and/or discourage some other entities that may enter their product domains is to be seen.

vii. FreeWire's location in the radius of 25 miles from the heart of Silicon Valley is one of the important factors for its early traction with investment firms and that fact should never be underestimated by the budding entrepreneurs in their planning.

viii. The application domains for FreeWire's products cover a large spectrum from automotive OEMs, entertainment industry, construction industry, to aviation and hence bestows a large market and hence a large revenue potential in USA, and all of them growing market with high potential further makes the product domains valuable to this company and its investors.

ix. This startup could also be in the EV charging startups in a later section but this startup has a unique product in that they have the mobile charger and hence it is placed here. Their product extension from mobile charger to conductive charger station comes naturally as they have a strong and proven technology base of charging through their mobile chargers.

5.6 BATTERY ENERGY STORAGE SYSTEMS

Energy storage has been in practice for a long time, and they can be classified into Electro mechanical, electro chemical, thermal, hydrogen pumped hydro, and compressed air types of energy storage systems. Battery energy storage is emerging as a leader in this field thanks to lithium-ion batteries. Energy stored in them comes from one of the following sources: electric power supply grid, solar energy installations, and wind energy farms. Then, the energy from the battery is sent to the power grid at times of its need and also used by the load in the homes and electric vehicles among other possible loads. The energy storage system working with solar in KWs of power has been studied in the previous section and also the key role of the power electronics and control in it. Utility-scale implementations of battery energy storage system (BESS) handle peak loads of the grid that cannot be sustained with its present and limited generation capacity, store its own generated energy at off-peak times, and also store the energy from other sources such as solar and wind. Because of the renewable energy sources that have grown in capacity at utility-scale, corresponding BESS are rising and in demand. Many startups see specific spots they can occupy in this domain and are emerging on the scene. The key subsystem is power electronic converters and their control but in the utility-scale BESS at MWs of power.

The power converters and their control are established products in the solid realm of established companies such as Siemens, Schneider, ABB, et al. and therefore, the startups carefully avoid getting into their production and source them with the big companies. By doing that, they avoid getting into manufacturing and also having to raise a large capital to bring them into the market. That is a key aspect of some of these startups at present but that does not mean and guarantee that the future may remain the same. Three startups are studied in this section.

The current market and future market scope are of importance to the startups. Some data is available as to the market value of BESS implemented and future demand and many sources such as the National Renewable Energy Laboratory (NREL) as part of the Department of Energy (DOE), USA does release studies and they are of value to interested readers. There has been a cumulative 536 MW BESS installation in the USA by 2017 and out of which the utility-scale installations are of 495 MW capacity, about 92% of the total. This alone shows the market range and the rest of the world is about 150% of the US market in this sector. In 2019, the US capacity grew by 45% (report from the Smart Electric Power Alliance). The prediction for the future is even much higher considering the fact that many utilities prefer to have the BESS so that they do not have to invest in new power stations, (in the process saving themselves a ton of money), just to handle the peak loads and emerging growth in their market. Some other BESS installation predictions are from 9 GW in 2019 to 1,095 GW for the next coming 21 years. Regardless, the market size is considerable even if the predictions do not hit the high end. In terms of real dollars, the global market for battery energy storage is about $13B by 2023.

5.6.1 Advanced Microgrid Solutions

Advanced Microgrid Solutions (now known as AMS and used hereafter) is founded in 2012 in San Francisco, CA, USA. Its founders are:

i. Susan Kennedy, Chief Executive Officer until the early part of 2020, and now Chairman of the Board of Directors: She holds a B.A. in Management from Saint Mary's College of California. She served in state and federal government and in visible positions as chief of staff to Governor Arnold Schwarzenegger, deputy chief of staff to Governor Gray Davis and communications director for US Senator Dianne Feinstein. She served on

California's Public Utilities Commission where she authored the largest energy efficiency program in utility history.

ii. Jackalyne Pfannenstiel: She worked 20 years for Pacific Gas and Electric Company and was the first woman promoted to corporate officer in the company. Served as Assistant Secretary to U.S. Navy and also as Chair of the California Energy Commission.

Currently, the top management is headed by Seyed Madaeni, and his team is spearheading into their software approach to power industry operation including all the renewable and battery storage systems.

Technology: They declare themselves as technology-agnostic and are in the business of:

i. Hybrid Electric Buildings (HEB): Control of clean energy storage in batteries from a fleet of hybrid-electric buildings for the owners' use and also supply the utility grid when demand is high. Examples: (i) Irvine Co has storage of batteries in 21 buildings in South California working with AMS to provide up to 10 MW for a maximum of 4 hours to help the Sothern California Edison (SCE) to balance the grid and earn revenue. (ii) As of 2019, AMS is managing 27 MW, 142 MWh fleet of batteries spread across SCE's operational region.

ii. Optimized Resource Management: Use of energy storage and analytics software to optimize cost, distributed energy resources (DER) including renewable energy sources and reliable backup power for large-scale commercial and industrial facilities through their DER software called Armada platform. This software works to extract financial value for customers and utilities.

iii. Intelligent Grid: Work with utilities and users to realize the smart grid through their software.

At the time of starting, they developed hybrid electric building projects which required a multimillion-dollar financing arrangement between various parties. They found that their market positioning becomes better only concentrating on the software to control and manage these facilities for energy storage, support utility loads when demand is high and maintenance of these facilities. One of their investors, Macquarie Capital, which is an investment bank and has $367B under its management globally, has pledged to support AMS's projects. AMS does design, finance, install and manage storage solutions for commercial, industrial and government buildings and these projects are also within their technology base.

Product: There is no product sold, but it consists of their software, project leadership and management. Note that the energy storage and flow of energy from utility to their grid and vice versa require battery, solar panels, and to power electronic converters with the capability of 100s of kW to tens of MW for their projects. They source them from established companies. DER management software

Armada is seen to be their major product for use by their team to work on projects. Their current focus is elegantly expressed in their graphic shown in Figure 5.20 as a wholesale market solution for actively managing assets such as solar and wind renewable energy with battery storage for optimal utility power supply. All with the eye on the financial aspects also from market prices of the power, its optimal dispatch, buy and sell decisions from asset holders to utility and vice versa for energy storage and working with independent system operators (ISOs). It is a complex operation and made feasible with their software forming their key product. They use AI-based approach for their software to enable it to make decisions based on the previous history and for continuous optimization of their approach and implementation.

Intellectual Property: Two granted US patents on "Method and apparatus for facilitating the operation of an on-site energy storage system to co-optimize battery dispatch".

Funding: $52.7M in two rounds from eight well-known investors and they are: Southern Company, Arnold Schwarzenegger, GE Ventures, Macquarie Capital, AGL Energy, Energy Impact Partners, DBL Investors, and Engie.

Observations:

i. The founders with extensive public exposure to state and federal governments in the domain of power and energy efficiency were able to see the market opportunity in rising renewable and battery-based energy technologies to form a virtual power plant (VPP). That led to the formation of AMS.

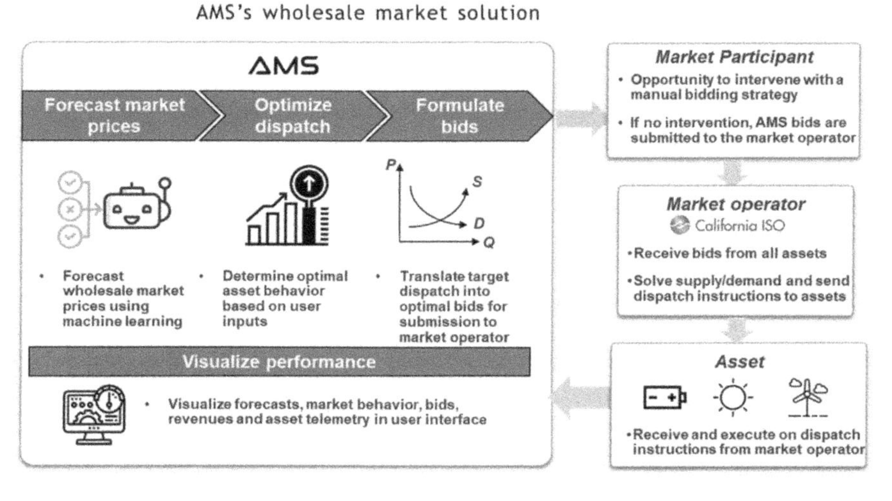

FIGURE 5.20 Wholesale market solution graphic. (Courtesy of American Microgrid Solutions.)

ii. An important corollary is that one does not need to be a leading electrical engineer but a keen observer of the industries in that arena with a high amount of interactions in that community can conceive startup ideas of this magnitude and even bigger ones as seen from Microsoft and Facebook.
iii. The idea behind a startup is alone not sufficient but to have the leading brains in that field of expertise is critical for building the startup. This company attracted two high-level Tesla Company executives and others with the appropriate background in battery technology, power and energy management that helped to move it in the technical arena.
iv. The founders' exposure to very powerful corporations in the energy sector, politicians and banks has been central to raising finance for this startup. The big hybrid electric buildings project could not have taken off without the $200M funds from Macquarie Capital and such backing comes to those only with high public/governmental exposure and influence is not to be overlooked by budding entrepreneurs. It also means that the startup requiring much capital has to bring a very well-connected person to be their CEO or on the board or other top positions.
v. In a short span of time, they have built projects of large power and energy in Southern California and through their operation for the past 3 years, they have demonstrated that the technology of battery-based energy storage helps utilities in times of their need and at the same time it is beneficial to the owners of the storage systems.
vi. The company management realized that building and bringing to operation of large projects and completing them makes a name but does not generate revenue after the projects are completed.
vii. They realized that the maintenance, continued monitoring, control and data management are all required even after the project completion and that is a revenue source for ever and ever is not to be underestimated. Therefore, they have put their focus of the company in software, analytics and data management. That change in strategy also attracted investors to fund $34M in Series B round for its operation and growth.
viii. They are growing with storage energy partnerships in May 2019 with big entities such as 38 North and John Hancock and its affiliates.
ix. This company has a very small number of patents at this time, 2019.
x. This startup demonstrates that one does not need innovations in power electronic converters to get into the renewable grids and VPPs, even though they form the big bulk of the products used in their deployment. A mix of strong electrical engineering background, the study of developing energy market sector and an inclination to entrepreneurship can lead some to similar ventures such as AMS is not a farfetched statement.

5.6.2 Greensmith Energy Management System

Greensmith Energy Management System (hereafter referred to as Greensmith) is founded in 2008 in Herndon, Virginia, USA.

The founder and CEO is John Jung. He obtained his MBA in Strategy and Finance and a Bachelor's degree in Economics from Western University. Before founding Greensmith he worked for Braxton Associates, A. T. Kearney and as managing director of Commerce One for varying periods specializing in corporate business strategies. From 2008 to 2018, he worked as CEO for Greensmith and after its acquisition also. Currently (from May 2019) he is the president and co-founder of BrightNight, LLC, San Francisco, CA which is based on solar energy projects of GW and above capacities.

Technology: Focus is mostly on battery energy storage systems from and for utility grids. The energy from the battery is used for supplying and load balancing the utility system. The technology base consists of solar or wind or hydro or thermal generation, battery and its chemistry and finally their integration with the utility via power electronic converters. The companies in this space do not or no need to have an independent technology base in any of these subsystem technologies but to have experience and expertise with choosing the right subsystem technologies for a given project and their integration via control software is where the strength lies. Greensmith's integration software is called GEMS, an acronym for Greensmith Energy Management Software. Right in the integration and control software is an opportunity for embedding intellectual property. The software platform performs optimization of energy storage, cost and return on investment. Also performs data collection such as charge/discharge rate, clean power generation, utility grid management data and offers the data to the utility control center for its use.

Products: They have no specific products for sale but projects that they assist customers to build. They have delivered 70 projects in 10 countries and one of the projects is a 180 MW of energy storage and the customers are utilities and independent power producers. An example is AltaGas Pomona Energy Facility, CA for 80 MWh of energy storage system shown in Figure 5.21.

Intellectual Property: One patent filed in USA in 2009 and published in 2012.

Funding: Raised $32.1M and investors include American Electric Power, Signatures Capital and others.

Exit: Acquired by Wartsila in May 2017 and the sale price was $170M. Greensmith's revenue at the time of acquisition was around $40M.

Observations:

i. Greensmith is one of the early startups in this sector business of integrating renewables and/or batteries and utility grid and providing analysis, management and appropriate subsystem technologies for implementation.
ii. All these startups are technology-agnostic as they evaluate, choose and integrate all the subsystems from the primary manufacturers.

FIGURE 5.21 An example of a battery energy storage facility, AltaGas Pomona Energy Facility. (Courtesy of Wärtsilä Energy Business.)

iii. It has developed projects of 80 MWh battery storage project and has been working with utilities and others.
iv. It has expanded the business to many countries with successful marketing.
v. The founder who also served as its CEO until it was acquired has financial and executive management experience in corporate strategies and business building. His articulation of the business and leadership has been one of the key factors for the success of this company is hard to miss.
vi. Raised reasonable funding for a company to grow and expand its operation in such a short time in this sector as the magnitude of the projects requires a fair amount of time for completion and takes a good amount of highly skilled workforce.
vii. The key product they developed with their project management and leadership is their control and integration software, GEMS.
viii. The company has very limited intellectual property (only one US patent application). It may be fair to guess that significant intellectual property not revealed to the public must be contained in their software GEMS. It is also safely ensconced as it is not for sale except that it is only embedded in the projects of the company.
ix. A successful exit has happened for Greensmith with a return of more than five times their investment funding in less than 8 years on the whole, thus rewarding the investors and founder very well. The acquirer has created a separate division with the same name and has been expanding the business.

x. This startup can be termed as a successful company and future entrepreneurs in this sector of business have a lot to learn from it.
xi. One of the key lessons that can be drawn from this company as well as from the sector in which it operates is that software integration and control of many subsystems opens an opportunity apart from technology innovations, say, in power electronic converters even though they play a major role in this business. Opportunities are many and particularly in applications than in basic innovations in the subsystems. Researchers tend to overlook this aspect not only in this engineering sector but in many others as they are more focused on some small part of the subsystems and not even the whole subsystem itself.

5.6.3 Stem

Stem, Inc. (hereafter referred to as Stem) has come from Powergetics, a startup founded in 2009 in San Francisco Bay Area, CA, USA. Powergetics transitioned to Stem, Inc. in 2012/2013 in the same geographic location.

Founder of Powergetics, Stacey Reineccius: He is a serial entrepreneur having a background in computer science with a B.A. from University of California, Berkeley and study in US army intelligence school. He is the founder of Powertree, Quicknet Technologies, Inc., and Fibercorp International, Inc. He is the source of founding patent and software of Powergetics and hence Stem also.

CEO (current) of Stem, John Carrington: 25 years of high-level and high-profile leadership experience in energy and industrial technology-related companies such as MiaSole, First Solar and General Electric. He has a B.S. in marketing and economics from University of Colorado.

CTO (current) of Stem, Larsh Johnson: Came with relevant experience from Siemens Digital Grid, and eMeter in consumer metering, demand response, grid automation, smart grid applications and analytics and data management. He is a co-founder of CellNet Data Systems in smart metering which is now a unit of Toshiba Company, Landis+Gyr. He has B.S. and M.S. degrees in mechanical engineering from Stanford University.

Technology: Its technology is very much similar to Greensmith, AMS and others in this domain except it strongly emphasizes the artificial intelligence part of its software for power management including battery storage to the grid and vice versa, with financial aspects to the fore. Their technology and business strategy include the following:

i. Reduce peak electrical usage that costs the users with a high penalty and free the stress on the utility and lower electric bills. Note that 60% of the utility bill is connected to the timing of energy use.
ii. Lower and/or eliminate the need for new generation facilities which is of immense interest to the utilities for not increasing their financial investments and sparing the public from environmental concerns.

iii. Energy optimization solutions combining the storage with cloud-based predictive analytics for speed of monitoring the changes and acting fast to maintain the low electric bill.
iv. Reshaping electric energy consumption vs. time to increase the profitability of commercial businesses.
v. A new model where the user does not pay for Stem's services upfront but from gained savings with their services.

Its technology was focused on homes and small businesses to start with and subsequently when it transitioned to Stem, large installations were included in its orbit of operations.

Products/Projects: Their central product is the software but only installed by them with the customers and also installation of the power generation projects with energy storage. Their foray into businesses with no upfront investment by customers is made possible by their ability to raise project finance capital of $350M plus. This has accelerated many project deployments. They have installed a total of 250 MWh of energy storage projects at more than 900 sites. Some also are with solar energy with battery storage. A sample of the projects is given below:

i. Solar carport and ground-mounted systems of 5.2 MW capacity with 630 kW energy storage system at the University of California, Merced, CA location and shown in Figure 5.22.
ii. Solar carport of 2.6 MW capacity with 1.3 MW storage system at Santa Rosa Junior College, CA.

Acquisition: Stem acquires Constant Power, Inc., Toronto, Canada in December 2018. It is a strategic and tactical acquisition to expand its business to Canada.

FIGURE 5.22 Solar plus energy storage project at UC, Merced, CA. (Courtesy of Stem, Inc.)

This purchase by Stem is encouraged by C$200M project finance from Ontario Teachers' Pension Fund specifically promoting Ontario market power projects of Stem.

Intellectual Property: 15 patents granted and 9 applications pending, all in the USA.

Funding: Total raised is $320M plus. The major investors are who is who in the community of investment and power companies such as Ontario Teachers' Pension Fund, BNP Paribus Private Equity, RWE, Magnesium Capital, Activate Capital Partners, Starwood Energy Group Global, GE, Mithril Capital Management, Mitsui Global Investment, Total Ventures, Angeleno Group, Greener Capital, Iberdrola and others.

Observations:

i. The original founder's vision and intellectual property formed the basis of the company. The market focus is changed when the company transitioned to Stem but not the technology base.
ii. This is one of the companies formed in the 2008 time frame when the activity in the battery energy storage and integration of renewables were starting to happen. That way it had a good lead with a small number of peers.
iii. The later management of Stem steered its marketing focus to business and commercial customers with MW capacity away from home front with a few kW capacity thus positioning the company to bigger market segments and higher earning potential.
iv. The current management has a strong background in the technology and implementation that they understand with a clear benefit focus to the customers in terms of saving and earning revenues as seen from their large projects.
v. Stem's business model for clients without requiring an upfront investment from them and still the project benefiting them opens up a significant advantage in the market place for its projects.
vi. The ability of the management to raise large equity capital is phenomenal and unmatched in this sector. That allows them to go beyond US markets such as in Canada and Japan.
vii. Finding investors capable and willing to invest a large sum (such as C$200M) in Canada and using that connection to setup an operation to expand in that country is a win-win for Stem.
viii. Stem is one of the big energy storage companies and is in the select group of a few such as AMS, Greensmith, etc. Its focus is mainly on business and commercial installations with a capacity of a few MW per project unlike its other competitors involved in tens of MW per project and nearing even 100 MW plus. Stem may as well cover this space in the future without much difficulty.
ix. They raised $350M project finance capital to fund projects. It is matched only by its peer AMS that was able to raise as much as $500M.

Such capability put these companies at the top of this sector's group making them unique in this sector. The corollary effect is that it discourages many to enter this domain even if they may be armed with superior ideas and software.

x. The company has knowledge of integrating various subsystems involved in the energy storage business, leading or providing the finance, and integrating the system with the utility and their system with cloud-based software. But they do not make the subsystems taking the bulk of the cost of the system and they all go to these large subsystem manufacturers and many of them are worth $100 B and even more. They are few on the global scene. What if they decide to integrate their subsystems and eliminate the middlemen such as Stem and its peers on large projects? That is a scenario that can come to help the successful exit for these companies (for example, as it happened to Greensmith even though its acquirer is not one of those global giants). For many of these large companies, the acquisition cost of a few billion dollars is within their easy reach and part of their everyday business.

xi. Given such a scenario in (x), it is interesting to watch this sector for future development.

xii. Also exciting will be to think ways of disrupting technology or software or control that may dawn on this sector and how it can be perceived and received by investment and engineering communities.

5.7 GRID EDGE POWER ELECTRONIC SYSTEMS

5.7.1 Introduction

The present evolution of the grid is influenced by:

i. trend to stay away from fossil fuel-based power generation to promote a clean environment,

ii. financial imperative to avoid as much as possible expensive investments in new power generation plants,

iii. mature technologies for large-scale implementations of wind and solar renewable energy,

iv. distributed energy resources such as rooftop solar systems pumping power to the grid,

v. battery energy storage availability from small to large power levels,

vi. emerging electric vehicles and their fast-charging demands putting peak load demand that is unlikely to be met by existing generation capability,

vii. opportunity to use the battery in the EV for vehicle to grid (V2G) and grid to vehicle (G2V) energy transfer for evening out the peak demands without having to invest in new power plants,

viii. to compensate for feeder voltage drops, reduce and compensate reactive power drawn from the grid, and improve voltage waveform quality, and

ix. smart metering and technologies for access to distributed energy resources (DER) to control power flow to the grid for the overall effective operation of the grid to the benefit of both grid and DER operation,

Technologies have developed on various topics listed above and most of them have been demonstrated, or on field trials and/or considered for integration with the grid. Capabilities such as methods to stabilize the grid in the presence of DERs and their uncertainties in its operation using computing power and information available from DERs, and the grid itself are yet to emerge. The communication and control capabilities from grid center to various subsystems and power electronic converters and their subsystems are yet to emerge for the grid operational center. If all of these features are built into, then it goes by the name of smart grid.

There are three important subtopics with a strong power electronics role that influence the emergence of the modern grid and ideal for the study of entrepreneurial activities. They are:

i. Grid edge-connected power electronic systems.
ii. Large-scale battery energy storage system (Section 5.6)
iii. Home solar energy management systems (Section 5.4)

Only item (i) is to be treated below as items (ii) and (iii) have been completed in earlier sections.

5.7.2 Technology of Grid Edge Control with Power Converters (Mainly Inverters)

When the voltage and reactive power control is exercised in a distribution system, the energy consumption by the customer is reduced anywhere from 1% to 4%. It also implies that the power saved is available to the utility to cater to other customers and hence save them the up-front investment for additional capacity that they may otherwise have to install. Currently, there are solutions to achieve them through tap changers and capacitor banks that seem to be popular with utilities. But there are more power electronic-based solutions with independent active power filters and also through the use of inverters and controllers (refer to Section 3.4.1 on active power filters) in the distributed energy resource (DER) systems themselves. Basics of power electronic active solutions to the grid problems have been there since the late 1960s in one form or other, and power electronic textbooks cover them. The trend is to advocate the use of these instead of the current utility solutions and such efforts reflected in the startups will be considered in the next section. Whether the utilities will embrace the newer approaches or adhere to older ones or go for technologies and implementations that may be more centralized or for a combination of centralized and some distributed solutions is open to question and debate within the engineering community.

5.7.3 Market Scope

Very few studies have come up forecasting the value of the smart grid and within that for this technology more commercially branded as Volt Volt ampere reactive Optimization (VVO). Its value is estimated in McKinsey projection to be equal to $43B annually from 2019 onward for the VVO to the utilities. This projection is not based on any one technology to accomplish this technical function. The startups may find comfort in the fact that this value is some indication of their market opportunity and hence to position themselves for a possible share of that market. The investment required to get to that value possibly will be a larger fraction of if not multiple times that and that factor may not be known at this time.

5.7.4 Startups

5.7.4.1 GridBridge

Grid Bridge was founded in 2012 in Raleigh, North Carolina, USA. Only two founders have been identified from open sources and they are:

i. Chad Eckhardt, CEO: He has a bachelor's in electrical engineering from Milwaukee School of Engineering and an MBA from Carnegie Mellon University. Experienced in electrical distribution and power quality markets when working for original equipment manufacturers in these fields.
ii. Alex Huang, Chief Scientist: He is a well-known professor in the ECE Department, North Carolina State University (NCSU), Raleigh, at the time of founding with a specialization in power devices and power electronics with current research on SiC power devices for solid-state transformer and other novel applications. He is not serving as a chief scientist anymore and currently with the University of Texas, Austin.

Technology: Power conversion systems to serve the needs of the emerging power grid with its associated renewables, electric vehicle to grid and grid to electric vehicle energy transfer while providing reliable operation with attendant features of voltage and power factor controls among other services. It is in the same category as many other startups such as Varentec, Gridco and others emerging across the globe. The theory behind the technology not only of this venture but its competitors are in textbooks in the early 1970s but the adoption in various ways with the newer devices and software control yield technologies, innovations and intellectual properties for all of them.

Technology License: GridBridge has licensed technology in the form of patents from NCSU and the details of them not available. NCSU research output is considerable in power electronic converters particularly using SiC devices for high-frequency switching, efficiency enhancement, high power density to yield the smallest packaging volume, and novel applications such as solid-state transformers. Whether this license is exclusive (could be most likely) to GridBridge is not known at this time.

Products: Considerable number of products have been engineered in a space of less than 5 years for the power companies and for individual users such as poultry farms, solar power systems, and restaurants. The technology is embedded in a product called Grid Energy Router (GER). Broadly it is to deliver voltage control, power factor at the grid edge, integrate renewable energy and energy storage systems, enhance efficiency at the user end by 5%–15%, assure balance in three-phase supply system. It is a power electronic converter system with software control so that it lends itself to remote control, a feature desired by the users and the power utilities as well. This device is installed at the secondary side of the distribution transformer in the grid. The products are available in three packages and they are:

i. Pole mounted GER for single-phase system with 120/240 V input, 50 kVA, efficiency of 99% at or near full load, reactive VA for compensation is ±5 kVAR, output voltage regulation when input variation is within ±10 Vac.
ii. Three-phase GER for 120/208 Vac, neutral required, rated power 150 kVA, et al. Its performance is shown from the datasheets in Figure 5.23. During 1 and 2, it is being regulated to the set point, whereas in 3, the regulation is bypassed which shows a poor state of operation. Figure 5.24 shows the current balancing per phase leg with excellent performance during the GER's operation.
iii. The products listed in (i) and (ii) are available with transformers also.

Intellectual Property: (i) Licensed from NCSU. (ii) Self-generated: One granted US patent and one patent application pending.

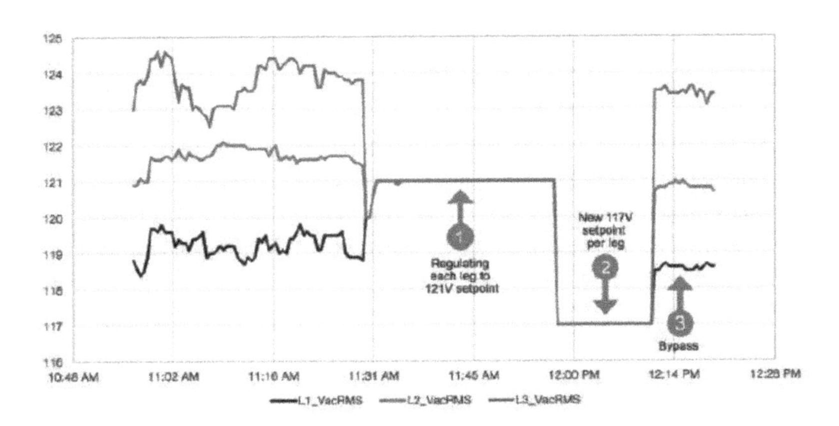

FIGURE 5.23 Three-phase system voltage balancing with GER. (Courtesy of GridBridge.)

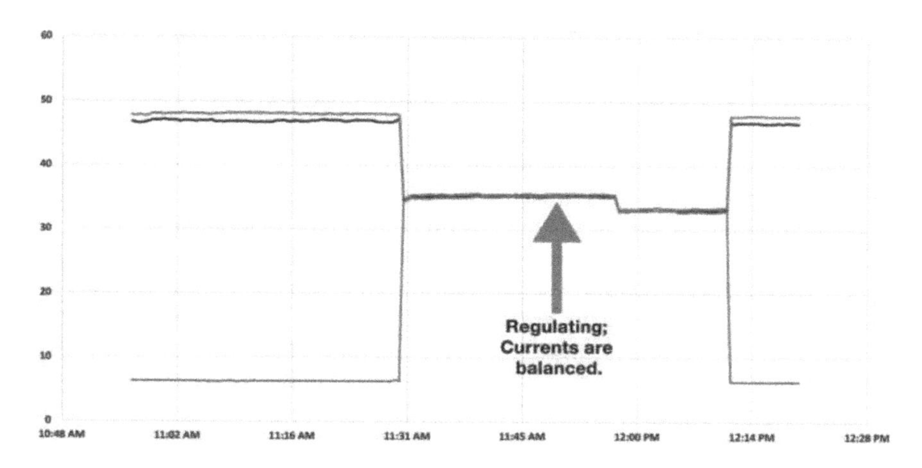

FIGURE 5.24 Source current balancing per leg performance. (Courtesy of GridBridge.)

Funding: The company mostly survived from federal and state grants and the former kind is more of research-oriented funding. Given its achievements, no other company in this domain has achieved so much with so little funding. The management deserves credit for it.

Exit: It was acquired by Electric Research and Manufacturing Cooperative, Inc. (ERMCO) with headquarter in Dyersburg, Tennessee. ERMCO is a subsidiary of Arkansas Electric Cooperatives, Inc. (AECI) in August 2017. AECI serves 500,000 homes, farms and businesses in Arkansas and surrounding areas. This is advantageous from the point of view that the principal can use GridBridge's products at the earliest and satisfy the customers in need of it and provides a greater adoption and testing ground for its technologies that are already developed and for its future products. The price paid for the acquisition is not known.

Observations:

i. Around 2010, many startups in power converter-based control of grids with various performance objectives came on to the scene. GridBridge is one of them with one founder having this kind of industry background and the other founder steeped in cutting edge research on some aspects of power control with modern devices and applications. Such a unique combination is a blessing for any startup!
ii. Technology licensing from NCSU and the connections that come with the research community including exposure of their graduate students for employment have been a remarkable advantage for GridBridge as for any startup for that matter.
iii. Startups in its domain have not built the products as GridBridge and taken them to power companies with such meager funding and in a short time of 4 years or so compared to its competitors. It is a unique

achievement for which the company's management and workforce can take immense credit.

iv. Funding is almost exclusively from research-based grants provided by federal agencies and state grants and that shows that it is a good idea for advanced research startups to take that route and it could be successful as in this case. Alignment with a university laboratory and faculty facilitates this mode of funding leading to product development is demonstrated by this startup as a viable means for developing the startup.

v. Their products have the best data sheets with adequate information as to their performance during continually varying power system conditions and in that regard, they have come up to the level of information sharing done by the power semiconductor companies both big and small. This is an important aspect to reach potential customers and showcasing their applications to different customers such as gas stations, poultry farms and the benefits the customers get out of them. Contrast this to their competitors' datasheets that are not very explicit in these aspects.

vi. The company had a successful exit in a very short time of 5 years since its inception. That alone shows its value in the market place. The acquirer has the potential to take its products to field adoption and beyond the testing stage. If this trend in grid control takes root, the acquirer is bound to have a high multiple of their investment recouped in a few years.

vii. Even though its own IP number is small, its tie-up with NCSU puts it at a very advantageous position both for licensing newer and evolving technologies and to take advantage of developing newer products with the help of the faculty and graduate students there.

viii. An ideal company for many graduate students to follow in its footsteps and design their research in a market-relevant way. The lesson that they can learn from this startup will go a long way for them to try their hands on their own ventures after their graduation.

5.7.4.2 Gridco Systems

Gridco Systems was founded in 2010 in Boston, MA, USA and its founder is Naimish Patel. He was also the co-founder of Sycamore Networks and served as its CTO for about 5.5 years. He has experience as Entrepreneur in Residence with General Catalyst partners prior to founding Gridco Systems. He served as the CEO of Gridco Systems from 2009 to 2017.

Technology: Software control of power electronic converters for dynamic, adaptive and precise management of voltage fluctuations, reactive power (volt-ampere) and hence power factor, and containing line faults among other functionalities required for emerging grids. In many ways, there is hardly much difference between this and its competitors such as GridBridge and Varentec as to the technology, products and the power utility customers and other users they all want and would like to have. The technologies as given by Gridco Systems under two different headings, from their information sheets and in their own words as some of them are trade-marked, are:

A. Power Regulation:
   i. In-line Power Regulators™ (IPR): IPRs are multifunction, utility-scale power electronics hardware systems that combine series-connected voltage control, shunt-connected current control, embedded sensing, and control logic to simultaneously provide voltage regulation, VAR compensation, harmonic compensation, power monitoring, and more.
   ii. Power Regulating Transformers™: PRTs are multifunction, utility-scale hardware systems that integrate distribution transformers and power electronics into one common footprint simultaneously providing power delivery, voltage regulation, VAR compensation, harmonic compensation, power monitoring, and more while further simplifying planning and operations.
   iii. Static VAR Compensators™: SVCs are lightweight, cost-effective modular hardware systems that combine a shunt-connected current source architecture, advanced algorithms, and fault-tolerant, self-balancing switching design to provide dynamic voltage boosting at minimal total install costs.
B. Intelligent Control:
   i. Distributed Grid Controllers™: DGCs are comprehensive computing platforms that provide coordination and control, data logging, local analytics, data networking, and communications
   ii. Advanced Software: Grid Management and Analytics Platform™ (GMAP) is a suite of software for providing remote monitoring and control, data collection, and seamless integration with other applications used in power systems.

These are precisely what everyone wants to accomplish in this technical area.

Products: Sample products are:

i. Low-voltage Pad-mounted 50 kVA in-line power regulator combines utility-scale power electronics and control to deliver load voltage regulation, VAR compensation, harmonic control and monitoring for use in residential, commercial and/or utility scale renewable integration. It is shown in Figure 5.25.
ii. Distributed Grid Controller (DGC): In brief, it provides computing platforms for coordination and control, power quality monitoring, data logging, communications for their power regulator and transformer products.

Intellectual Property: Three granted patents and three applications pending, all in USA.

Funding: Raised $54M from 2010 to 2017. The investors are North Bridge Partners, General Catalyst, Lux Capital, and strategic investor Maschinenfabrik Reinhausen GmbH (MR) which is one of the global leaders in power transformer regulation, and others.

FIGURE 5.25 Gridco system's 50 kVA in-line power regulator. (Courtesy of Gridco Systems.)

Pilot Projects: With the utility companies Duke Energy, Hawaiian Electric, California's Sacramento Municipal Utility District, and Canada's greater Sudbury Hydro in Ontario, Gridco Systems proved its products capabilities.

Litigation with Varentec: Varentec, one of the closest competitors, sued for infringement of its patent and got a verdict in its favor in the middle of 2017. The details may be found from the reference section and it may be of high interest to inventors so that they can get to know the intricacies of patent, its value and how companies have to be cautious of IP related lawsuits.

Exit: Company closed in December 2017 for various reasons. One assigned reason by its CEO is that the market for the products did not open up. There was a huge expectation that renewable energy integration will be very high in terms of the quantum of power and that utility companies will not be able to handle the challenges that come with it. First, the large power flow from renewables did not materialize as expected and the utilities have handled the power system so far without adopting the products marketed by Gridco Systems and its competitors too.

Observations:

i. Gridco Systems was started by a person who has been in technology management but not with equal experience or exposure to power electronics. This has not affected its vision and product developments.
ii. Its vision is phenomenal in that it tried to combine software control of power electronic hardware to meet objectives for power regulation and intelligent control of the grid itself with desirable features.

iii. In spite of (i), the company grew fast and built a line of products that passed the tests with users including some utilities.
iv. Its intellectual property in terms of number of patents and patent applications is in line with its competitors.
v. Vision for how it could serve the power systems industry, articulated comprehensively in public and through their websites, is unmatched in this business sector.
vi. It attracted a huge investment of $54 M unmatched by any of its competitors as of date in about 6 years.
vii. Strategic investor (Maschinenfabrik Reinhausen GmbH) invested with a special interest in its technology which is a recognition any startup greatly looks forward to.
viii. It had sufficient big utility companies in USA that were interested in testing the products and that is a big achievement for any startup and hence a step in the right direction for it.
ix. The lawsuit by Varentec seems to have weakened the company as lawsuits always do to entities. The lawsuits always invariably take the management's time and distract them from day to day activities of growing the company, drain the monetary resources, create an uncertain atmosphere within the companies, make the present and future investors nervous, and the potential customers to shy away from any possible deal.
x. Further, the market did not open up to use the products and that could put a lot of pressure on the company in terms of raising capital to survive and build the company in the absence of product revenues. Therefore, this situation also did not seem to help the company.
xi. In spite of a huge number of positives and achievements, Gridco Systems closed its operation at the end of 2017 and its assets were to be sold.
xii. A number of lessons could be learned from this company. One of them stands out and that is litigation aspects. Budding entrepreneurs have to do a large amount of work to build intellectual property that cannot be called in to question and legal aspects of it has to be airtight. Unlike other countries, it is much easier to resort to litigation in USA and get a fairly faster resolution of issues, even though it is a highly expensive process and the startups operating within financial and other constraints have are at a greater disadvantage taking this route. The flip side of that is the protection for a company's intellectual property is also weak or nonexistent in countries where the legal system is weak and judgments cannot be obtained in a reasonable time.

5.7.4.3 Varentec

Varentec is founded in 2009 and located in Santa Clara, CA, USA. Its founders, their roles and their brief bio are given below:

i. Andrew Dillon, Founder, VP Business: Obtained his BS in Electrical and Computer Engineering, Univ. of Illinois, Urbana Champaign and MS in communication technology from the University of Texas, Austin.

Founder of four companies including Varentec. He served Varentec until 2015.

ii. Dr. Deepak Divan, Co-founder: Professor in Georgia Tech's ECE Department, well-known power electronics researcher, author of 250 papers +, inventor and/or co-inventor of 50 granted patents and pending applications, founder of two other companies. He is a Fellow IEEE and member of the National Academy of Engineering and honored with prizes from IEEE including Newell Award.

iii. Mehrdad Hamadani, VP, Engineering and Operations, Co-Founder: He holds a BS in Electrical Engineering from the University of Pittsburgh, and MSEE from West Virginia University, and holds four patents. Prior to Varentec, he has a long and successful career as an Engineering leader in a number of technology and energy-centric organizations, including Mitsubishi Semiconductor America, CoSine Communications, Aruba Networks, Foundry Networks, and Brocade.

Its current CEO is Guillaume Dufossé and holds an MSc in Electrical Engineering from Supélec, France and MBA from Harvard Business School. He has over 20 years of experience in the energy industry ALSTOM, AREVA and Schneider Electric, SmartGrid & Grid Automation business units, SunPower Corp all in senior executive positions. Guillaume is a French citizen and lived for over 12 years in USA, China and Europe.

Technology: Control and power electronic converters applied to overcome the problems of the emerging power system and it is very similar to its competitors in the base technologies and products. Suggested to refer, to avoid repetition here, to the sections on its competitors GridBridge and Gridco Systems to get a broader picture of the general technologies behind.

Products: There are two products available to customers and they are:

i. Edge of Network Grid Optimization (ENGO): Intended to regulate voltage and reactive volt-ampere control and the unit is rated for 10 kVAR. It can perform on-demand voltage control and is installed on the customer side, i.e., the secondary side of the distribution transformer of the utility system. The pole-mounted and pad-mounted ENGOs are shown in Figure 5.26.

ii. Grid Edge Management System (GEMS): It is a web-based software platform for device control and data visualization so that voltage can be tracked for diagnostic and maintenance purposes.

ENGO and GEMS working together provide voltage regulation at the user end, peak power demand control, reactive volt-ampere (VAR) control and support solar integration to the grid or possibly any other power source. It is more helpful to the utility as it enhances its operation to help deliver quality power to the consumers as it is required to certain standards and also may be of some help to the users.

FIGURE 5.26 ENGO mounted on the pole and pad. (Courtesy of Varentec.)

Intellectual Property: 12 granted US patents and 8 US pending applications at this time. This is a reasonable quantity for a company of this size in its domain. The co-founders and technical team members are proven innovators contributing to the IP portfolio.

Lawsuit: Varentec launched a patent infringement suit against Gridco Systems in 2016 and it obtained a judgment in its favor in 2017 midyear.

Funding: Raised $47.1M in a span of 11 years and its principal investors are: Khosla Ventures, Bill Gates of Microsoft Corporation, 3M Company and WindSail Capital Group.

Customers: Two orders for sizeable units have come recently and they are:

i. Hawaiian Electric Company (HECo) intends to deploy ENGO and GEMS on a distribution circuit with a high penetration of solar PV units in 2016.
ii. Xcel Energy of Colorado has plans to install 2000 units of ENGO in Denver city and announced it in 2018.

It is not only a sizeable order for a startup like Varentec but may serve a bigger purpose than the sales by getting to establish their equipment performance and reliability in the field deployed in larger numbers. The results are critical to Varentec and many of the larger potential customers may be closely following the end result.

Observations:

i. The company's founders have well-balanced experience that can be most needed for a team, i.e., one in research but more focused on applications, two in founding companies, one in power system with equipment manufacturers and power utility and all of these experiences are absolute essential ingredients of success for launch and operation of the company. This is one of the startups having a concentrated combination of such talent pool among founders.
ii. It is also very fortunate in that there are not many startups in this area and it is left with one another viable startup GridBridge as its competitor on the east coast of USA. The most powerful competitor, Gridco Systems, has ceased operation recently and much to the advantage of Varentec.
iii. Varentec has been funded by the most well-known and powerful venture capital Khosla Ventures, headed by legendary Vinod Khosla, Bill Gates of global fame, and 3M one of the largest US companies. This is the most formidable combination of investors that not only can pour capital but provide advice, introduction to decision-makers in potential customer companies, and law firms and all of that unmatched by any other startup in this field. This is the biggest weapon in the armory of Varentec.
iv. It has, for time being, a limited product line of 10 kVAR capacity and that can go only for limited application and power utilities may need bigger than that. It is expected that when that need arises, it is a possibility that technology may be licensed to larger equipment manufacturers to make and sell.
v. They have raised almost a sizeable capital but not big by other standards of software or some power device companies. Unless orders come and revenue starts to flow in, they may have to raise capital, and at that time (which they did with WindSail Capital Group recently), the present investors will consider other alternatives such as an exit among others.
vi. The intellectual property position is big compared to its competitors but not at a size that many companies such as Siemens, ABB, GE, Mitsubishi, Toshiba and others will pause to think and worry about at this stage. But Varentec has the capacity to ramp up their generation efforts to amass more intellectual property. This may aid the startup for higher rewards during exit.
vii. Mostly benefits to utilities have been demonstrated in this domain with the technologies by all of them. Therefore, the user community of power system companies has to adopt the products in time to come hopefully. For that to happen, the user companies have to be convinced that there is a return on their investment in these products and for that they need how long these products will work once installed and what their failure rates are and will be and how they function under severe conditions.

There seems to be not enough published material to assess that for the general public and engineering community at large.

viii. GridBridge has demonstrated that certain individual users, such as poultry farms, gas stations etc., save as much 5%–15% energy with the use of their equipment. A similar study from Varentec and other companies in the domain and their publication will go a long way to gain the attention of user communities.

ix. If the deployment is going to be only more beneficial to utilities than to individual users, then the market will be determined by them. In such a situation, the big power system manufacturers may persuade them to deploy system-level large equipment to do all the functions with centralized control and at a lower cost.

x. Unless many distributed users can afford to buy these products, which is quite unlikely at this time, then the mass market, as in the case of mobile phones, may not open up in the near future for these products.

xi. The utilities, unless forced by regulators to strictly adhere to the standards and to improve the standards itself for power delivery, may not get into action as that may require a huge amount of capital for them to adopt this technology and which may increase the electricity bills to homeowners and that may be resisted. All these pose challenges for startups in the area of grid modernization to work in this environment including for Varentec. How these are going to be resolved is of crucial importance for budding entrepreneurs to learn from them.

5.7.4.4 Smart Wires

Smart Wires, Inc. (hereafter referred to as Smart Wires) is founded in San Francisco Bay Area, CA, USA in 2010.

Its founder is Woody Gibson who also served as its founding CEO. He is a prolific founder and is very active in the energy and storage technology-related startups. He is also the founder of Stalwart Power, Inc., an energy storage company, cofounder of Zenergy Power, Inc., and Zenergy Power PLC. He has a degree in marine engineering and an executive MBA from Pepperdine University.

Technology: It is based on directing and controlling the power flow in transmission lines and balancing power flow between various transmission systems with passive and power electronic converter systems. Its importance is much more to the rising integration of renewable energy sources with the conventional utility power systems. The concept of Smart Wires is attributed by the current Vice President of Smart Wires Dr, Frank Kreikebaum to Prof. Deepak Divan of Georgia Tech and note that the attributor studied with him. The present company came from Smart Wire Grid which seemed to have intellectual property licensed from Georgia Tech and not much more details available in the public domain on it.

Products: A brief description of the technology and the company's products follow. The technology in the form of one of their most recent products, Smart Valve, is given from their sources. Its concept in schematic form is shown

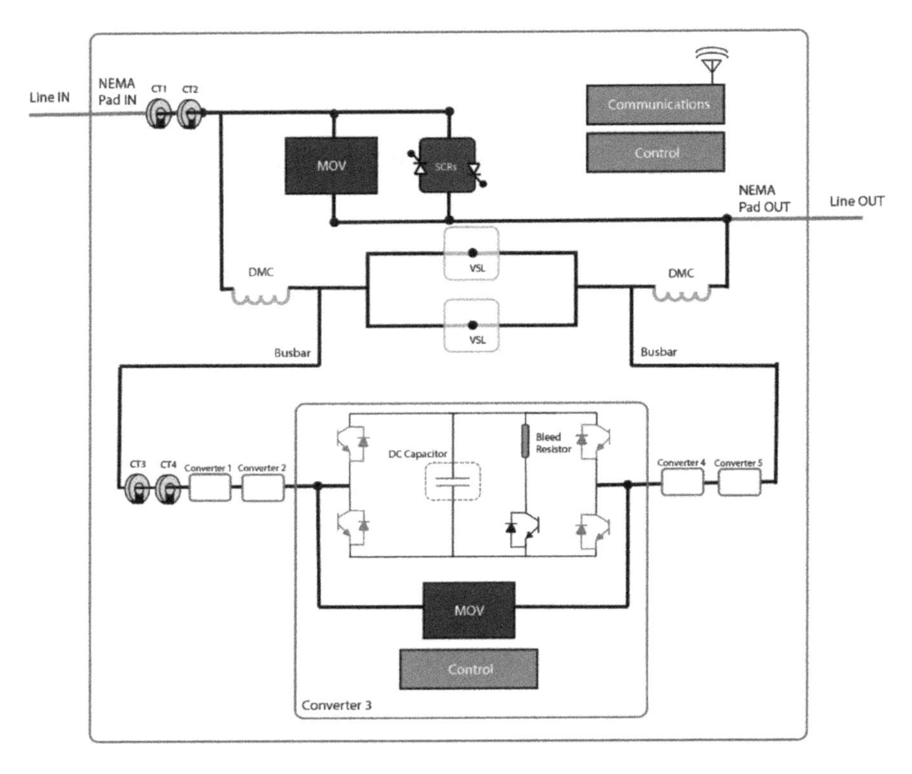

FIGURE 5.27 Smart valve schematic. (Courtesy of Smart Wires, Inc.)

in Figure 5.27. The device is connected directly to the transmission/distribution line. The device can be connected or disconnected with a set up including vacuum series link (VCL) acting as a breaker, connected in series with differential mode chokes (DMCs) on either side of them and with SCRs to serve as backup for isolation of the line and metal oxide varistors (MOVs) to control the transients in parallel to this setup. Current transformers in the device give measurement of the line currents. The line current is routed through a set of voltage source converters (five shown in the figure) and connected in parallel to the VSLs and in series with the DMCs. The voltage source converters control the phase of the voltage and hence resulting in the reactance reflected to the line. That changes the power flow in the line and hence in its control. The control and communications network part of the device gives access for control from the power systems operator as well. Because of these features, it is named as a smart valve. A smart valve and its installation are shown in Figure 5.28.

They also have technology with transformer coupling of the line with their device. The line goes through its system forming the primary of the transformer and the control system similar to the smart valve is connected through the secondary winding coupling it to the line. It is easy to install, much similar to the

FIGURE 5.28 Smart valve installation in the power system. (Courtesy of Smart Wires, Inc.)

direct line connected smart valve also. The secondary of the transformer when it is open, the primary sees only the magnetizing reactance of the transformer and hence presents a high reactance to the line, thereby curbing power flow. When the secondary of the transformer is shorted/bypassed, the reactance introduced in the secondary of the transformer to the transmission line is practically zero and the power is allowed to flow in the transmission line. The transformer coupling is not currently used in their products and can be seen from the smart valve discussed earlier.

Their products include the one directly mounted on the line known as Power Line Guardian and the one deployed on a platform or ground is the Power Guardian in its installation environment, and they are shown in Figure 5.29.

Intellectual Property: 11 granted patents and 2 pending applications, all US. There may be licensed technology from Georgia Tech even though it is not found on their website.

Funding: $103.4M raised in 5 rounds. Its investors are Purple Venture Partners, Alumni Ventures group, Total Access Fund, RiverVest Venture Partners, and Arnerich Massena & Associates.

Status: They have been installing their products with select utilities and undergoing evaluation and study. They have opened a European sales and engineering office in Dublin, Ireland, and working with distribution agents in South America. It is being listed as one of the top 50 companies for growth in Silicon Valley.

Observations:

i. Smart Wires current management (Executive Chairman, CEO and CTO) has a good background in power utilities and renewable energy. They have brought immense connections to the utilities and a record of raising finance.

(a) (b)

FIGURE 5.29 Power line guardian (a) and installation of power guardian (b). (Courtesy of Smart Wires, Inc.)

ii. They were able to take the concept to installation and testing in a fairly short time, and their product developments with their deployment are at the forefront of their activities now.
iii. Many studies have been initiated by them that promote their products for major utilities and they are worth perusal by anyone interested in this area of grid control.
iv. There is an overwhelming emphasis on the power-sharing between various transmission lines, particularly in light of rising renewable energy connection to the power utilities and resulting congestion necessitating or likely to necessitate the modernization of the power transmission network. With that comes the opportunity for Smart Wires and related companies to introduce their products. This is a good scenario to consider.
v. But what if the following scenario rises? (i) Some older power plants are retired as they will be thus making room for the renewables in their places where the power flow can be controlled from power converters required to work with renewables. (ii) There has been a decrease in demand for electric power except in Texas for the past few years. (iii) Power consumption may not increase in the future as more and more energy-efficient appliances, motor drive systems and LED lighting come into wide use. For example, the lighting power consumption in USA is around 15% of the total power consumption, and with the LED lighting, it may come down to 3%–4%, thus relieving the transmission system congestion significantly. The wide adoption of variable speed motor drives will increase system efficiency. The energy consumed by fixed speed motors (mostly) is to the tune of 40% of the total energy consumption. If an additional

slice of that energy, say 10%, is taken off by high energy-efficient operation, then the congestion in the system will decrease significantly. That leaves the question as to what happens to the prospects of companies specializing in such problems as the startup and others.

vi. Utilities are highly familiar with solutions to the congestion problem using phase-shifting transformers or flexible ac transmission systems (FACTS) devices and many have been installed already for some time. They may stand in the way of newer proposed methods on a larger and higher voltages and higher power scale.
vii. The momentum is there for Smart Wires side as shown by their management's ability to raise a large amount of funds from investors and draw keen interest from utilities for select test installations which may pave the way for system-wide deployment as well.
viii. Engineering solution and its sophistication in subsequent products seem to be less complex as many of the control concepts are familiar but the specific application target deserves study and reflection by budding entrepreneurs.
ix. The company went through some management changes around 2014 and 2015 but has put together a team that could be the envy of any startup. Team building is another aspect that entrepreneurs, particularly engineers, can learn from Smart Wires for their own future use and growth.

5.7.4.5 Envelio

Envelio is founded in 2017 in Cologne (Koln in German), Nordrhein-Westfalen, Germany. It is a spin-off from RWTH Aachen University. Its founders are all doctoral degree holders from the same university with specialization in distributed energy systems research and they are:

i. Dr. Smon Koopmann, Chief Executive Officer,
ii. Dr. Fabian Potratz, Chief Technology Officer,
iii. Dr. Philipp Goergens, Head of Software Department
iv. Dr. Moritz Cramer, Head of Customer Relations
v. Dr. Philipp Erlinghagen, Head of IT operations.

Technology: Application of analysis and optimization techniques based on artificial intelligence from the founders' theses for the study of emerging utility with the challenges of integrating distributed renewable sources and providing analysis, assistance in efficient and cost-effective operation of utility systems. This service and tool are assuming a significant role with the rising renewable energy in the form of solar and wind coming into the utility grid. These energy source installations have operating and financial aspects such as: (i) uncertain and uneven generation throughout the day stressing the grid to cope with them and (ii) investment recovery necessitating the maximization of revenues by coordinating their energy sale to the grid. Further, these energy sources, particularly the solar ones, are more distributed requiring much more analysis for integration and effective use particularly if there is energy storage with batteries charged mostly by solar renewable energy. The focus of startup's technology is to address these issues through their products with their software and targeted for the grid planners, managers and operators.

Products: Software under the name of intelligent grid platform (IGP) with three subdivisions in them for: (i) data quality with a focus on data inspection, transparency and analysis of measurement data; (ii) planning for a connection request, grid study, and analysis; and (iii) operations with online monitoring, grid forecast and congestion management.

Intellectual Property: Information is not available at this time.

Funding: €1.2M (around $1.33 m) raised, the bulk of it from High Tech Grunderfonds (HTGF) and Demeter Partners, and a small amount through a grant from the European Union Executive Agency.

Observations:

i. The founders who also constitute the senior leadership in the company are all electrical engineers specialized in distributed energy systems and grid operation among other related topics. Doctoral thesis work is plenty around the globe, but very little emerges into the entrepreneurship domain, and in this aspect, the founders have made a contribution to the university researchers in general. Further, five of them joining together to venture out from the same group of researchers with a common understanding of the startup's base technology and having a long and good working relationship prior to formation is an asset to any and all founders. Numbers give strength, make it possible for the distribution of workload and day to day responsibilities, and emotional support - something to emulate as it reduces the strain and stress involved in the startup process for the founders.
ii. The company has a cohesive leadership team. At some point, sooner than later, it may have to broaden to include former executives and senior leaders from utilities to establish viable access to these companies for the startup's products and services. This is a fairly common practice among US startups can be seen from many startups studied in this chapter.
iii. Many services are required to handle successfully the various distributed energy sources coming into grid integration and particularly the solar renewable energy sources. Most of the ways of transferring the stored or direct solar energy to grids have been achieved with power electronic converters and their control and many technologies and products have been standardized but the grid and its control are hardly scratched. That does not mean that existing software and/or their combinations are inadequate to provide the solution but they individually may not address all the issues surrounding the planning, control and operation of the grid faced with distributed energy sources. This is where the opportunity exists for this startup and others.

iv. The tendency to go the software route for solutions is predominant in every sector of the industry and electric grids are not an exception to this rule is being seen. Many more startups have arisen in various aspects of the grid and their operability, etc. in recent times. The mostly hardware-based companies such as energy storage and home energy management systems startups, for example, AMS, Stem, Greensmith, AlphaESS, and ElectrIQ Power are more focused on their system software demonstrating the key role of the software. From this point of view, the Envelio founders' focus on software is keeping with the trend in the market place.

v. Envelio has been hardly 2 years old, and they have survived with very limited funds. It will be good to watch how the company expands through its own revenue from operations or raising funds from investors. It is not common to see huge investments from venture capital companies going to startups in Europe unlike the situation in USA. Given the market size of USA, Envelio may start its subsidiary there and in that case both the leadership team expansion and high levels of fundraising may be achieved.

vi. Examples of their software's role from planning to operation either by simulation or preferably with customer's projects may enhance the understanding of this company's products and services and that is some degree not available to interested engineers and possibly customers.

vii. The specialization and preference to serve European clients are natural at this time, given the location of the startup and familiarity of the founders with the regulations there. Is there a plan to address other markets such as USA by the startup? If so, will there be efforts either from Envelio or in collaboration with entrepreneurs outside of Europe to make it suitable for use in other countries? If the latter is the case, there are further opportunities for both Envelio and aspiring entrepreneurs from other countries, mainly from USA, China and India to work together to make that happen. Interested readers should explore such an avenue to get their feet wet in entrepreneurship!

5.7.4.6 Faraday Grid

Faraday Grid (hereafter referred to as FG) was founded in 2016 in Edinburgh, Scotland (United Kingdom). The founders and their current roles with their backgrounds where possible are given in the following:

i. Andrew Scobie, CEO: He holds a Bachelor of Economics from the Australian National University (ANU), a Master of Design from University of Technology, Sydney, and is conducting Ph.D. research in Economics at RMIT in Melbourne. He has founded and directed multiple technology start-ups and chairs Exigen, a systems design consulting business.

ii. Mathew Williams, CTO: He has a Bachelor of Engineering degree from the University of Queensland. He is the lead engineer developing the Faraday Exchanger IP, and is the co-author of the Faraday Exchanger and Faraday Grid patents. He is also Managing Director of systems design engineering consultancy Exigen.
iii. Jason Cordner: (Not in FG site but listed in Pitchbook website)
iv. Jacqueline Porch: (Not in FG site but listed in Pitchbook website)

Background brief to the company's technology: Power system (or known as power grid) is faced for the past 25 years with the rising influx of renewable energy from wind and solar and switching loads from large motor drives to smaller but millions of switch-mode apparatus in the form of chargers for laptops, solar microinverters, battery chargers, uninterruptible power supplies. Effects of them on the power grid in the form of harmonics, voltage fluctuations, frequency variations, and higher demand on reactive power affect the power quality to the users and also the control of the power system itself. Distributed power electronic solutions have been in the study for about 30 years, and the interest has picked up in recent times mainly from the investors. Power electronic apparatus and its control features as part of the solution to many of the problems in the power systems are embodied in Faraday Grid startup.

Technology and Products: FG's technology in brief is derived from their only one pending patent application as they studiously avoid discussing it anywhere else and therefore it is based on limited information. Hopefully, much more detailed information may be released by the company in the future. It is given below:

- The device consists of a transformer with an air gap (they call it virtual air gap) having three windings: (i) primary winding is connected to the power system grid; (ii) secondary winding is connected to the load, i.e., homes, factories, etc.; and (iii) control winding that is fed from the output of an inverter.
- The transformer primary excitation establishes a mutual flux in the core connecting the other secondary and control windings. If there is no excitation given to the control winding, then the established mutual flux induces a voltage in the secondary winding and a current is established in it by the load. The secondary current is reflected on to the primary current which is the normal operation of a two winding transformer.
- Here, the third winding called control winding is injected with power signals to correct the power quality issues on the primary and/or secondary side system in the form of their voltage and current magnitudes, harmonics and hence in their waveforms, and power factor. If no correction is required, note that the control winding will not be excited. The control winding communicates with the primary and secondary windings through the mutual flux that is impacted by the control winding current.

- The system has a feedback control based on desired variables for control and the final resulting control signal is amplified to get power level signal from a full-bridge single-/three-phase inverter's output and it is where the power electronics plays a part in the system control.
- This device is named Faraday Exchanger. Note that it is not on the edge of the power grid as in many other solutions and it is in the system itself. For each individual or a group of users, such an equipment for a single and/or three phases is being promoted.
- Faraday Exchangers are the products of this startup. Note that the transformers and inverters are central to them.

Patents: One application pending.

Funding:

i. Adam Neumann, the co-founder of WeWork and a billionaire, has invested $32.5M in 2019.
ii. Series C funding by a large VC invested $5.27M in 2018.
iii. Innovate UK Grant amounted to $1M in 2018.

Exit: The company closed its doors in mid-2019 for various reasons, a month after a large commitment of funding was received. Efforts were underway to acquire the property and restart by one of the founders according to press reports.

Observations:

i. Power system solution is attempted using both passive (transformer) and active components, which is not the approach followed by others in this field, even though similar techniques are familiar in the field.
ii. The approach has the advantage of transformer isolation for the utility and the user. It is desirable in that it is easy to isolate any one party when the problem is not on their side and keep their systems functioning. Note that transformers are very reliable in practice.
iii. The solution may be expensive because of the transformers in the products.
iv. The inverter and its control are very well known, and its reliability is fairly assured.
v. The device Faraday Exchanger has been tested at very low power levels and there is hardly any publication describing its operation at higher power levels. Most of them are available on their own website with very little engineering details of the system.
vi. Cost-effectiveness study for the system is yet to be made or if it is already made, it is yet to be made public and this will determine its adoption vis-à-vis their competitors' system solutions.
vii. There is tremendous publicity effort from the company with events in UK and USA. A foothold is being established in USA with a division in Washington, DC.

viii. The reaction of the power utilities will be known in the future and can be gauged by their adoption.
ix. The level of funding is very high compared to other companies involved in similar efforts given the time of the duration it has been in existence, their patent position and customer base. The fact that it is able to attract a high net worth individual investor shows that they may not lack funding to grow and scale up much bigger should they prove themselves with the customers.
x. Entrepreneurs interested in this area will find it compelling to follow this company's developments very closely and find ways to differentiate themselves from FG as well as other competitors. They will come to the conclusion that there is a lot of space for many more viable startups but require heavy funding and investors backing. Note that the founder or founders, according to a report, plan to come up with another startup along a similar line.

5.8 CHARGING OF EVs

Many small electronic devices such as cell phones, laptops, wearable devices in the category of exercise monitors have their batteries to be charged periodically and until recently the only recourse to that end is through a chorded plug-in power supply. The way to remove the plug-in power supply requirement is through wireless power charging and that opens up greater convenience to the device users as they are not tied by the task of plugging, seeking particular location and access to the plug wherever they are, and freeing up space taken by the chorded power supply with plug-in the bag while traveling away from home and most of all the certainty of having fully charged device at all times assuming such wireless chargers are prevalent in public places and also in homes and offices.

With the introduction of electric two-, three- and four-wheel vehicles in the market place, the need for high-power chargers both with and without the wireless feature has become urgent. The combined market for them in the face of expected EV expansion within the next decade is considered to be very big that will make it the largest of the power electronics and motor drives application market segments.

Introduction to charging for EV applications (without wireless power transfer) and wireless power transfer presented in essentials is given in the next section. Then startups in the wireless charging of small devices and EVs are presented in the subsequent sections.

5.8.1 Introduction to Power Transfer Schemes for Charging

Power transfer for charging is in two modes: (i) conductive meaning that the charging is done by the wired connection from the grid, and (ii) wireless power transfer where the load (i.e., battery and regulated dc source feeding that) is supplied wireless from the grid source via a system of transformer and power converters. Each of these is described here briefly.

5.8.1.1 Conductive Power Transfer for Charging

The entire system with its power electronic converters and control are all connected by wires and there is no isolation between the input side or part of the input side and output side and that is termed as conductive power transfer. Charging is one of the applications for such a system. Various methods of conductive charging are in practice and they are:

i. Conductive: No new technology required for home use so long as the battery system comes with its own ac to dc converter and control so that it can accept the ac grid input from a wall plug, as shown in Figure 5.30 and may be termed as direct ac to ac conductive system, say, Scheme 1. Note that minimal or almost no extra cost is involved in this mode of operation which will be embraced for low power EVs such as two, three and even smaller four wheelers. It requires attention from the user for remembering to plug in the cable for charging and the ac source may not be available at all places where the vehicle may have to be parked and they are the disadvantages of this scheme. Note that the power transfer is only in one direction in this scheme as indicated by the arrow over the plug-in cable. Many users may not need or want the bidirectional power flow in their system because of the additional cost of providing this feature in their power electronics and control systems.
ii. Conductive with Dc Output to Match with the Battery: A modification of Scheme 1 where the output of the conductive charging is dc to directly charge the battery. This approach named Scheme 2 is shown in Figure 5.31. Scheme 2 does not depend on the ac to dc converter and dc to dc converter that is required on the vehicle for converting the ac input as in Scheme 1. This is a big advantage as it reduces the burden on the vehicle but has the disadvantage that charging at places where only ac

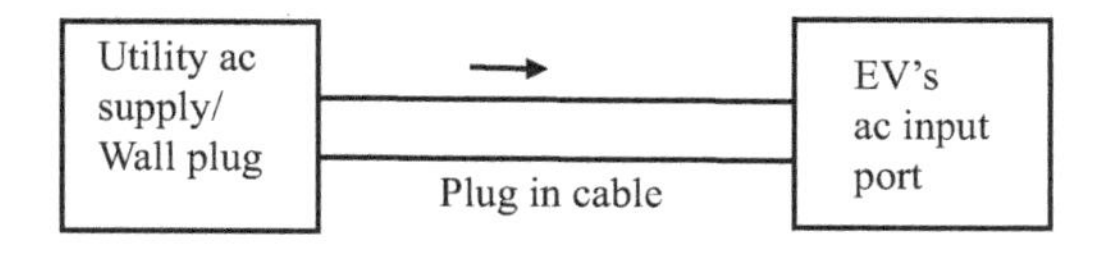

FIGURE 5.30 Direct ac to ac system for charging (Scheme 1).

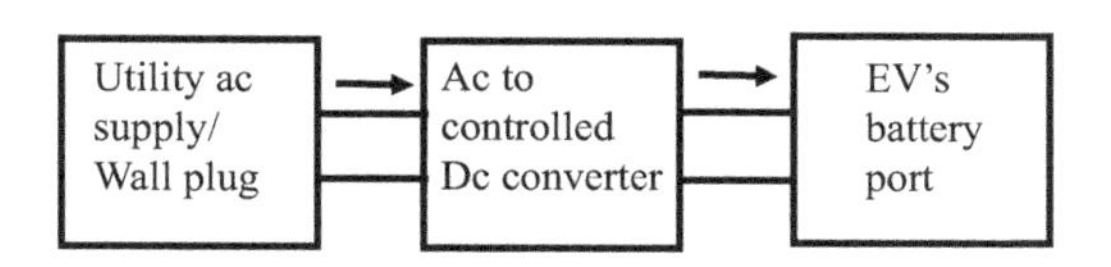

FIGURE 5.31 Conductive power transfer with dc charging (Scheme 2).

source is available and neighborhoods where Scheme 2 charging stations are not available. For home use, this scheme is acceptable. The power flow from ac source to the power converter with ac and controlled dc output and from that to the battery port is as shown by the arrows and they are all in one direction.

iii. Bidirectional Conductive Power Transfer and Charging: Bidirectional power transfer from ac source to the battery and vice versa shown in Figure 5.32, named Scheme 3, is a modification of Scheme 2 by adding power electronic converter, in the form of an inverter, from the battery to ac source. This is a feature known as V2G and G2V where V means vehicle and G means grid (ac source) system. It is being promoted where charging of the battery is accomplished when the cost of electricity from the grid is the lowest and the energy stored in the battery is sent to the grid when the grid can pay a much higher rate for the power than when the battery was charged. This is known as time of use charge by the utility grid. The reason for floating this scheme is to assist the utility during its expected peak loads caused by EVs when they come in significant numbers in the future to replace the gas/petrol cars. Power system studies are projecting that such peak loads will happen and one way to tackle that is to use this energy source including solar and wind energy sources without having to upgrade their power generating and system capabilities. The disadvantage of the Scheme 3 is its higher cost to the consumer if located in individual homes and that is where many of the conductive power transfer charging systems are targeted.

iv. Programmable Conductive Bidirectional Power Transfer and Charging: Scheme 3 requires programmability for individual user interface particularly for the time of use function and for monitoring system functioning and analytics at the tip of the finger to go with the trend of the market in all other sectors. Laptop and mobile interfaces together with remote access and control features on the equipment are almost a must in the evolving environment and its incorporation is shown in Figure 5.33 by adding on the controller part in the block diagram itself. This scheme for further reference is denoted as Scheme 4. The controller requirements on the system and its hardware are transfer of key variable readings

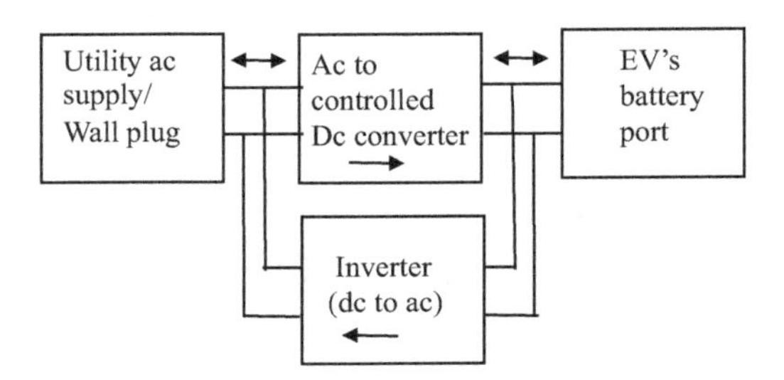

FIGURE 5.32 Conductive bidirectional power transfer (Scheme 3).

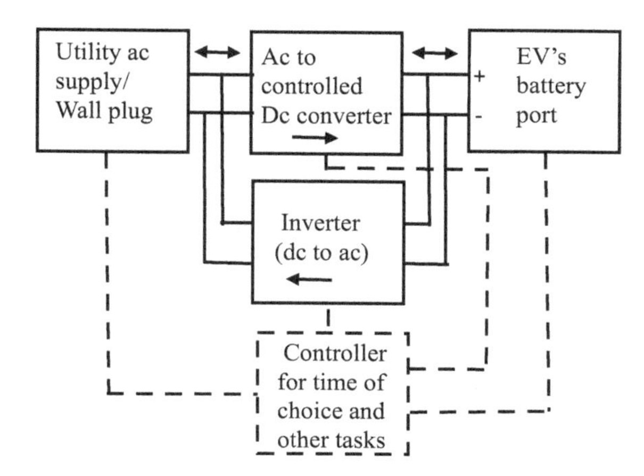

FIGURE 5.33 Conductive bidirectional power transfer with time of choice and other task-based user-friendly controller device (Scheme 4).

from the system to the controller such as charging current, voltage, input power, time and the cost of electricity at this time, variation of the cost over the day and access to the grid controller and/or user so that they can initiate automated charging of the battery at the lowest demand time and hence at the lowest price and to draw power from the battery when the demand on the grid is high. These advantages of the system can be augmented by adding security measures on the controller so that it is never under attack or malfunction.

v. Automated Conductive Power Transfer for Charging: All of the conductive schemes from 1–4 require to manually plug in the charger to the battery port or EV port (receptacles or receiver plug) of the EV. It could be automated so that the vehicle user does not have to take care of this chore except that to make sure that the vehicle is parked within a specified space. In that space, the conductive power transfer charger cable is buried from the utility ac side receptacle, and its output point is connected to a moving platform which has a moving arm capable of raising and moving its arm and at the end of the arm is a cover enclosing the receptacle and shown in Figure 5.34 as Scheme 5. Once the vehicle is stationed, the platform can move to find the dc port or ac port for the vehicle that is in the underbelly of the vehicle. Then the arm can raise, comes in alignment with the receiving side receptacle (receptor plug) in the vehicle, then slide the cover and plug in. The arm can slide back once the charging is complete back to its position and lower itself and lie flat on the platform for the rest of the time. Charging has to be protected against water coming into the garage and the ac power supply has to be disabled during the times and such safety measures have to be in place for its operation and use. Startups have been working on this scheme as well as other Schemes 2–4 as well.

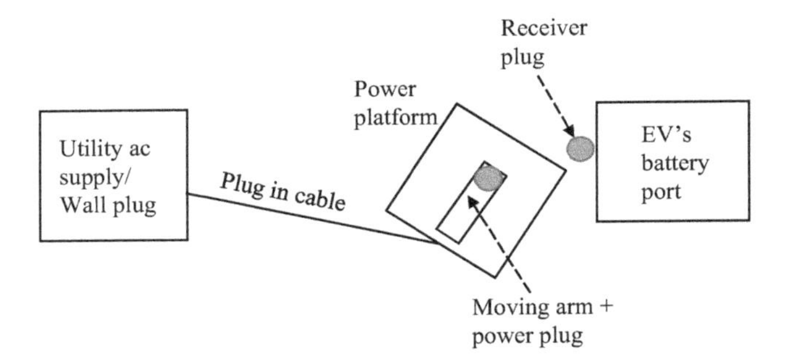

FIGURE 5.34 Conductive power charging scheme with automation for connecting to EV (Scheme 5).

5.8.1.2 Wireless Power Transfer for Charging

Wireless power transfer has been in existence from the days of Tesla's experiments. Two types of wireless power transfer emerged and they are: (i) radiative power transfer at radio frequencies for long distance and for near distances, (ii) inductive power transfer and (iii) capacitive power transfer. Applications in category (i) are not considered in this section and study. Only power transfer schemes employing (ii) and (iii) are considered as the emerging applications are starting to embrace them and also the schemes involve largely power electronics and control.

Wireless inductive power transfer (IPT): It is through the means of a transformer without the core, i.e., with the air core having the primary (also referred to as transmitting) and secondary (also referred to as receiving) coils. When the primary coil is excited with an alternating current a magnetic field is created which also encompasses the secondary coil which induces an emf resulting in a current through the secondary and its load that is battery pack over various power conversion stages. In a regular iron core transformer, the mutual coupling between the primary and secondary is around 0.98 and above with resulting low leakage of the flux in the coils. They contribute to high efficient power transfer. In the case of wireless power transfer, the coils are separated to the order of tens of mms and depending on applications such as EV charging, it may be as much as 10–20 cms. Under such circumstances of separation and not having an iron core adds on to the deterioration of mutual coupling between the coils to the tune of 0.1–0.4, in general, and ending with the efficiency that cannot match regular transformers. The coil shapes are rectangular, circular, elliptical and other shapes and are designed with minimum resistance by using Litz wires and for reducing the leakage flux with ferrites in the back of the coil base in both primary and secondary. The leakage inductances are significant in these coils.

The wireless power transfer system then consists of primary and secondary coils separated in space but facing each other. The power conversion unit requires an input and its output eventually supplies the battery pack and the schematic to do that is given in Figure 5.35. The primary is excited with a high frequency (in the range of 60–120 kHz) ac input which is supplied from an inverter that is

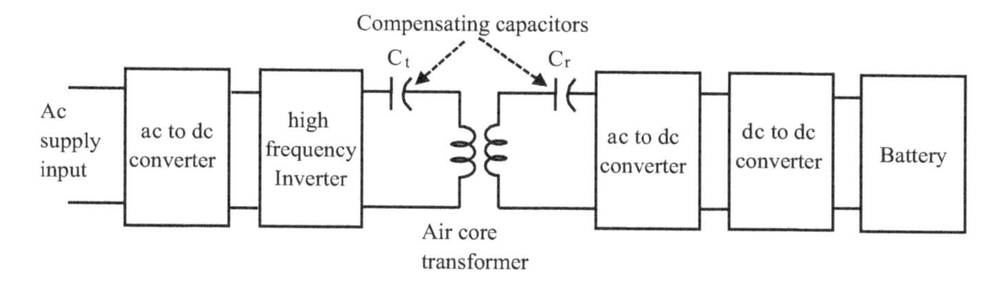

FIGURE 5.35 Inductive power charger schematic.

fed with a controlled dc that in turn is obtained from the grid ac supply (say, a wall plug for powers lower than 3 kW depending on the input ac line voltage). Various converters used in this system can be found in Chapter 2 and also some of their implementations with wide bandgap devices in Chapter 4. The secondary coil is connected to a rectifier whose output is then controlled via a dc to dc converter to supply the given battery pack.

Compensation (or matching) Network: Between the primary coil and high-frequency inverter and secondary coil and ac to dc converter, capacitors C_t and C_r are placed, respectively. They are to compensate for the leakage and to provide resonance so that maximum power transfer happens between the primary and secondary coils. Early systems did not have these compensating capacitors resulting in lower efficiency. Only one type of compensation using capacitors in series with the coils is illustrated and is denoted as CC compensation. There are many ways of compensating the network with capacitors such as: (i) C_t C_r series (already discussed), (ii) C_t series and C_r in parallel, (iii) C_t in parallel and C_r in series and (iv) C_t and C_r both in parallel to the coils. LCC compensation on both sides where L is an inductor and its combinations are being considered and some in development. The simplest series CC type will be considered to illustrate the functioning in the following.

Resonant Operation: This includes the compensating network which facilitates the resonance in the circuit by canceling the impedance due to leakage inductances in the coils. That amounts to reducing the leakage in the coils and that is one of the advantages of the compensation. It also enhances the operational efficiency of the coils. The optimal layout of the coils can ensure the minimum variation of these inductances and the mutual inductance when the position between the primary and secondary coils is different from the ideal design position. That mismatch between the coil placement will happen in practice as it is very difficult for the drivers to bring the vehicle underbelly secondary coil plate to be in exact alignment with the primary coil plate with an accuracy of a few tens of millimeters. Therefore, it is very important from the efficiency point of view to minimize the sensitivity of the coil position between each other. Also in the process of design, optimization is carried out to the extent that leakage inductances have minimal changes with regard to position and magnitude of the current. This entire system of the magnetic coils and its operation is called by the name of magnetic resonance (coined by MIT researchers who went on to a startup, WiTricity, more of it later) when the chosen frequency brings resonance in the primary as well as secondary coil sides of the network. Regardless, the basic principle is the same with the transformer excited in the primary with an alternating current inducing an emf in the secondary coil for feeding a load, and hence, the majority of researchers and engineers use the term inductive power transfer and charging.

Resonance condition and its impact on the system: Consider the system shown in Figure 5.40 and its simplest equivalent circuit given in Figure 5.36. The transformer primary and secondary resistances are R_t and R_r, leakage inductances are L_{tl} and L_{rl}, and series compensating capacitors are C_t and C_r, respectively, their mutual inductance is M, and the load resistance R_L which is the equivalent of battery load and the intermediate converters' equivalent resistance contributed by the switching devices. The voltage equations for the primary and secondary side are written as:

$$V = \left[R_t + j\omega(L_{tl} + M) - \frac{j}{\omega C_t}\right] I_t + j\omega M I_r \tag{5.1}$$

$$0 = j\omega M I_t + \left[R_r + R_L + j\omega(L_{rl} + M) - \frac{j}{\omega C_r}\right] I_r \tag{5.2}$$

where ω is the angular frequency in rad/s, and the self-inductances of the primary and secondary side coils are :

$$L_t = L_{tl} + M \tag{5.3}$$

$$L_r = L_{rl} + M \tag{5.4}$$

and for resonance, the conditions are:

$$C_t = \frac{1}{(\omega^2 L_t)} \tag{5.5}$$

$$C_r = \frac{1}{(\omega^2 L_r)} \tag{5.6}$$

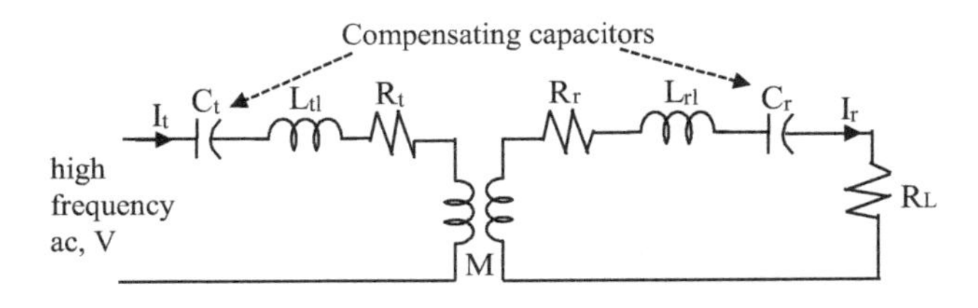

FIGURE 5.36 Equivalent circuit of the transformer with load for inductive power transfer system.

Substituting these conditions into the system equations result in the primary and secondary side currents as

$$I_t = \frac{V(R_r + R_L)}{R_t(R_r + R_L) + \omega^2 M^2} \tag{5.7}$$

$$I_r = -\frac{jV(\omega M)}{R_t(R_r + R_L) + \omega^2 M^2} \tag{5.8}$$

Using these currents, the input and output power are calculated from which the efficiency is obtained as:

$$\begin{aligned} \text{Efficiency} &= \frac{\text{Output power}}{\text{Input power}} = \frac{R_L I_r^2}{R_t I_t^2 + (R_L + R_r) I_r^2} \\ &= \frac{R_L (\omega M)^2}{(R_r + R_L)\left[R_t(R_r + R_L) + \left((\omega M)^2\right)\right]} \\ &= \frac{R_L}{R_r + R_L}\left\{\frac{1}{\frac{1}{Q_t}\left(\frac{1}{Q_r} + \frac{1}{Q_L}\right) + 1}\right\} \end{aligned} \tag{5.9}$$

where Q_r, Q_t and Q_L are quality factors, respectively, of the receiver (secondary) coil, transmitter (primary) coil, and load with respect to mutual impedance of the transformer and defined for a conventional series RLC circuit, and derived for the system under consideration as:

$$Q_r = \frac{\omega M}{R_r}, Q_t = \frac{\omega M}{R_t}, Q_L = \frac{\omega M}{R_L} \tag{5.10}$$

Some comments are in order based on efficiency:

i. Higher the quality factors in the circuit, higher will be the system efficiency.
ii. Higher quality factor means here that the coil resistances have to be very small which is what all electromagnetics designers do and therefore, that practice is applied in this system also excepting that they go to Litz wire to accomplish it and note that Litz wire is expensive compared to the regular copper wire used in electrical transformers.
iii. Load resistance varies depending on the level of charge in the battery and that is beyond the magnetic designer's control and therefore, there is not much that could be done when it comes to the quality factor of the load.

iv. Mutual inductance of the transformer also varies depending on the relative placement of the coils and operational current and that needs to be optimized so that it remains maximum for all possible operating conditions in the system.
v. Efficiency is affected mostly by mutual inductance and operating frequency.
vi. Efficiency derivation is based on resonant operation given by equations (5.5) and (5.6). For a given compensating network say, with C_t and C_r, there is only one resonant frequency for each side of the coil network assuming constant L_t and L_r. But the inductances change with coil positions and to maintain resonance then the excitation frequency on each side has to be controlled since the compensating capacitors have fixed values. This is the way resonance is achieved dynamically in wireless chargers to maximize efficiency.

Efficiencies achieved are in the range of 85%–93% for power up to 7 kW, and much higher efficiency is being targeted. Because of a number of converters (as much as four) involved in the scheme, the maximum efficiency is limited to around 96% assuming that the individual converters are all 99% efficient which is a stretch in many cases. Then, the magnetic coil efficiency has to be taken into account given its varying characteristics such as misalignment between them, and an additional challenge of resonant operation at all times which is not guaranteed with the parameter variations of the coils and load as well. A maximum of 95% efficiency will be more than gladly acceptable in the market and that is a bit far at this time.

There is a large amount of product development activity with most of the established automotive companies around the world in conductive as well as inductive power charging with compensation. Many startups have come into being mostly from 2000 onwards and the scope for more is there.

Capacitive Wireless Power Transfer and Charging: An alternative to conductive and inductive power transfer has recently surfaced using capacitive power transfer. Charge transfer occurs between two plates, say when an electric field is applied. The power transfer scheme is very much similar to that of the inductive power transfer scheme shown in Figure 5.35 excepting that instead of the magnetic transformer, two plates on both sides of the current path are introduced in the scheme and that is shown in Figure 5.37. A sample compensation network for this scheme is shown in Figure 5.38. Note that the compensation network contains inductors as there are no inductors in the capacitive plates and additional parallel capacitors are introduced. Simple series inductors will work as series capacitor compensation worked in the inductive power transfer systems. There are some advantages to the capacitive power transfer system as it is not sensitive to metals around the receiving end of the plate, and electric field is considered to be made safe. It is yet to be demonstrated at power levels desirable for EV charging at this time and at efficiencies to match that of the wireless inductive power transfer schemes. The system itself is in the research phase and hence caution must be exercised to consider this system's negatives with a different non-judgment perspective at this time as it is in the R&D phase, and it may change in the future.

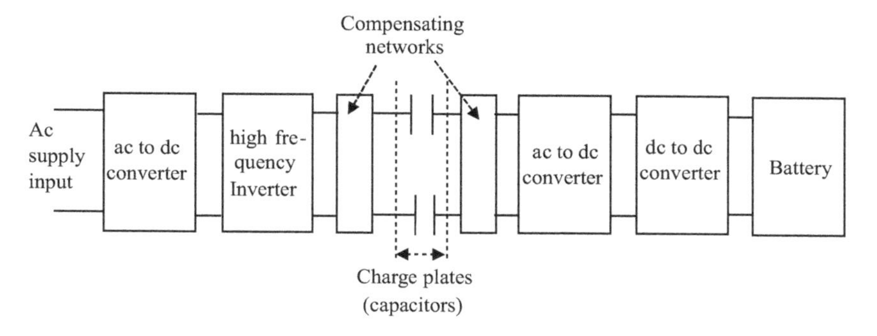

FIGURE 5.37 Capacitive power transfer system schematic for charging.

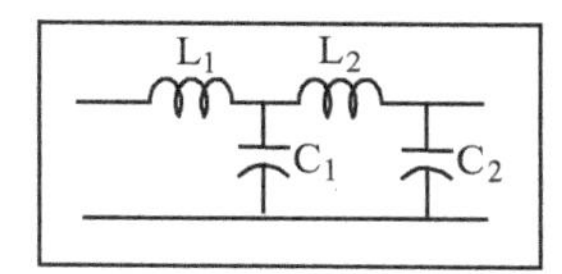

FIGURE 5.38 A sample compensating network for capacitive power transfer charging system.

5.8.1.3 Market Potential for EV Chargers

Many countries including China and India have declared that there will only be electric cars for sale from 2030 and the rest of advanced countries are trending toward a sizeable market ranging from 15% to 30% penetration of their market with electric and hybrid electric cars and vehicles by 2030. Various projections are being made as to the number of EVs that will be on-road at that time and assuming about 100M EVs and 40–50 M of hybrid electric vehicles around the globe seems to be a very acceptable number amongst the business analysts. That number may be credible given China and to some extent India may contribute to significant numbers as well as many other governments are committing to supporting the EV for a cleaner environment. Given these numbers, it may be possible to project the charging infrastructure market if the information is available as to how these vehicles are going to be charged and that may be based on the power level requirement, and access to the power from the grid. The commercial classification of the chargers given below goes a long way to an estimate (or projection) of the chargers and their placements.

Commercial Classification of the Chargers: The following is the adopted practice as of now for classifying the chargers:

i. Level 1 (or known also as ac Charging): This is an ac to ac port in the EV which has a rectifier and inverter to provide the dc to charge the battery (as shown in Scheme 1 of the conductive power transfer scheme) and the ac source is the standard home outlet in USA which is 120 V. The power that can be drawn from it is limited to 1.9 kW.

ii. Level 2: Similar to the Level 1 scheme but has the ac source of 240 V and maximum power to be drawn is limited to 20 kW at this level.
iii. DC Fast Charging (DCFC): Here the input goes directly to the battery port of the EV from the charging circuit such as shown in conductive schemes from two to five and having power capability of 25–350 kW for fast charging of EVs. This is expected to be in public charging stations near the highways to cater to EVs on long-distance travel.
iv. Wireless Chargers: Inductive types of all kinds with power capability of up to 11 kW, mostly for home use are in this category.

This classification helps to identify the markets for each and every category. Individual users residing in private homes (that will be about 80% of EV users in USA, while in China, that will be very small) will take to categories (i), (ii) and (iv) and commercial charging stations will go the route of (iii) in addition to stations with multiple chargers of category (ii) as well. Estimates of the market for all kinds of charging installations at this time indicate $100B by 2030. Based on the national EV sales projection, it is possible to estimate the annual market for the chargers.

5.8.2 Startups

The startups considered are in three wireless categories: (i) non-EV categories mainly intended for cell phones, laptops, lights, wearables, etc., (ii) inductive charging systems for EVs, and (iii) magnetic resonance charging systems for EVs. Another startup related to the business of charging stations is also included first in this section because of its pivotal role in the existing and evolving EV world, and it may use all the charging technologies in its implementations. The order of the startups' line up in this section is different may be noted.

5.8.2.1 ChargePoint

ChargePoint (hereafter referred to as CP) was founded in 2007 in San Francisco, CA, USA. It started out under the name of Coulomb Technologies and then transitioned to its current name in 2012. Its founders are:

i. Richard Lowenthal: He served as the company's Chief Technology Officer. Prior to founding ChargePoint, he was instrumental in the startup of Lightera, Pipal Systems and Procket Networks. He served in executive roles at Cisco, StrataCom, Stardent Computers and Convergent Technologies. He is a former Mayor of Cupertino, California, and is involved in nonprofit organizations. He holds a BS in Electrical Engineering from UC Berkeley and is co-inventor of many patents.
ii. Praveen K. Mandal: He started six successful companies and they are: Pipal Systems (served as CEO, acquired by Riverstone Networks), ChargePoint (served as President), the Emerging Technologies Group at SGI (served as Senior VP & General Manager) and 2 Know Services

(serving as Chairman). He served in Riverstone Networks (VP of R&D), Lucent Technologies (VP of R&D for Carrier Ethernet Solutions) and the aforementioned SGI (SVP of R&D and GM of Emerging Technologies Group). He is a co-inventor on ten awarded patents and holds a BSCE from Santa Clara University.

iii. Harjinder S. Bhade: He received his degree in Computer Science from Ohio State and his MBA from the University of Phoenix. He served as Vice President until 2015 in software engineering for ChargePoint. He also was a co-founder and director of software at Pipal systems. His other employers were Pluirs, Bell-Northern Research/Nortel, and also Zeitnet. He has also worked as the head engineer at Lucent Technologies and Riverstone Network.

iv. Dave Baxter: Serves as VP, Hardware engineering. He was Vice President of Engineering at 3Com Corporation and Adept Technology, where he developed multiple generations of assembly robots and flexible automation controllers and as Vice President of Engineering for Jetstream Communications. Dave received a B.S. from the Engineering College at Cornell University.

v. Tom Tormey: VP, Product Management from 2008 to 2015. Authored/co-authored 15 granted patents for CP. Served as VP, Engineering in Echelon and VP, Sales and Marketing in Design Tech. Has a BSEE from UC, Irvine and MSEE from UC Berkeley, CA.

Technology: It consists of conductive charging with proprietary design, hardware and software. They are not claiming products with wireless chargers as of this time and that does not mean that they may not have that technology for future deployment. Their technologies cover a wide spectrum of aspects and features associated with electric vehicle chargers and some of the topics are cited in the intellectual property section.

Products: They have products ranging from home use to charging station with the capability to provide for one vehicle or two vehicles from a charging set up. Further, the power that they can supply varies from 16 A for home charging (with power levels of 3 kW and above) to 62.5 kW for fast charging for varied applications. Their future product for fast charging of larger EVs can go to 400 kW level. Their products come with very much software and user-friendly controls for payment to the utility and to charging stations, depending on where the charger is located. For example, they provide charging station availability, charge scheduling and energy usage data to EV drivers among other software network solutions to users. Their product line is very large and interested readers are recommended to their website for further information. Their basic products for: (i) home EV charging station, (ii) DC fast commercial charging station, and (iii) commercial charging station for specific uses, respectively, are shown in Figure 5.39a–c.

Intellectual Property: 43 issued US patents, and most of the key patents are authored and co-authored by the founders. The patents cover proprietary technologies related to, for example in: (i) authorization in a networked EV

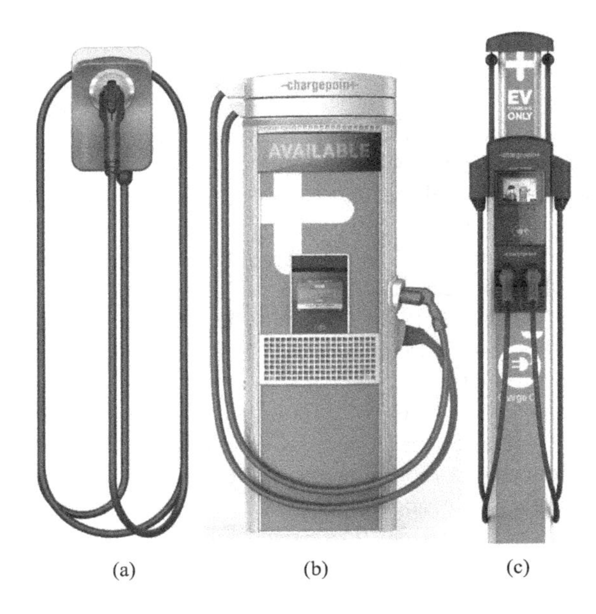

FIGURE 5.39 ChargePoint's charging stations: (a) home flex charger, (b) CPE 250 charging station (DC fast) and (c) CT4000 charging station (level 2). (Courtesy of ChargePoint.)

charging system, (ii) power modules for chargers and power cubes, (iii) dual EV charging station, (iv) detecting and responding to unexpected EV charging disconnections, (v) managing electric current allocation between charging equipment for charging EVs, (vi) connector for an EV, (vii) EV charging station with movable swing arms, (viii) EV charging station load management in a residence, (ix) street light mounted network-controlled charge transfer device for EVs, etc.

Funding: ChargePoint has raised a total of $532.2M in funding in more than ten rounds. Their latest series H round alone secured $240M in November, 2018. Some major investors are American Electric Power, Chevron Technology Ventures, Canada Pension Plan Investment Board, Daimler Trucks & Buses, GIC, BMW i Ventures, Siemens, Quantum Energy Partners and many others over the years.

Acquisition: ChargePoint has acquired Kisensum in June 2018. Kisensum was founded in 2014 in San Francisco Bay area, CA to successfully integrate energy storage to the grid through software control allowing the user control over their charging and energy management to their advantage. The same may be applied to fleet charging control which is at the heart of the business interest of CP and hence the acquisition.

Current status: CP has exceeded 100,000 charging places in its network in September 2019. The dominance can be seen in comparison with Tesla's 1,342 charging stations though they are only for Tesla's EVs. CP has made a commitment to introduce 2.5M charger spots by 2025 around the globe to match the needs of the EVs at that time. They have about $40M revenue at present, and with the forecasted expansion of EVs on the road, it can only grow by leaps and bounds.

Observations:

i. The founding of CP in 2007 required a great belief and confidence in the emerging market for which there were signs but with the battery prices as they were at that time, it was a long shot. The founders made that leap of faith and positioned themselves to the conductive technology (that is eschewing wireless feature for charging) to make it happen with the technology that was quite feasible at that time.

ii. The plan did not stop with providing chargers for home use but taking them to public charging spots, charging stations, fleets, businesses and all kinds of opportunities. Such a plan in a dream is not hard to come by but going after its execution is the hallmark of this company's founders and their leadership.

iii. The founders are all versed in the technology relevant to EV charging and its control in both the hardware and software aspects and their working together to bring out the products early to the market place has been phenomenal for CP in attracting attention mainly from investors and users.

iv. Most of the founders were successful entrepreneurs and exited those startups also successfully. That kind of experience has been well used in the selection of the business sector to aim for and drawing its plans and going after their execution in a big way. Their previous experience endowed them with a higher confidence level from investors resulting in attracting huge funding as and when they required to build and expand CP. Note that not all startups have such an advantage to start with!

v. The founders, all of them, had higher levels of corporate executive experience prior to founding CP only added and worked to the benefit of CP is unquestionable as seen from their development today.

vi. The vision for CP is more of historic importance. It must be a lesson to all the power electronics companies because the technology for CP came from mostly power electronics field and cutting edge research knowledge of it which they all had in their possession but did not have the vision and/or the leadership to pursue it and hence did not gain the market and revenue stream. It is a classic case of a disruptive idea in this field not necessarily in technology but more in business.

vii. The name Coulomb Technologies won't have much impact on or relevance to non- electrical engineering and physics audience and its change to ChargePoint is timely and meaningful to one and all is a stroke of marketing acumen displayed by the company. Attention to such details, however minute, paves the way for success.

viii. The intellectual property of CP is very limited given its size and ambition but that is likely to be sufficient to maintain their visibility and dominance in the market.
ix. The intellectual property of CP as seen from their patents could have been made by a group of power electronics professionals and researchers without any doubt in any part of the world is something that budding entrepreneurs have to come to grip with and realize that alone is not sufficient for a startup and its success. The leadership and vision play a bigger role and resulting fundraising and execution are the key elements for success.
x. Wired power transfer for charging has a number of issues such as: safety issue arising out of rain and storm water entering the charging stations, mishaps with the cords by human traffic, and most of all the cost of labor to assist drivers in public charging stations may persuade many stations to go for wireless charging alternative, if not now but in the future. How this trend will grow, and if it grows, how it will affect CP and its strategies are of interest to watch and learn.
xi. The conductive technology of CP is great for most of the intended market space. Whether CP has the technology for wireless power transfer is not clear at this stage. If that becomes a major threat to CP's products, it is quite possible that they may have strategies to cope up with. The obvious way out is to build their own development and research in the wireless power transfer field and get the resulting intellectual property and/or to acquire a startup in that field at a nominal price and luckily many such startups exist already and many more will sprout too in the future. Some such potential targets may be in the following sections (who knows!).
xii. CP is no more in the startup stage but in the big business domain. The day for initial public offering (IPO) of their stock to the public may not be far off and most likely it will be one of the few startups in this domain to be a multibillion-dollar company.

5.8.2.2 Wiferion

Wiferion transitioned in 2018 from the startup Blue Inductive founded in 2016. Its founders have worked together for a number of years in Fraunhofer Institute for Solar Energy (Fraunhofer ISE) in Freiburg Area (the largest research and development solar energy organization in Europe), Germany, prior to founding Blue Inductive. They have specialized in power electronics and control. They with their past and current roles in the startup are:

i. Benriah Goeldi: CTO then and currently Head of Software Development
ii. Florain Reiners: CEO then and currently Managing Director
iii. Johannes Mayer: CFO then and currently Managing Director
iv. Johannes Tritschler: CTO then and currently Head of Hardware Development

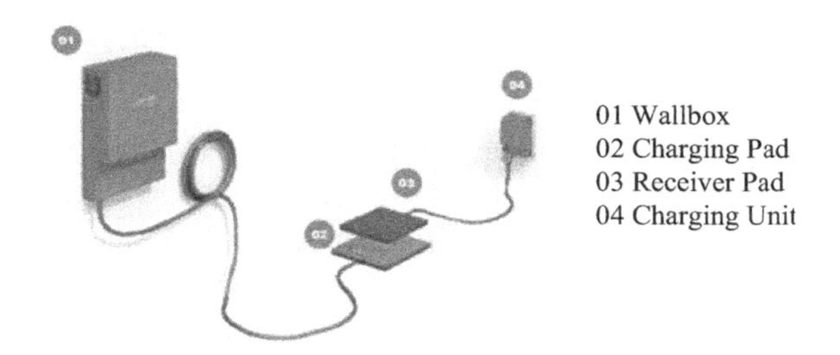

FIGURE 5.40 Wiferion's etaLINK 12,000 inductive wireless charging system. (Courtesy of Wiferion.)

Technology: Wireless inductive power transfer (IPT) system.

Products: The company targets charging of robots and industrial vehicles with built-in intelligence for these applications. One of the products, under the name of etaLink 12,000 has the ratings of: Charging voltage of 15–120 V dc with charging currents of 400–100 A, respectively, 12 kW power max. The product is shown in Figure 5.40. They intend to expand their product lines to on the road EVs and at higher power levels.

Intellectual property: One granted patent and two pending applications with the US patent office at this time. Two founders (Johannes Tritschler and Benriah Goeldi) are listed as the inventors in the patents apart from another from Fraunhofer ISE. The patent applications have to do with: (i) control of the IPT system (Publication number 20190148979), and (ii) multi-input multi-output converter for ac to dc conversion process (Publication number 20190149060). The granted patent (US Patent 10367363, Device for charging an energy store) deals with some configurations for accomplishing ac and dc inputs for charging and with or without motor windings, say in the case of hybrid EVs.

Funding: Starting with a seed fund from 2016 and investments from VCs, it has attracted a large investment recently from a major investor, Nordic Alpha Partners to position it for the international market catering up to 12 kW IPT charging systems.

Observations:

i. This is a startup founded by electrical engineers with expertise in power electronics and control with significant experience in research and development in Fraunhofer ISE, Germany. From this angle, it is a positive and encouraging sign for emerging entrepreneurs from this domain.
ii. Important to notice that after education, the founders also got grounded in experience that allowed them to learn of industrial practice, identifying emerging trends and markets together with how to meet those market demands with innovations that can stand against formidable behemoths such as WiTricity and large automakers having their own product lines

and large intellectual property portfolios. This is a good sign for emerging entrepreneurs encouraging them to innovate regardless of the players in the market space.

iii. The startup has limited intellectual property, but nevertheless, it may cover all their product lines for now. With increasing products from them for newer market applications, they can augment their intellectual property portfolio by increasing their engineering staff afforded by a large investor that they have recently attracted. They seem to be on the roll.

iv. Their market focus on robots and industrial vehicles including autonomous guided vehicles is a great marketing approach when not many were talking about at that time and choosing that application also allowed them to develop IPT chargers at lower power levels, i.e., max 3 kW level in a shorter time. This is the result of considerable marketing, management and engineering talents for a startup with seed funds and it shows the founders' breadth of knowledge in various domains in this decision process.

v. It has brought out products up to 3 kW charging capacity within a short time of 2 years and that is an achievement for any startup. Their aim to increase the rating of their product line will eventually cover EVs which will be the largest market existing and to emerge in the future as well.

vi. The performance of their product with an efficiency of 90% and above (and the highest efficiency is not mentioned) for misalignments (that are yet to be given clearly) and from a separation of coil plates with 20 cm is good at this time. Emerging standards will help them to fine-tune, if necessary, to even better performance. Note that some other manufacturers are also claiming 90%–94% efficiency in wireless energy transfer but at higher power levels but not at 3 kW level. With time, all of that will become public knowledge and some haziness in that aspect will clear away.

vii. Note the marketing aspect in changing the name from Blue Inductive to Wiferion for this startup. With the former name, there is nothing to indicate wireless operation of the products but with the latter name wireless is now sticking out in the first part of the word itself as in the case of WiTricity. Such minor adoption may make it easier for potential clients to find some connection with Wiferion's product offerings. This is a subtle change and it shows the management's attention to very minute but important details. The name itself may become very important as it is the property of the startup when there is an exit and the potential buyer is not only interested in the products but sometime in the name as well.

5.8.2.3 Momentum Dynamics

Momentum Dynamics Corp. (MDC) is founded in 2009 and is located in Malvern, PA, USA. Its founders are:

i. Andrew Daga: He is the CEO and Chairman of the Board of MDC. He has degrees in Civil Engineering and Architecture from Spring Garden College and has studied space engineering too. He has varied experience in these fields at the executive level and is one of the inventors of the MDC's technology.
ii. Jonathan Sawyer: He is VP, Product Design at MDC. He has degrees from Yale University and Pratt Institute in industrial design and is an expert in visualization. He has high-level management experience in his specialization with companies over 20 years before joining MDC.
iii. Dr. Bruce R. Long: He was the VP of Research and Development at MDC. He received his doctoral degree in electrical engineering from Pennsylvania State University and has been a product designer in his career. He is the lead inventor for MDC in all their patents.

Technology: MDC's core technology is based on wireless inductive power transfer (WIPT). They have developed methods, for example, for:

i. Adjusting the reactance in the system so that resonance in the power circuit could be maintained when there is a misalignment between the transmitter and receiver coil placements which is an important control feature for efficient operation (Patent number: 10319517, Method of and apparatus for generating an adjustable reactance).
ii. A scheme and its implementation in the WIPT to detect coil alignment error so that it can be corrected during positioning of the vehicle (and by that the receiver disc) over the transmitter disc and this procedure is critical for the efficient operation of the system (Patent number: 10193400, Method of and apparatus for detecting coil alignment error in wireless inductive power transmission).

Such measurement and control techniques to aid proper operation of the WIPT have been developed and this may give them proprietary implementations while protecting themselves from counterclaims from their competitors should they arise. The efficiency claims in their implementations are >90%.

Products: They have tested their prototypes from 50 to 200 kW range for charging the electric buses in a few installations. Some are: (i) a 50 kW wireless charging system at the *The Mall* in Columbia (Maryland), and (ii) a 200 kW Wireless Charging System is embedded in the pavement at Link Transit's Columbia Station in Wenatchee, Washington State, shown in Figure 5.41. Also shown in Figure 5.42 is the underbelly receiver panel installation on the bus. It is being claimed that the 200 kW charger is capable of 450 kW output. Note that these high-powered charging systems enable fast charging which is a must for highway travelers in buses and trucks, etc. By doing the quick charging available and affordable, the batteries can be frequently charged in a short time when the buses stop to unload and load passengers in a bus terminal which takes only 2 minutes and less. The amount of battery capacity can be reduced as the battery needs to supply energy

FIGURE 5.41 MDC's charging station on the pavement. (Courtesy of Momentum Dynamics Corp.)

FIGURE 5.42 Receiver panels for their bus system. (Courtesy of Momentum Dynamics Corp.)

only to cover the travel distance of 1–3 miles and about. In such cases, batteries can be charged at all terminals without stopping the flow of the bus journey and time schedule currently practiced with gas engine-based buses. Because of the lower battery capacity in such buses, considerable initial saving will result in lowering the cost of the EV buses. Their approach is branded by them as wireless power stations to replace the fuel stations of today, and there will be no difference between their systems in terms of time and comfort. They are yet to list available products on their website which may take a while. Most of their expected customers are known to be large companies in the applications sector they specialize in.

Intellectual Property: Six granted US patents and the lead inventor for all the patents is Bruce Long with fellow co-inventor Andrew Daga.

Funding: They have raised $32.9M and the bulk ($26.5M) of it in the last round in 2019. The major investor in the recent round of financing is the Volvo group's venture capital. The leadership plans to raise more funds to take the company products into the market.

Alliances: They have put their system in BYD's electric bus and getting financing backup from Volvo also opens up their technology applications to its company's vehicles.

Observations:

i. Founders are experienced in different fields, but one of them is in the electronics and control areas, another one in software-related expertise and the other one bringing management and marketing experience apart from applications knowledge. This gives MDC a solid leadership and a good start.

ii. The founding of the company was around when EVs were slowly making their entry into a small market space and not many forecasters and automakers were excitingly positive at that time. This gave them time to form their own ideas, develop their intellectual property, file for patents, raise initial rounds of finance to support these activities and find a market niche where their technology development will reward MDC. The leadership seems to have excelled at this.

iii. Their focus on large EVs such as buses, vans, construction equipment, and commercial vehicles has put MDC as one of the few players in the world in this technology implementations. There is very little competition say in USA such as WAVE IPT in Utah (this is in the EV bus charging market), HEVO power in NY (this one more in the passenger electric car charging market), to mention a few. Note that WiTricity is not making much of a claim in these applications though their technology licensees may be building on to them.

iv. It is important to note that their focus does not preclude passenger EVs when it comes to their fast charging market application and therefore their technology and future products may well be positioned for this market in what they brand as 'electric fuel stations'.

v. Their fundraising level has risen significantly in early 2019 to match partially their plans and it could be attributed in no small measure due to their earliest focus on the EV bus market with successful implementation and continued operation in Washington and Maryland states. This is a critical point of learning for budding entrepreneurs.

vi. Their system efficiency is claimed to be >90% (just like their competitors) in this field. But a specific profile of the efficiency vs. misalignment parameters is not available. That makes it difficult to compare their product performance vis-à-vis their competitors and for objective study by third parties such as academic researchers. This aspect is not very specific to them as it equally applies to their competitors also. They may benefit from when such performance information is made available for the general public, provided their system comes on top.

vii. The marketing plans may be worth watching for the interested reader as to what order they may go for the following application segments: (i) charging big EVs such as buses, (ii) fast charging of passenger electric cars in electric fuel stations, and (iii) car chargers for home use.

viii. They are getting into ramping of their operation both in product manufacturing and marketing and it will be worth following them to see how fast and how much they can get into the market space and resulting revenue stream. This will give interested readers to know the time requirement for a startup of this size to achieve its ambitions in this sector of business.

ix. International marketing is very critical to the charging company sector as the majority of the market will be in China and Europe than in USA. MDC seems to have started working with companies like BYD in China and possibly with Volvo through the investment they obtained from it recently. These international markets in all likelihood generate more and faster revenue than the US market and therefore this is critical not only to MDC but also to companies and startups in this sector.

5.8.2.4 WiTricity

WiTricity is a spinoff from MIT in 2007. It is located in Watertown, MA, USA. The founders are:

i. Dr. Marin Soljačić: He is Professor in Physics, MIT, MA and came up with the concept of resonating coils on both the transmitter and resonator end at the same or near-equal frequency for high efficient power transfer between them.

ii. Dr. Andre Kurs: Graduated from Princeton with A.B., and from MIT with a Ph.D. in Physics.

iii. Dr. Peter Fisher: Professor of Physics, MIT.

iv. Dr. John D. Joannopoulos: Francis Wright Davis Chair and director of Institute for Soldier Nanotechnologies (also in the department of Physics).

v. Dr. Aristeidis Karalis: MIT graduate in electrical engineering.

Technology: Resonance control on both transmitter and receiver coil sides for high-efficiency wireless power transfer constitutes their technology as described under the technology section in 5.8.1.2. The major claims of the technology and its implementation are high-efficiency operation with a great tolerance for the mismatch in the alignment positions of transmitter and receiver coils.

The latter aspect is a very important factor given the mismatch where the vehicle in its underbelly with the receiver coil and the charging plate on the ground with its transmitter coil that will come in to play in practical wireless EV charging due to parking inaccuracy of the vehicle by either the driver or the driving autonomous control system. To additionally fine-tune the resonance in the circuits, WiTricity resorts to tunable matching networks in addition to the impedance matching networks. Further fine-tuning is possible with the frequency control that may compensate for the alignment mismatch within the standards governing their use.

Products: Products range from low power wireless chargers to EV chargers at power levels from 3.6 to 11 kW, as shown in Figure 5.43. Many of its products have been licensed worldwide to many companies including leading automakers. Their EV charging products have been under testing with many automakers' laboratories since 2018. They are also into expanding their product for V2G and G2V applications to stabilize the grid in times of peak load by transferring power from battery to grid and vice versa when grid power is cheap. A study by them and Honda has indicated that >90% efficiency is possible for V2G operation. Whether the startup will go the whole hog of licensing their technologies and products or they get into manufacturing and sales to serve the markets directly is not clear at this time. From their licensing history, it is quite likely that may be their strategy.

Intellectual property: Around 875 patents and 600 of them are granted worldwide.

Acquisition: Qualcomm Halo's wireless power transfer technology for EVs was acquired in February 2019. It was seen as a powerful competitor to WiTricity not only in its implementations and product opportunities, but Qualcomm Halo also has a large portfolio of intellectual property.

Licensing: It has licensed its technology to companies in more than 20 plus countries around the globe, and there are not many startups that can boast of this singular achievement.

FIGURE 5.43 Wireless EV charger. (Courtesy of WiTricity.)

Funding: Overall $60M has been raised, but it started having revenue from 2010 onward through licensing of its technologies that netted about $65M till 2018. That may have reduced the necessity for additional investment funding.

Observations:

i. A key part for the company's technology came from four founders with a physics background and only one of the founders is from an electrical engineering background. This is unusual in that most of the similar innovations were driven by founders and engineers with specialization in the electrical field.

ii. The proof of concept was enabled and demonstrated at MIT for low power that subsequently led to the founding of the company and the attention it received from investors and companies. In no small measure, a fair amount of credit goes to MIT.

iii. The technology has many claims and one is efficiency for 11 kW charging from 90% to 93% efficiency. Its efficiency in the face of the mismatch in the alignment between the charging plate and the car parking is not very clear but even one taking it to be equal to the lowest efficiency mentioned, i.e., 90% may not be satisfactory for some users. Clarity on this and improved operational efficiency would be very important for application considerations.

iv. The misalignment comes from the parking of the vehicle given human errors and/or autonomous driving device errors and an estimate of maximum efficiency achievable for maximum misalignment within a car garage is something that is not available in the open literature and that may go a long way in assuaging the concerns of the users as well to compare and contrast other competing products and to choose finally one that fits their requirements. This particular service has to be hopefully made available by this company as well as its competitors in the near future.

v. Fundraising does not seem to have been a challenge to this startup and many multinationals such as Intel Capital (principal is a chipmaker), Toyota (automaker), Foxconn (electronics assembler and integrator) and Schlumberger (oil exploring company) support it. The key component and system manufacturers and users of EV chargers have participated in WiTricity's development from its early days.

vi. Its intellectual property portfolio is the envy of any startup and even large established companies. With very limited fundraising and revenues (a total of around $125M), this portfolio must have cost them a significant portion of these funds. It is a wonder to see how they have managed to build such a big portfolio of patents and it is something that entrepreneurs have to learn from this classic example. The management deserves credit for such an important step in building up this company.

vii. Its acquisition of Qualcomm Halo charging technologies is one of the best and probably the most paying step from the management in its positioning to be without a powerful rival at this time and that may further consolidate its market position.

viii. Its early presence in the wireless charging scene has enabled its popularity partly assisted by its ability to distinguish itself from others' technology by branding their own with the term "magnetic resonance". This is an important factor in its sales pitch, aura and identity apart from its technology claims. Mostly, others were doing practically the same (with their impedance matching network on both sides of the coils and operating them in resonant mode) but not emphasizing the quality factor in the circuits for high-efficiency power transfer. The concept of resonance and its use in power electronics applications is well known.

ix. Whether the startup will go the whole hog of licensing their technologies and products or they get into manufacturing and sales to serve the markets directly is not clear at this time. From their licensing history, it is quite likely that may be their strategy.

x. WiTricity has drawn a lot of attention and following from automakers worldwide and when they adopt its system and/or technology, its revenues can propel this startup to a billion-dollar company. That will be a rarity for a company that is not a pure software product based which is mostly the norm in USA.

xi. As the EV charging market grows internationally, it is quite likely that in the EV charger market there could be a hectic competition that may lead to legal wrangling (more like warfare) among many of these companies with a principal role to this startup. By then it may be loaded with licensing revenues that it may take up on the competitors!

5.9 LOW-POWER PLUG-IN AND WIRELESS CHARGERS

Chargers at low power are prevalent in almost all small electronics gadgets to cell phones and laptops. As smaller as they are, the user feels better to accommodate them in their travel package and handbags. For example, the laptop chargers came from a handful of volume to nearly one-fourth of its size and even that is losing its charm in the evolving market place. Smaller in weight and volume are most preferred in the current climate as technologies have evolved to satisfy that need. There are two broad classes of solutions that are available:

i. Wired known also plug-in type or referred to as conductive in the EV charging: They are the most predominant in today's market. The weight and volume of the package will be dramatically reduced if the switching frequency of the power converters can be in the MHz range (as against switching frequencies of <20 kHz with conventional power devices) with the result that the inductors and capacitor ratings will be reduced significantly. Recent developments in the wide bandgap power devices, their commercial availability and their use in high-frequency switching of the power converters, in the MHz range are one of the strong driving forces. Then, there is the understanding emerging out of research laboratories to minimize losses during switching with resonant switching, using the layout of the PCB itself to engineer small inductors go a long way toward the objective of realizing small size chargers. All this technology comes with plug in the utility source end and the other end of the charger is connected to the appliances. One such advanced development resulted in a startup, FINSix, and it is considered in this section. Many such companies will come into play and supply one or a few big market appliance manufacturers and they may not have to enter the market place on their own. Such a possibility exists for them in this domain.

ii. Wireless chargers: Similar to large wireless EV-type chargers, but they are for small power with a maximum being aimed for 500 W in the appliances market. There are three types: inductive charger, inductive or magnetic resonant charger and RF radiative charger. The first and second kinds have entered the market place mainly for lighting, phone charging and other smaller appliances. The charging pad has to be close to the appliance in a fixed position in the wireless inductive type whereas the resonant version will have some tolerance of misalignment between the appliance and the charging pad and the RF charging does not have that constraint even though it is yet to come into the market in a noticeable way. One startup that tries to use all these three wireless technologies for their product lines, PowerSphyr, is given in this section. There are a large number of startups in this domain, though the big players like Apple and Samsung are not sitting quietly and their products are in the market. All startups face competition from behemoths and that is a fact of life. But the other side of the coin, because of the small investment required and short development time for products in this domain, startups can spring overnight and release products that may surprise the established players.

The market size for these products is estimated to be a few billions to grow more with the wireless chargers likely to grab a larger share of the market with time, at least in western countries. No clear market figures are available in the open for various appliance chargers.

5.9.1 FINSix

FINSix is a power electronics MIT spinout founded in 2010 in Boston, MA and then moved its headquarter to Menlo Park, CA. Its founders are as follows:

i. Vanessa Green, CEO: Holds an M.Eng. in Civil and Environmental Engineering degree from MIT, an MBA from MIT Sloan, and a BA in Environmental Studies from Dartmouth College. Prior to FINSix founding, Vanessa served as manager at TECOM Investments in Dubai, UAE, and a consultant with the Monitor Group in New York, NY.

ii. Anthony Sagneri, CTO: He holds Ph.D. and S.M. degrees in Electrical Engineering from MIT, and BSEE from Rensselaer Polytechnic Institute. He is a graduate of the U.S. Air Force Intelligence Officer Course. He served in Air Force as an officer for 5 years prior to joining MIT.

Currently, the interim CEO is Ken Morikawa, and there is no mention of a change of hands in the top leadership on their website.

Technology: It is power conversion at very high frequency (VHF) in the 30-100 MHz range. At these frequencies, the inductor and capacitor required in power conversion circuits go to tens/hundreds of nano and pico levels, respectively. That enables physical magnetic core-based transformers and inductances to go to air core structures which can be weaved into the printed circuit board itself thus saving space, weight and cost. The device switching losses increase at VHF and that is countered by the resonant circuit in the converter. Consider a dc/dc power conversion system to consist of a resonant inverter which then is followed by a rectifier, known as a Class-φ2 converter. It converts a dc into an ac with the resonant inverter and then it is rectified to yield output dc. The control of the resonant inverter gives the desired magnitude and shape control of the output with the least amount of losses.

Product: An ac/dc with a dc to dc conversion through the Class –φ2 converter for laptop charger and simultaneous USB charger with it capable of 65 W has been the most significant product and it is known as DART. It has 90% plus efficiency and the unit is shown in Figure 5.44. Claims are: Universal ac input 100-240 V ac, 50/60 Hz, 65 W power, 2.75″×1.1″ dimensions, volume of 3 in^3, and naturally cooled. This product is available in the market.

Intellectual Property: Six granted US patents.

Funding: $21.6M. Investors are Venrock, Launch Angels, Astia Angels, Green D Ventures and others.

Alliances: (i) Strategic partnership in 2018 with Toyota Industries Corporation, a major company of Toyota Group to develop miniaturized power electronics modules for hybrid vehicles. (ii) Lenovo collaboration in 2016 to develop the smallest charger for their ThinkPad laptops.

Current Status: There is no more access to its website from late 2018 or early 2019.

Observations:

i. FINSix is an MIT spinout with licensed IP developed by one of the cofounders with others there. Contrary to expectations that many such spinouts would arise, it is not very common given the challenges facing the graduates in startups, the major being fund raising. Hope the trend of FINSix may become pervasive.

ii. The cofounder comes with VHF research and development experience which is a great asset for a startup and as well from one of the

FIGURE 5.44 FINSix's laptop charger with an additional USB charger, DART.

most reputed research laboratories in the world where most of the work around is only in this area facilitating learning beyond his own work.

iii. VHF-based power conversion to bring about miniaturization in various applications including laptop chargers is the logical step from where chargers were in early 2010 with high-efficiency battery chargers but at a lower frequency with passive components taking considerable space in them. Opening for VHF products in this line was waiting for such products as the users of the expensive laptops and phones can afford to pay for the convenience of size reduction in them for portability and convenience.

iv. The ingenuity of combining the charger for a laptop with a USB supply that is used for phone in one single package is to be lauded as it satisfies the customer needs. This played an important role in the successful launch of this product line and greatly helped to establish FINSix in the market place.
v. The diversification opportunities are there in many other product lines apart from the current product line and they may be working on them.
vi. The research and development alliances with Lenovo and Toyota Industries Corporation guarantee the continued existence and growth and success of this venture. It may free them from raising capital for their survival and growth.
vii. There seems to be a change in the CEO that has come after 8 years of its founding. It has come with additional positive growth opportunity for it in the form of tie-up with Toyota Industries Corporation.
viii. Their IP position in numbers may be small but it is quite strong enough to have provided the basis for a successful product. The continued augmentation is a must in this line as many are likely to enter or have entered this VHF-based power conversion arena.
ix. They have raised a medium amount of capital compared to the number of years of their existence, in spite of the fact that they did not seem to have any difficulty from the start with investors. That shows: (i) the management is cost-conscious and frugal which is a big attraction for investors, (ii) the technical team wants to go one step at a time with their product lines and their success, (iii) they have generated revenue from sales since 2015 that could have partly offset their operating and development expenses, and (iv) the management may not have been keen to dilute the shares of the founders and original investors. All of them put the management in a very good light.
x. The alliances with the big giants and the products that they can launch with them will determine their future and their independence as well.
xi. This is one of the very attractive candidates for acquisition by the power electronic giant companies surrounding them.
xii. Important events to watch in this company's future and to learn from are: (i) their new product lines, (ii) the reaction of their competitors, and (iii) possible exit through acquisition.

5.9.2 PowerSphyr

PowerSphyr, Inc. (referred to as PS in this section) was founded in April 2016 in Danville, CA. USA. Its founders are:

i. Neil Ganz, Chairman of the Board: A seasoned executive and entrepreneur in tire sales and its manufacturing for nearly two decades. He obtained his B.S. in business administration from American University.

ii. Will Wright, CEO: Serial entrepreneur and has an education in mechanical engineering from College of DuPage. He has multiple areas of interest from small unmanned aircraft systems, industrial design and wireless power transfer and control and is a co-inventor of PS's patents.

Technologies: Inductive, resonant and RF power transfer for charging up to 500 W range.

Products: Power pad which is the transmitter platform and the receiver to be built into the phone together with the battery constitute the product line application in regard to mobile phones, shown in Figure 5.45. Instead of the phones, many other products can also be attached to the receiver pad with coils and battery for their charging. These are the types of products that PS and its competitors provide. Products that are being promoted from PS are:

i. Dual More Charging Pad: Targeted for phones to light industrial charging applications with the additional feature of fast charging built into them. Primarily these products use inductive charging technology. Their products also cover battery management system with their software and it is projected to expand the battery lifecycle by 20%–40%
ii. Dual Mode Charging Pad: This product is to serve similar applications but uses the resonant inductive technology which is very similar to inductive technology with resonance of the circuit on the transmitter and receiver sides built into for high-efficiency power transfer, more tolerance for misalignment between the transmitter and receiver coils as a result of arbitrary positional placement of the devices for charging.

It is not clear whether the products are available in the open market as of this time.

Market: PowerSphyr's estimate of the potential wireless charger market in its segment was $2.3B in 2019. For all types of wireless charger products, a market of $37.2B by 2022 is forecast by Allied research.

Acquisition: Gill Electronics, Grand Rapids, MI, in 2017

FIGURE 5.45 Charging pad product. (Courtesy of PowerSphyr, Inc.)

Competitors: Startups such as Chargifi, Air Charge, Ubeam et al., and big global companies such as Apple, Samsung and others.

Partnerships: Works with well-known wide bandgap semiconductor device companies EPC and GaN Systems and also ON semiconductors. The nature of these partnerships is not revealed except that PS may use their components and some of the controllers for the devices in their products.

Intellectual Property: It has two patents granted in USA. Both are on the intelligent multimode wireless power system (Patent numbers: 10411543 and 10069328), the first one more on the intelligent control aspects and the second one on the power circuit and its realization including necessary control. The essence of their IP may be here and interested readers are encouraged to go through them thoroughly to learn of the technology and to compare them with others' IPs. The first author of the patents is David Meng and the coinventor is Will Wright.

Funding: Raised $4.6M within the short time of its existence, and the major investor is Faurecia. A bit about Faurecia will assist in understanding where PS' products may have a preferential market and the extent of it too. Faurecia is a global automotive company with a presence in 35 countries and covers the following sectors: automotive seating, interior systems and clean mobility. Its revenue is €17.5 billion (approximately $19.25B) in 2018.

Observations:

i. PS's founders are from business, entrepreneurial and industrial design backgrounds. That enabled them to come to an area of product with the help of good inventors such as David Meng and professionals in the area of RF power transfer and power electronics to start this venture.
ii. The level of intellectual property as reflected in their patents is very small for a company in this domain. Needless to say, they may be sufficient for them to enter the field and with time they generate more patents.
iii. The level of funding has to go higher for them to enter the market and to secure a place in it. It might be sufficient for them to get to this stage.
iv. They have secured a probable high user, Faurecia, of their technology products in the form of a lead investor in the automotive sector. If that works out with SP's products and the requirements of this investor, the chances of safe and protected market availability will enable this startup not only to survive but also to do extremely well in the market.
v. The startup is in a market sector where the competition from similar startups as well as from the existing giants is formidable. They are at a stage where they are making a presence in the conferences and trade shows relating to their field trying to advertise their technology and get written about them in the trade magazines and meeting potential customers.
vi. Their focus on the software aspect of the battery management control is following very similar trends in the energy storage systems market and from low to very high power levels of many MWs, as seen from the earlier sections in this chapter. Their claim of extending the lifecycle of

batteries by 20%-40% in these small power batteries in SP's products can be very useful to users for their products and may make a way for them in the market. Note that the battery because of this may last the lifecycle of the phones resulting in a little bit of saving to the customers and thus help in partially paying for the PS's charging pads.

vii. SP's ability to organize the technology and product development in such a short time of less than 3 years is a good indicator of how a small group of inventors and founders can enter this field and try to make a mark. It is a plus as well as a minus as that fact may propel similar activities from various groups and individuals around the world!

viii. Interested entrepreneurs and students have a lot to learn from following this company's march and its competitors from other startups to see the evolution of their technology products, marketing, revenue, ability to hold their own in the face of big giants, and finally their success in the market place.

5.10 CONCLUSION

A few fields of dominant power electronics application have been illustrated, and the resulting entrepreneurial activity in a limited way has been studied. Some key observations are summarized below:

i. Textbook and to a smaller degree research knowledge in power electronics are essential to understand the subtle innovations demanded in these emerging applications.

ii. University spin-offs in entrepreneurial activity are limited. It is understandable given that power electronics and control in most of the undergraduate curriculum is absent in most of the universities and also because of very limited graduate programs in this arena in USA.

iii. The university spin-offs that happened are by doctoral scholars and their supervisors in most instances. This may be a factor that budding entrepreneurs in this field may be better off heading to some of these university campuses.

iv. The more one learns of the application challenges and methods adopted to solve them leaves the impression that the previous solutions as old as few decades can be adopted with the facilities afforded by prevalent modern technology practices such as software control, communication with phones and tablets, et al., emerging high-frequency switching techniques in power electronics, use of wide bandgap devices, physical size reduction and compact packaging. Then it implies that knowledge in all these practices is required to realize the desired products, and hence people with these expertise or aptitudes are a pre-requisite for a successful startup in this domain.

v. Very few startups from outside USA have been included in this chapter. It is primarily that information for chosen format is generally not available. However, it is hoped that interested readers residing in those countries may have access to the sources in the local languages as well as through contacts and word of mouth.

vi. The application fields are emerging and will continue to emerge and that entails more entrepreneurial activity to bring innovations to the market place is shown from the sectors under study in this chapter.

vii. The more that is learned of the applications and startups' activities in them must leave the impression that many other solutions in the core technology, user-friendliness and usability aspects, cost reduction (that may involve a much more rigorous study) are all certainly possible and are waiting to be discovered and invented. That may propel more entrepreneurships in the future.

viii. Limited applications of wide bandgap devices in power electronics have come to the fore and that too mostly due to the startups in those devices. Many applications will benefit pursuing such devices' use in power electronic applications and that is an avenue yet to be fully exploited and exhausted in the entrepreneurial activities.

5.11 DISCUSSION QUESTIONS

1. Hardware-based single-chip controllers for select converter operation have been in the domain of large corporations. Discuss about the barriers for startups to shine in this area.
2. New converters (not the textbook types) adopted for emerging applications are in the increase. If so, discuss it with evidence from academic studies and market.
3. A large number of converter topologies emerge from various research laboratories and academic institutions. What are the barriers to adopting them in emerging applications?
4. Discuss the steps to overcome the barriers discussed in question 3?
5. What do you think is the best way to learn about the power converter circuits for emerging applications: Industrial internship, graduate studies at a well-known university, with a recognized professor in that domain, private self-study and research, and/or forming a team that is not connected with work, degree program and working together? Discuss the merits and demerits of each option.
6. Are there other options to achieve the objective in question 5?
7. If you have to exercise one option from five and six, list that one and your reasons for selecting that.
8. Discuss that in a team inside the class and note the choice variations that emerge from each and every one.
9. There are not many startups in the solid-state breaker arena. What may be the reasons for it?

Note: Questions from 10 to 31 are on the subtopic of home solar and home energy management systems

10. Microgrids with solar energy have been in use for nearly three decades. Many may have done projects in this field. List the converters that may form the subsystems in them.
11. Discuss how complex or easy it would be to build these converters with undergraduate course knowledge.
12. Will it require integration knowledge of the solar panels, converters and control that is outside of the classroom instructions?
13. If so, how to acquire it?
14. With that will any electrical engineer get into the business of working, making, selling and servicing these systems? If so, what are the requirements to start that effort?
15. Discuss the range of technologies that are available through startups and differentiate between them.
16. Why is that many of these technologies are targeted for markets in East Asia, Africa and to a certain extent in Australia?
17. List technologies that are appropriate/attractive for western markets.
18. What would be the reason that they are not attractive to rural communities in Africa and East Asia?
19. Energy storage is integral to solar homes and why is that?
20. How much effort is being put into the software part of the system as against the hardware in these systems? Is it too much or too little?
21. Why software aspects are being promoted in some startups (and many existing startups also go after that)?
22. Access via phones and computers is considered very important in many of these applications in western countries and that may also get into other countries as well. Is that really necessary and if so, why?
23. Can undergraduates in electrical engineering develop software-related control, integration and information sharing in these systems in a short time? For both affirmative and negative responses, discuss the reasons.
24. How could they acquire the knowledge to do the software? Discuss the methods.
25. Why is that large investments are flowing into these startups?
26. USA is not the leading country in the number of startups in this sector. Discuss if there are reasons regarding the market or the home solar system may not be able to fully or significantly supply the energy needs (heating in winter and cooling in summer) of a home?
27. The constraint of 26 may not be there in other countries. List them and discuss whether the opportunities are greater there than in your home country.
28. What is the status of this technology sector and market where you happen to live? If there are students from more than one country, then each can discuss it in the open to educate others.
29. Many startups have similar interests as charities in this particular field. Why is that?

30. Many have started with the charitable notion. But they all are becoming big businesses in some parts of the globe such as in Africa for the present. This leaves room for startups with pure public service goals. Discuss the possibility of such startups to come up with their own technologies.
31. Will you think of doing that and if not, why?

    Note: Questions from 32 to 38 are on the topic of energy storage systems

32. Discuss the relevance of high-power energy storage systems in your country of residence.
33. Will this increase the use of renewable energy sources such as solar and wind and to what extent in the coming decade?
34. Cost is an important factor in their implementations and is that justifiable for all markets?
35. Any new power electronic technology absolutely essential for their implementations? Give the reasons for both yes and no answers.
36. How will the use of wide bandgap devices impact power electronic converter subsystems in these large installations?
37. Discuss why wide bandgap devices may not be used in them in near term or even longer term, from your point of view.
38. The startups in this domain are not focused on making the hardware part but more in the software side. The reason is that they may not have the wherewithal to accomplish initially given the high power level of these installations and/or they may not want to compete with the big established companies that supply the hardware. Is this a good strategy or a necessity for the startups?

    Note: Questions from 39 to 45 are on mobile power platforms.

39. Sparsely populated northern territory in Australia is a good target site for mobile power platforms. List where they may be employed apart from the example given in the text as well as here.
40. Tourist vans, buses, recreation vehicles (RVs) can be made with solar power-based mobile platforms so that they can be taken for camping and nature vacations. Will they come into them or already fitted with them? Discuss your familiarity and also what needs to be done to make them happen, say, in your own future RV?
41. Could the applications listed in question 40 open a big market for solar mobile power platform? Discuss the feasibility, cost and market scope of that.
42. Assuming these applications happen and will have a market, list the subsystems required to make them happen.
43. Is there a market for them in East Asia and Africa where the cost is a major barrier? List the potential opportunities.
44. Are there other alternatives to mobile power platforms for some of the applications cited earlier? And if so, discuss.
45. Emergency applications such as during and after natural disasters and relief work could use the mobile power platforms. Do they need

additional features in the mobile power platform as seen earlier and if so, discuss them?

Note: Questions from 46 to 57 are on the topic of power grid with power electronics

46. Discuss the future power system in regard to power electronic converters in them.
47. Distributed interventions with power electronic converters are being promoted by recent startups. Discuss the merits and demerits of that approach.
48. Discuss the opportunity for high power centralized intervention and its merits and demerits.
49. If centralized intervention comes into play, can most of the startups catch up to that paradigm shift?
50. From the discussed startups, discuss the likelihood of their success in the market place.
51. Elaborate on question 50 by identifying the most likely to succeed startups, from your point of view, and present your reasons.
52. Success or failure will be governed by the large utilities and the government regulation within which they must work. Given that, discuss the hurdles that a startup may have to face in that environment.
53. Time to blossom in such an environment will be long. Given that, the startups may have to access to deep pocket investors to fund them or forced to become part and parcel of the large utilities. Discuss this statement and present your conclusion with reasons.
54. From the technologies required for the startups in this domain, identify the converters and controllers required.
55. Discuss from the identification whether much more startups can come into being from existing knowledge from the 1970s in these subjects.
56. Identify that for any startup discussed, how its technology is distinct from existing knowledge base up to the year 2000 in books and research papers.
57. Present your discussion to the class and debate all points of view.

Note: Questions from 58 to 70 are on the topic of charging electric vehicles

58. Discuss the upper power levels that can be reached by conductive and wireless inductive, inductive resonant, and RF technologies.
59. Discuss the applications of the technologies, from question 58, to other than the listed ones in this book chapter.
60. A single application-based startup in these lines will be easier to launch and get to market to the earliest. Discuss the pros and cons of such an approach.
61. Can a small team of electrical engineering students in this course join together and build a low power product using any one of the technologies for a target application within a semester? Discuss the plan, funding requirement, equipment and time to accomplish it.

62. Following on 61, can a team of graduate students achieve that within a course of three credits in a semester? Discuss also the challenges to such an approach and give a plan of action to achieve it.
63. Questions 61 and 62 follow up: How to go about identifying an application? Will you require a marketing student or anyone interested to help the team or the engineering team can do it? Discuss various options and advocate what is the best option available to you and your team within the university confines.
64. Conductive (i.e., plug-in) charging will dominate in home applications until the wireless charging can make the price comparable, installation easier, and efficiency to match. Discuss how each of these requirements will be met at least for this market sector.
65. Where in the market will wireless charging come to dominate? Discuss various possibilities and their likelihood of it happening.
66. Questions 64 and 65 follow up: Discuss them for various countries and regions of your interest.
67. Will wireless charging predominate in China for cars? Note that individual residences are not prevalent there as compared to USA. Discuss this and the likely method to prevail given this and other information that you may come across.
68. There are standards for wireless charging and charging in general for EVs. Collect the information on them for discussion and presentation to the class.
69. Will concern for health affect the wireless charging products market in general?
70. Will the enforcement of standards on them completely eliminate people's concern regarding the health and safety of the wireless charging methods?

    Note: Questions from 71 to 76 are on small chargers

71. Immense applications for small chargers are in the market place, mostly satisfied by plug-in types but not necessarily by the small-sized chargers. Identify some applications that may use small-sized chargers which the market may not be supplying at this time.
72. Discuss the distinct features of small plug-in and wireless chargers that are being promoted.
73. The advantage of the deployment of a wireless charger at home does not seem to be very high. Discuss about this.
74. Discuss the cost difference between the wireless chargers as against the small size plug-in chargers in specific cases.
75. One advantage of this field is that many startups can come overnight and capture a small market at least with very minimum resources at its hand. Discuss whether this is true or false and outline your reasons and facts behind such a conclusion.
76. Is it possible to develop such chargers as part of an undergraduate project spread over a year or master's thesis spread over one semester or

two semesters or doctoral work spread over more than 2 years? Discuss the approaches and examine whether anyone approach is viable for your own case.

5.12 EXERCISE PROBLEMS

1. By doing the following exercise problems, you will gain in-depth system knowledge and performance of various startups related to solar homes with energy storage aspects. Note that each exercise problem may take considerable time and each one may be considered as a separate problem for grading and exercise purposes. The exercise problems require the development of block diagrams, models for each and every subsystem, converter, and loads and then their mathematical modeling, developing Matlab and/or Simulink programs and simulating the system and getting the dynamic and steady-state results for analysis and learning. You may use the Matlab programs from Chapter 3 and from the problems and projects assigned in Chapter 3. If necessary, develop additional programs to develop system simulations and plots with many of the key system variables as a function of time, for the following:
   - i. A solar-powered home with solar panels (assuming average daylight, sunny hours, etc.), maximum power point tracker, and a dc to dc step down converter with a load consisting of lights and a fan.
   - ii. Extension of (i) with an inverter to be connected to the power grid.
   - iii. The problems with (i) and (ii) give solar panel to ac grid connection. Connect the load to dc output, i.e., prior to the inverter and a control panel that could read the dc load current, voltage and power as well as the feed into the grid.
   - iv. Simulate the reactive power flow with the inverter to the grid in system (iii).
   - v. Model the system that is capable of supplying dc loads in the home from the solar power and when it is not there, take power from the grid and rectify and feed it to the home loads. Develop the block diagram before proceeding to the modeling and simulation aspects.
   - vi. Introduce safety factors in the form of isolation circuit breakers between the grid and dc bus feeding the home loads and any other point that you would consider important to isolate.
   - vii. Introduce a dc to dc converter and battery for energy storage into the solar home system. Develop a block diagram for that with no connection to the grid at all but only having a dc bus supplying the home loads. Model and simulate the system.
   - viii. Add to (vii), ac power grid link to charge the battery when solar power is unavailable through a rectifier that is feeding to the input of a dc to dc converter on the battery input side. Develop a suitable way to connect the outputs of the solar panel as well as the dc output of the rectified ac grid system so that they will form a common input

to the dc to dc converter to regulate the battery input current and voltage.

2. Model and simulate a utility-scale energy storage system. The following steps are required to do that and each one may be an independent exercise problem:
   i. Develop a broad schematic block diagram of the system you may have interest.
   ii. Build suitable models for each of the subsystems and the MATLAB/Simulink models for them.
   iii. Integrate these subsystem models into the system under consideration and run the simulation for various operating conditions such as changing loads, changing ac grid voltages and frequency.
   iv. Integrate utility-scale solar power into this system along with all others as given in (iii) and simulate the system for varying loads and intentionally introduced disturbances in the ac grid supply.
   v. Control coordination between various input sources, battery state of charge, varying loads, time changing electricity rates and optimal operation of the system itself. Develop a broad picture of them and build a model and simulate such a system to the best of your understanding.
3. Here, only grid-related power electronics control is considered. The startups studied under this topic are much similar to one another in some respect and basics have been dealt with in power electronics text books and some in Chapter 3 and in this chapter as well. Based on that, the following exercise problems are taken up:
   i. Consider a feeder having one or more converters for controlling the reactive power and hence the power factor in the system near the load point. Model the inverter, load, and the ac feeder (which is not an ideal source).
   ii. Develop MATLAB/Simulink models and programs for the dynamic simulation of the system.
   iii. Consider reasonable disturbances in the grid voltage and frequency fluctuations and simulate the system so that the performance enhancement can be achieved with the power electronic system.
   iv. Up to now, this amounts to the simulation of GridBridge, Gridco and Varentec systems. The simulation can be extended to include other factors that are being mentioned in them such as for a three-phase system the deviation from the ideal three-phase systems and the practical system is almost small to be ignored with the use of inverter system of (i). Simulate that aspect of the system.
   v. SmartWires is a different system from the system described in (iv). It may be modeled for power flow control with their system as described in this chapter. Modeling and simulation of the system with MATLAB/Simulink is encouraged for further understanding.

4. Conductive and various methods of wireless charging are considered for simulation here and the following exercise problems are considered for modeling and dynamic simulation:
   i. Model the subsystems from the ac input via dc link to the battery in the vehicle and dynamically simulate a plug-in type EV charger for 3 kW power capability.
   ii. Model a system that can do both the grid to vehicle (G2V) and vehicle to grid (V2G) in problem (i).
   iii. Simulate the system in (ii) and incorporate time of use charges of the utility into the optimum operation of the battery energy storage system.
   iv. A higher power level is required for fast charging of the EV battery. Upgrade the system in (i) to 350 kW for that purpose and simulate it. Use appropriate three-phase ac voltage inputs.
5. Repeat the appropriate exercise problem format given in four for inductive charger systems with and without resonance built into control. Similar steps given in problem 4 may be followed. The information required on the transmitter and receiver coils and their inductances and resistances may have to be assumed or obtained by finite element simulations and/or transformer design equations learned from other courses. Sometimes, the required information may be given in publications and/or theses.
6. Wireless chargers for small power transfer modeling and simulation require very precise modeling of the coils on both sides and their performance as inductive and/or inductive resonant transformer. Students are encouraged to learn finite element methods and/or software that will generate the data, some of which are available online and free.
7. Model and simulate the plug-in charger for low power such as for laptop but using very small capacitors and small inductors in the circuit. They can be minimized in their physical size by using high frequency, and the resulting small values required in the inductor can be realized on the circuit board itself which reduces the overall size of the total system. Different designs and values of them as a function of switching frequency may be tried and such optimization has to be included in your modeling and simulation.

5.13 CLASS PROJECTS

Some may require collaboration amongst many teams. A suggestion to instructor and students: It may be very time efficient if the exercise problems and discussion questions are selected to focus on one of the applications that will be pursued in the project.

1. Determine the interest in industries around you for technologies illustrated in this chapter by general surveying, conducting face to face interviews with company management, marketing leadership among

them, and where they foresee market development or require assistance from the teams to design, develop, or study to get them into a particular market sector. The team must have electrical engineers and can be assisted by business students if that is allowed or preferred by the instructor.

2. Project 1 should have multiple teams doing the study on various topics discussed in this chapter.
3. Presentation of their study results and based on discussion, narrow down the number of technology projects to one or two items for further development by mutual consent. If teams do not want to merge, they can continue to pursue as per the following projects.
4. For the selected technology and application, build a full-fledged business plan for its pursuit.
5. Select an implementation (circuit, control, battery, if necessary) plan with schematic diagrams, design of the circuits, modeling, dynamic simulation of the system (may use results from the previous simulation if available through exercise problem solutions in the previous section), design of the controller on a printed circuit board, contacting vendors to make the PCB and for getting the components, populate the PCB with components, testing, making the converter and testing, and final assembly of the subsystems for system testing. Costing is an important factor and that needs to be done for various quantities in production and that will require further collaboration with local industries who will manufacture these products. Here because of so much work to be done, it will be ideal if many teams can join in this effort with each contributing to some aspect of the project and that may make the project realizable within a semester (or two, if necessary and allowed in the curriculum).
6. Final project report and presentation to the class and invited audience from the industry who are all interested in the project and its outcome, as well as investors big and angel kind.
7. If a large number of teams are available, they can pursue different applications of the technologies that are studied in this chapter resulting in strong tie-up with local industries and opening up opportunities for a startup and/or employment.
8. Most importantly, the teams given time have to look into testing requirements specific to the application and collect information and develop a schedule for testing.

REFERENCES

Powervation Ltd.

1. www.powervation.com.
2. https://www.bloomberg.com/research/stocks/private/snapshot.asp?privcapId=29806254.

3. http://www.4thlevelventures.ie/news.html.
4. https://craft.co/powervation.
5. https://www.tuugo.info/SiteViewer/0340001887202?url=https%3A%2F%2Fwww.independent.ie%2Fbusiness%2Firish-computer-chip-company-powervation-sells-for-64m-31397883.html.
6. https://www.hotfrog.com/business/powervation/products-786890.
7. https://www.prnewswire.com/news-releases/rohm-semiconductor-acquires-powervation-300117576.html.

Atom Power

1. https://www.atompower.com.
2. https://www.linkedin.com/in/ryanjkennedy.
3. https://ca.linkedin.com/in/denis-kouroussis-82045013.
4. https://www.linkedin.com/pulse/atom-switch-now-ul489-listed-its-official-denis-kouroussis?articleId=6536612511951659008#comments-6536612511951659008&trk=public_profile_post.
5. https://www.linkedin.com/pulse/one-small-step-atom-power-giant-electrical-denis-kouroussis?articleId=6175289929560576000#comments-6175289929560576000&trk=public_profile_post.
6. https://www.crunchbase.com/person/denis-kouroussis#section-jobs.
7. https://www.globenewswire.com/news-release/2019/05/21/1833447/0/en/Atom-Power-Introduces-First-Ever-Digital-Circuit-Breaker.html.
8. https://www.ktvn.com/story/40596454/atom-powers-silicon-carbide-semiconductor-module-now-ul-recognized.
9. https://spectrum.ieee.org/energywise/energy/the-smarter-grid/atom-power-is-launching-the-era-of-digital-circuit-breakers.
10. https://patents.justia.com/inventor/denis-kouroussis.
11. https://patents.justia.com/assignee/atom-power-inc.

Solar Storage Home Energy System

1. https://energy.mit.edu/wp-content/uploads/2015/05/MITEI-The-Future-of-Solar-Energy.pdf.
2. https://www.energy.gov/eere/solar/articles/solar-plus-storage-101.
3. https://www.energysage.com/solar/solar-energy-storage/what-do-solar-batteries-cost/.
4. https://www.energy-storage.news/blogs/energy-storage-system-integrators-six-of-the-best.
5. R. Fu, T. Remo and R. Margolis, "2018 U.S. utility-scale photovoltaics-plus-energy storage system costs benchmark", https://www.nrel.gov/docs/fy19osti/71714.pdf.
6. https://www.solarreviews.com/blog/is-solar-battery-storage-worth-it-given-current-solar-battery-cost.
7. Washington State University, "Solar electric system design, operation and installation an overview for builders in the U.S. Pacific Northwest", Extension energy program, October 2019, http://www.energy.wsu.edu/documents/solarpvforbuildersoct2009.pdf.
8. Solar United Neighbors, "Battery storage for homeowners", https://www.solarunitedneighbors.org/wp-content/uploads/2018/11/Solar-United-Neighbors-Battery-Guide.pdf.

9. A. Stock, P. Stock and V. Sahajwalla (Climate Council of Australia), "Powerful potential: Battery storage for renewable energy and electric cars", 2015, https://www.climatecouncil.org.au/uploads/ebdfcdf89a6ce85c4c19a5f6a78989d7.pdf.
10. H.A. Samad, S.R. Khandker, M. Asaduzzaman and M. Yunus, "The benefits of solar home systems: An analysis from Bangladesh", The world bank, Development research group, Agriculture and rural development team, December 2013, https://openknowledge.worldbank.org/bitstream/handle/10986/16939/WPS6724.pdf?sequence=1&isAllowed=y.

Solar Energy Capacity and Market

1. https://www.ibef.org/industry/renewable-energy-presentation.
2. http://www.solarpowereurope.org/wp-content/uploads/2019/05/SolarPower-Europe-Global-Market-Outlook-2019-2023.pdf.
3. https://www.imarcgroup.com/home-energy-management-systems-market.
4. https://www.pbs.org/newshour/science/the-state-of-the-u-s-solar-industry-5-questions-answered.
5. https://www.imarcgroup.com/home-energy-management-systems-market.

Mera Gao Power

1. http://meragaopower.com/team.html.
2. http://gcep.stanford.edu/pdfs/NikhilJaisinghani_IndiaWorkshop.pdf.
3. http://smartgrid-for-india.blogspot.com/2012/05/mera-gao-microgrid-power-among-10.html.
4. https://vol11.cases.som.yale.edu/mera-gao-power/introduction/home.
5. https://vol11.cases.som.yale.edu/mera-gao-power/mgps-business-model/construction.
6. http://energyaccess.org/news/recent-news/mini-grids-around-the-world/.
7. http://www.regainparadise.org/clean-energy/company-mera-gao-power.html.
8. http://meragaopower.com/pressrelease1.html.
9. https://economictimes.indiatimes.com/company/mera-gao-micro-grid-power-private-limited-/U40102DL2010PTC205337.
10. https://pitchbook.com/profiles/company/162748-90.
11. https://www.owler.com/company/meragaopower.

Me SOLshare

1. https://energypedia.info/images/1/14/Factsheet_ME_SOLshare_Ltd.pdf.
2. https://unfccc.int/climate-action/momentum-for-change/ict-solutions/solshare.
3. https://www.crunchbase.com/organization/me-solshare#section-related-hubs.
4. http://nymag.com/developing/2018/10/solshare-mini-solar-power-grids-energy-bangladesh.html.
5. https://challenges.openideo.com/challenge/bridgebuilder2/review/me-solshare-ltd-bridging-planet-and-prosperity-through-rural-electrification-and-smart-grids.
6. https://www.powerforall.org/news-media/news-and-announcements/new-p2p-grid-energy-access-emerges-bangladesh.
7. https://www.fastcompany.com/90241777/this-startup-lets-villagers-create-mini-power-grids-for-their-neighbors.

8. https://medium.com/le-lab/solshare-the-startup-for-sharing-solar-energy-in-bangladesh-en-377002b21716.
9. https://www.dhakatribune.com/bangladesh/2018/10/01/solshare-raises-1-66m-series-a-to-build-the-future-of-energy-on-a-global-scale.
10. https://iixglobal.com/trailblazers-interview-series-solshare/.
11. https://bd.linkedin.com/in/sebastian-groh-80549241.
12. https://energypedia.info/wiki/Blockchain_Techologies_For_the_Energy_Access_Sector.

Cygni Energy

1. https://www.cygni.com/.
2. https://microgridknowledge.com/wp-content/uploads/2019/05/Cygni-Energy-Microgrid-2019-.pdf.
3. https://www.slideshare.net/BrajakishoreBasantar/cygni-energy-pvt-ltd.
4. http://eshop.cygni.com/solar-solutions.html?___SID=U&ajaxcatalog=true&limit=36.
5. https://www.cygni.com/products/48v-dc-appliances/.
6. https://www.vccircle.com/endiya-indusind-bank-pump-6-4-mn-in-solar-power-solutions-firm-cygni-energy.
7. https://yourstory.com/2018/08/iit-m-incubated-startup-cygni-energy-raises-6-4-m.

Zola Electric

1. https://zolaelectric.com/.
2. http://powerhousemt.org/profile/erica-mackey/.
3. https://www.linkedin.com/company/zolaelectric?trk=public_profile_position_group_header.
4. https://www.linkedin.com/in/xavierhelgesen/.
5. https://www.greenbiz.com/article/solarcitys-co-founders-charge-african-venture-dedicated-improving-energy-access.
6. https://unfccc.int/climate-action/momentum-for-change/financing-for-climate-friendly/off-grid-electric.
7. https://fortune.com/2016/02/09/off-grid-electric-oxford-energy-solar-east-africa/.
8. https://www.greentechmedia.com/articles/read/zola-electric-smart-storage-infinity-rive.
9. https://www.greentechmedia.com/articles/read/55-million-investment-sets-new-record-for-off-grid-service-companies.
10. https://pv-magazine-usa.com/press-releases/zola-electric-closes-20-million-in-debt-financing/.
11. https://www.crunchbase.com/organization/off-grid-electric#section-funding-rounds.
12. https://pitchbook.com/profiles/company/62070-94.
13. https://www.greentechmedia.com/articles/read/investment-in-off-grid-energy-access-2018.
14. https://www.gogla.org/sites/default/files/resource_docs/global_off-grid_solar_market_report_h1_2018-opt.pdf.

Alpha ESS

1. https://www.alpha-ess.com.

2. https://www.ibesalliance.org/fileadmin/content/download/presentation/ees___Power2Drive_Forum_2018/MI_13.00_Alpha_ESS_LinFREIGABE.pdf.
3. Company Introduction ALPHA ENERGY STORAGE SOLUTION CO., LTD, June, 2015 (Note: google as it is and presentation becomes available).
4. https://www.linkedin.com/in/yuan-thomas-5243069?trk=org-employees_mini-profile_cta.
5. https://i3connect.com/company/alpha-ess.
6. https://worldwide.espacenet.com/searchResults?submitted=true&locale=en_EP&DB=EPODOC&ST=advanced&TI=&AB=&PN=&AP=&PR=&PD=&PA=Alpha+ESS&IN=&CPC=&IC=&Submit=Search.
7. https://www.enfsolar.com/directory/component/57377/alphaess.
8. http://midnight-sun.co.za/wp-content/uploads/2019/03/Smile5-Inv.pdf.

ElectrIQ Power

1. https://electriqpower.com/.
2. https://angel.co/chadwick-manning.
3. https://www.linkedin.com/in/jim-lovewell-31352a8.
4. https://www.solarpowerworldonline.com/2019/08/6-thoughts-on-the-current-state-of-the-residential-energy-storage-market-in-california/?trk=organization-update-content_share-video-embed_share-article_title.
5. https://www.altenergymag.com/article/2017/09/complete-control-over-your-energy/27068.
6. https://www.greentechmedia.com/articles/read/electriq-launches-smart-home-battery-system-for-8999#gs.yjzqke.
7. https://www.greentechmedia.com/articles/read/electriq-offers-battery-solution-for-homes#gs.yk0cyc.
8. https://www.energycentral.com/c/iu/electriq-still-alive-and-now-shipping-fast-moving-residential-energy-storage.
9. https://www.marketwatch.com/press-release/electriq-power-secures-series-seed-financing-2018-09-24.
10. https://www.crunchbase.com/organization/electriq-power#section-competitors-revenue-by-owler.
11. https://pitchbook.com/profiles/company/135635-77.
12. https://patents.justia.com/assignee/electriq-power-inc.

Geli

1. https://geli.net/.
2. https://www.greentechmedia.com/articles/read/geli-makes-storage-project-design-software-available-for-free#gs.zjutu6.
3. https://www.greentechmedia.com/articles/read/geli-raises-5-million-to-take-battery-control-software-wider-stage#gs.zjsw63.
4. http://www.prweb.com/releases/2017/04/prweb14249633.htm.
5. https://www.crunchbase.com/organization/geli#section-recent-news-activity.
6. https://my.nps.edu/web/eag/january-11-2019.
7. https://www.forbes.com/sites/peterdetwiler/2015/12/07/geli-bringing-smarts-to-the-smart-grid/#5046693e5013.
8. https://www.tabuchiamerica.com/our-products/eibs-intelligent-battery-system.
9. https://patents.justia.com/assignee/growing-energy-labs-inc.

Sunverge Energy

1. http://www.sunverge.com/.
2. https://www.energysage.com/supplier/21876/sunverge-energy/.
3. https://cleantechnica.com/2015/02/17/interview-sunverge-ceo-co-founder-ken-munson/.
4. https://www.linkedin.com/in/kennethmunson.
5. https://www.linkedin.com/in/dean-sanders-b96b3a16.
6. https://www.solarpowerworldonline.com/2019/01/puget-sound-energy-selects-sunverge-for-grid-scale-solar-storage-demonstration/.
7. https://www.prnewswire.com/news-releases/con-edison-selects-sunverge-for-the-smart-home-rate-project-300775915.html.
8. https://www.marketwatch.com/press-release/alectra-expands-business-relationship-with-sunverge-energy-2018-09-24.
9. https://www.smart-energy.com/industry-sectors/energy-grid-management/sunverge-combines-residential-tou-solar-storage-smart-home/.
10. https://solarbuildermag.com/batteries/ac-vs-dc-coupling-solar-systems-with-storage/.
11. https://patents.justia.com/assignee/sunverge-energy-inc.
12. https://www.crunchbase.com/organization/sunverge-energy-inc#section-overview.
13. https://www.renewableenergyworld.com/articles/2019/08/sunverge-raises-11-million-for-utility-distributed-energy-resource-controls.html#gref.

Bboxx

1. https://www.bboxx.co.uk.
2. https://uk.linkedin.com/in/chrisbakerbrian.
3. https://www.linkedin.com/in/mansoor-hamayun-9b47b41b.
4. https://uk.linkedin.com/in/laurent-van-houcke-70820b17.
5. https://www.forbes.com/sites/philipsalter/2018/06/04/bboxx-is-delivering-off-grid-energy-to-the-developing-world/#4909f58d3f4a.
6. https://energypedia.info/wiki/BBOXX.
7. https://energypedia.info/wiki/BB120_Solar_Kit.
8. https://www.sustainablebusiness.com/2013/11/khosla-invests-in-bboxx-20-million-solar-kits-by-2020-52002/.
9. https://cleantechnica.com/2016/09/12/off-grid-solar-company-bboxx-closes-20-million-investment/.
10. https://www.bboxx.co.uk/bboxx-closes-50-million-series-d-funding-round-led-by-mitsubishi-corporation/.
11. https://www.energycentral.com/c/gr/mitsubishi-bets-bboxx.
12. https://www.crunchbase.com/organization/bboxx.

Multicon Solar AG

1. https://solarcontainer.info/.
2. https://www.developmentaid.org/#!/organizations/view/85431/multicon-ag-co-kg.
3. https://www.solarserver.de/uploads/tx_essolpagegenerator/2947__20171122133906-94b552-FILE_1.pdf.
4. https://issuu.com/multiconsolar.

5. https://www.pes.eu.com/renewable-news/german-armed-forces-purchase-mobile-solar-containers-equipped-with-high-power-aleo-modules/.
6. https://www.solarpowerworldonline.com/2018/01/solar-industry-responding-increasing-intensity-natural-disasters/.
7. https://soledion.com/pdf/multicon-solar-trailer.pdf.
8. https://www.flickr.com/photos/120538782@N04/16687033286/in/photostream/.

FreeWire

1. https://freewiretech.com/.
2. https://www.bloomberg.com/profile/company/1538636D:US.
3. https://www.crunchbase.com/organization/freewire-technologies.
4. https://blogs.haas.berkeley.edu/the-berkeley-mba/evening-and-weekend-mba-startup-jolts-electric-vehicle-charging.
5. https://freewiretech.com/team/.
6. https://www.linkedin.com/in/lkothari.
7. https://www.linkedin.com/in/sanatkamal.
8. https://www.linkedin.com/in/sameer-mehdiratta-9baa273.
9. https://www.linkedin.com/in/arcady.
10. https://chargedevs.com/features/freewire-deploys-battery-systems-to-increase-the-scalability-of-ev-infrastructure/.
11. https://patents.justia.com/assignee/freewire-technologies-inc.
12. https://freewiretech.com/news/freewire-technologies-raises-15-million-series-a-financing/.
13. https://electrek.co/2018/10/24/volvo-mobile-charging-startup-freewire/.
14. https://www.cnet.com/roadshow/news/volvo-freewire-investment-mobile-ev-fast-charging/.

Energy Storage Systems and Companies

1. http://css.umich.edu/factsheets/us-grid-energy-storage-factsheet.
2. https://www.scientificamerican.com/article/utility-scale-energy-storage-will-enable-a-renewable-grid/.
3. https://www.greentechmedia.com/articles/read/research-5-companies-postioned-to-succeed-in-grid-scale-energy-storage.
4. https://energyacuity.com/blog/2019-top-40-most-viewed-energy-storage-companies/.
5. https://www.greenbiz.com/blog/2014/04/21/how-10-innovative-companies-are-giving-energy-storage-jolt.
6. https://cleantechnica.com/2015/01/15/27-battery-storage-companies-watch/.
7. https://energystorage.org/why-energy-storage/technologies/.
8. https://solarmagazine.com/three-large-scale-energy-storage-technologies-all-out-renewable-energy-transition/.

Market of Utility Scale Battery Energy Storage Systems

1. https://www.greenbiz.com/article/utilities-are-starting-invest-big-batteries-instead-building-new-power-plants.

2. https://www.tdworld.com/energy-storage/report-offers-forecasts-utility-scale-energy-storage-systems.
3. https://pv-magazine-usa.com/2019/08/01/utility-scale-rather-than-behind-the-meter-batteries-will-drive-energy-storage-take-up/.
4. https://www.pv-magazine.com/2019/07/22/australia-singapore-power-link-worlds-biggest-solarstorage-project-get-government-backing/.
5. https://www.pv-magazine.com/2019/05/06/battery-storage-market-will-be-worth-13-billion-by-2023/.
6. https://www.utilitydive.com/news/us-energy-storage-market-grows-by-nearly-45-interconnects-over-760-mwh-in/560297/.
7. https://www.power-eng.com/2019/07/10/eia-utility-scale-battery-capacity-growing-ten-fold-over-10-years/#gref.
8. https://energyacuity.com/blog/2019-top-energy-storage-companies/.

Advanced Microgrid Solutions (AMS)

1. https://advmicrogrid.com/.
2. https://advmicrogrid.com/team.html.
3. https://advmicrogrid.com/projects.html.
4. https://www.greentechmedia.com/articles/read/advanced-microgrid-solutions-and-irvine-co-s-hybrid-electric-building-fleet#gs.Ht_HfVk.
5. https://www.prnewswire.com/news-releases/advanced-microgrid-solutions-raises-34-million-in-series-b-energy-leaders-see-der-optimization-as-major-growth-market-300485024.html.
6. https://advancedmicrogridsystems.com/microgrids/.
7. https://www.ny-best.org/page/advanced-microgrid-solutions-member-spotlight.
8. https://electricenergyonline.com/article/energy/category/solar/142/767577/advanced-microgrid-solutionsams-announces-solar-and-storage-partnership-with-38-degrees-north-and-john-hancock.html.
9. https://www.greentechmedia.com/articles/read/advanced-microgrid-solutions-breaks-2-gigawatt-hours-in-grid-services.
10. https://www.energy-storage.news/news/infrastructure-fund-susi-invests-big-in-macquarie-ams-california-portfolio.
11. https://www.globenewswire.com/news-release/2018/05/02/1494986/0/en/Black-Veatch-EPC-Services-Help-Advanced-Microgrid-Solutions-Meet-California-Energy-Storage-Goals.html.
12. https://www.crunchbase.com/organization/advanced-microgrid-solutions#section-overview.
13. https://patents.justia.com/assignee/advanced-microgrid-solutions-inc.

Greensmith

1. https://briefs.techconnect.org/wp-content/volumes/Nanotech2013v2/pdf/706.pdf.
2. https://pv-magazine-usa.com/2017/01/27/greensmith-announce-grid-connection-on-20-mw-storage-system/.
3. https://renewablesnow.com/news/greensmith-energy-gets-20-mw-storage-order-in-california-536794/.

4. https://www.solarpowerworldonline.com/2018/04/sungrow-and-greensmith-to-deliver-solarstorage-to-massachusetts/.
5. https://www.forbes.com/sites/peterdetwiler/2014/05/12/greensmith-provides-intelligent-solutions-to-energy-storage-market/#5a2ff56e77a7.
6. https://www.bizjournals.com/washington/blog/techflash/2015/12/can-an-18-3-million-funding-round-take-greensmith.html.
7. https://www.crunchbase.com/organization/greensmith-energy-management-systems#section-company-tech-stack-by-siftery.
8. https://www.powermag.com/press-releases/greensmith-energys-texas-waves-energy-storage-project-now-operational/.
9. https://www.linkedin.com/in/john-jung-99690a3.
10. https://www.greensmithenergy.com/wp-content/uploads/2019/03/greensmith-energy-whitepaper-future-proofing-energy-storage.pdf.
11. https://www.renewableenergymagazine.com/energy_saving/greensmith-and-aep-launch-hybrid-energy-storage-20171020.
12. https://patents.justia.com/assignee/greensmith-energy-management-systems-l-l-c.

Stem

1. https://www.greentechmedia.com/articles/read/stealthy-energy-storage-firm-powergetics-wins#gs.ydzubo.
2. https://gigaom.com/2013/12/02/startup-stem-raises-funds-for-energy-data-analytics-machine-learning-smarter-batteries/.
3. https://truveria.com/view/Stacey-Reineccius-DcFfCBFD.
4. https://www.stem.com/.
5. https://www.greentechmedia.com/articles/read/stem-a-reinvented-storage-startup-leverages-batteries-and-the-cloud#gs.ycbh6e.
6. https://www.peakload.org/assets/17thSpring/2e.Owens.Stem.pdf.
7. https://www.globenewswire.com/news-release/2019/01/07/1681709/0/en/Stem-Announces-Expansion-of-Business-Operations.html.
8. https://www.greentechmedia.com/articles/read/stem-lands-80-million-to-fuel-its-growing-behind-the-meter-battery-business#gs.ybtneq.
9. https://www.crunchbase.com/organization/stem#section-recent-news-activity.
10. https://www.bloomberg.com/profile/company/0087451Z:US.
11. https://patents.justia.com/assignee/powergetics-inc.
12. https://patents.justia.com/assignee/stem-inc.
13. https://www.stem.com/case-studies/?spg_category%5B%5D=solar&profile=16056.
14. https://www.stem.com/case-studies/?spg_category%5B%5D=solar&profile=16058.

Future Grid

1. US Department of Energy, Smart grid system 2018 report to congress, November 2018.
2. https://www.greentechmedia.com/articles/read/what-is-the-grid-edge.
3. https://www.greentechmedia.com/articles/read/what-2018-grid-edge-trends-reveal-about-2019.
4. https://www.greentechmedia.com/articles/read/grid-edge-20-the-top-companies-disrupting-the-u-s-electric-market.

5. World Economic Forum, "The future of electricity: New technologies transforming the grid edge". In: *Thought* Leaders Speak Out: *Key* Trends Driving Change *in the* Electric Power Industry, L. Wood and R. Marritz (Editors). Institute for Electric Innovation, The Edison Foundation, Washington, DC, 2015.

Value and Modern Grid Technology

1. A. Booth, M. Greene and H. Tai, "US smart grid value at stake: The 130 billion dollar question", www.mckinsey.com, https://www.mckinsey.com/~/media/mckinsey/dotcom/client_service/EPNG/PDFs/McK%20on%20smart%20grids/MoSG_130billionQuestion_VF.aspxhttps://www.mckinsey.com/~/media/mckinsey/dotcom/client_service/EPNG/PDFs/McK%20on%20smart%20grids/MoSG_130billionQuestion_VF.aspx.
2. Y. Xue, M. Starke, J. Dong, M. Olama, T. Kuruganti, J. Taft and M. Shankar, "On a future for smart inverters with integrated system functions", *9th IEEE International Symposium on Power Electronics for Distributed Generation Systems*, 2018.
3. W.U. Tareena, S. Mekhilefa, M. Seyedmahmoudianb and B. Horanb, "Active power filter (APF) for mitigation of power quality issues in grid integration of wind and photovoltaic energy conversion system", *Renewable and Sustainable Energy Reviews*, vol. 70, pp. 635–655, 2017.
4. "Technologies for advanced volt/var control implementation: Integration of advanced metering data", http://www.ieee-pes.org/presentations/gm2014/PESGM2014P-002524.pdf.
5. D. Divan and P. Kandula, "Power electronics: An enabler for the future grid", *CPSS Transactions on Power Electronics and Applications*, vol. 1, no. 1, pp. 57–65, 2016.
6. D. Divan, R. Moghe and A. Prasai, "Power electronics at the grid edge: The key to unlocking value from the smart grid", *IEEE Power Electronics Magazine*, pp. 16–22, December 2014.
7. R. Moghe, et al., "Grid edge control: A new approach for volt/var optimization," *Proceedings of IEEE PES Transmission and Distribution Conference*, Dallas, TX, pp. 1–5, 2016.

GridBridge

1. http://www.grid-bridge.com/.
2. https://www.crunchbase.com/organization/gridbridge#section-overview.
3. https://www.bloomberg.com/profile/company/1102155D:US?cic_redirect=3.
4. https://pitchbook.com/profiles/company/89355-52.
5. https://www.wraltechwire.com/2014/05/20/watch-out-for-gridbridge-it-could-be-triangles-next-startup-powerhouse/.
6. https://www.solarpowerworldonline.com/2015/02/next-solar-inverter-distribution-transformer/.
7. https://patents.justia.com/assignee/gridbridge-inc.
8. http://www.grid-bridge.com/wp-content/uploads/2018/05/ERMCO-GBI-Poletop-Grid-Energy-Router-1-10-18-3.pdf.
9. https://electricenergyonline.com/article/energy/category/Mergers%20&%20Acquisitions/58/650477/ARKANSAS-ELECTRIC-COOPERATIVESERMCO-acquires-GridBridge-Corporation-to-add-advanced-power-management-options.html.

10. https://www.greentechmedia.com/articles/read/gridbridge-a-new-contender-in-grid-power-electronics.

Gridco Systems

1. https://www.crunchbase.com/organization/gridco#section-overview.
2. https://pitchbook.com/profiles/company/54289-63.
3. http://gridcosystems.com/wp-content/uploads/downloads/2016/02/Gridco-Systems-emPower-Solution_Overview.pdf.
4. https://www.globenewswire.com/news-release/2016/10/06/1254200/0/en/Gridco-Systems-Closes-12-Million-Financing-to-Grow-Global-Footprint.html.
5. https://www.greentechmedia.com/articles/read/gridco-shuts-down-its-digital-grid-controls-business.
6. https://www.prnewswire.com/news-releases/gridco-systems-introduces-industrys-first-integrated-power-regulating-transformer-for-der-integration-and-energy-efficiency-cvrvvo-300217400.html.
7. https://www.hawaiianelectric.com/hawaiian-electric-company-deploys-gridco-systems-technology-to-help-increase-pv-hosting-capacity-of-distribution-grid-leverage-installed-asset-base.
8. https://www.bizjournals.com/boston/blog/techflash/2013/11/gridco-raises-10m-ahead-of-public.html.
9. https://www.energycentral.com/c/iu/gridco-systems-empowertm-solution-adds-amazon-web-services-hosting-option.
10. https://www.marketwatch.com/press-release/gridco-systems-recognized-with-awards-from-greentech-media-smart-grid-today-and-smart-grid-news-2015-06-04.
11. https://www.m-t-g.com/gridco-systems-introduces-new-static-var-compensator.
12. https://patents.justia.com/assignee/gridco-inc.
13. https://www.leagle.com/decision/infdco20170607a97.
14. http://www.greenpatentblog.com/wp-content/uploads/2016/05/Varentec-Inc.-v.-GridCo-Inc.-et-al.pdf.

Varentec

1. http://varentec.com/.
2. https://angel.co/andrew-dillon-1.
3. http://varentec.com/applications/.
4. https://pitchbook.com/profiles/company/14339-35.
5. https://www.crunchbase.com/organization/varentec#section-funding-rounds.
6. https://patents.justia.com/assignee/varentec-inc.
7. https://law.justia.com/cases/federal/district-courts/delaware/dedce/1:2016cv00217/59145/109/.
8. https://blogs.duanemorris.com/greenip/2017/11/22/varentec-dodges-gridco-requests-for-inter-partes-review/.
9. http://varentec.com/applications/.
10. https://www.greentechmedia.com/articles/read/varentecs-power-electronics-to-tame-hawaiis-solar-rich-distribution-gr.
11. https://www.greentechmedia.com/articles/read/varentecs-big-deal-with-xcel-energy-a-tipping-point-for-grid-edge-power-ele.

Smart Wires

1. https://www.smartwires.com/.
2. https://www.linkedin.com/in/woody-gibson-18a3581.
3. https://patents.justia.com/assignee/smart-wires-inc.
4. https://www.greentechmedia.com/articles/read/smart-wires-raises-30-8m-for-transmission-power-routing.
5. https://www.greentechmedia.com/articles/read/smart-wire-grid-wins-first-customer-for-its-power-flow-control.
6. https://www.smartwires.com/2012/06/13/smart-wire-grid-may-hold-key-to-power-transmission-woes/.
7. https://gigaom.com/2013/02/26/a-startup-quietly-delivers-smart-wires-to-big-power-players/.
8. https://www.crunchbase.com/organization/smart-wire-grid#section-overview.
9. https://www.freeingenergy.com/what-are-the-biggest-trends-that-have-the-potential-to-disrupt-the-power-industry/.
10. http://needtoknow.nas.edu/energy/energy-efficiency/lighting/.
11. http://www.irjabs.com/files_site/paperlist/r_555_121228124614.pdf.

Envelio

1. J. See, W. Carr and S.E. Collier, "Real time distribution analysis for electric utilities", https://www.smartgrid.gov/files/Real_Time_Distribution_Analysis_for_Electric_Utilities_200812.pdf.
2. https://web.ornl.gov/sci/electricity/facilities/physical-software/.
3. https://www.accenture-insights.nl/en-us/articles/smart-grid-simulation-software.
4. https://envelio.de/en/.
5. https://high-tech-gruenderfonds.de/en/smart-grid-startup-envelio-raises-1-million-euro-from-high-tech-grunderfonds-and-demeter/.
6. http://www.rwth-aachen.de/cms/root/Die-RWTH/Profil/Gruenderhochschule/Gewinner-des-ersten-Spin-off-Awards/Gewinner-des-Spin-off-Award/~okyn/envelio-GmbH/?lidx=1.
7. https://impakter.com/envelio/.
8. https://app.dealroom.co/companies/envelio.
9. https://www.crunchbase.com/organization/envelio#section-overview.

Faraday Grid

1. www.faradaygrid.com.
2. https://www.faradaygrid.com/media/2018/6/21/a-journey-through-the-world-of-technology-patents-with-faradays-founder-and-cto-matthew-williams.
3. https://www.greentechmedia.com/articles/read/faraday-grid-investment-wework-founder.
4. https://www.greentechmedia.com/articles/read/faraday-grids-spending-spree-begins.
5. https://www.businesswire.com/news/home/20190328005099/en/Faraday-Grid-Launches-U.S.-Presents-Energy-System.
6. https://www.bloomberg.com/profile/company/1529517D:LN.
7. https://pitchbook.com/profiles/company/178303-60.

8. https://www.smartcitiesworld.net/news/news/smart-energy-faraday-grid-launches-in-the-us-4021.
9. https://www.worldenergy.org/news-and-media/news/faraday-grid-in-new-partnership-with-world-energy-council/.
10. https://www.freedm.ncsu.edu/wp-content/uploads/2019/04/Glavaski-Faraday-Grid.pdf.
11. https://theenergyst.com/faraday-grid-we-come-from-the-future-to-stabilise-the-power-system/.
12. https://dfml.co.uk/emergentstaging/home/board/.
13. https://www.current-news.co.uk/blogs/inside-faraday-grid-the-firm-hoping-to-enable-the-internet-of-energy.
14. https://www.greentechmedia.com/articles/read/transformer-technology-hopeful-faraday-grid-goes-bust.

Wireless Power Transfer and EV Charging

1. C. Mi, "Development of an extremely efficient wireless EV charger", http://web.stanford.edu/group/peec/cgi-bin/docs/events/2014/10-24-14%20Mi.pdf.
2. C.T. Rim and C. Mi, *Wireless Power Transfer for Electric Vehicles and Mobile Devices*, Wiley - IEEE, Hoboken, NJ, 2017.
3. G.A. Covic and J.T. Boys, "Inductive power transfer," *Proceedings of IEEE*, vol. 101, no. 6, pp. 1276–1289, 2013.
4. J. Shin, et al., "Design and implementation of shaped magnetic-resonance based wireless power transfer system for roadway-powered moving electric vehicles," *IEEE Transactions on Industrial Electronics*, vol. 61, no. 3, pp. 1179–1192, 2014.
5. M. Bertoluzzo, K.N. Mude and G. Buja, "Preliminary investigation on contactless energy transfer for electric vehicle battery recharging," *In Proceedings of IEEE ICIIS*, pp. 1–6, 2012.
6. G. Buja, M. Bertoluzzo and K.N. Mude, "Design and experimentation of WPT charger for electric city car", *IEEE Transactions on Industrial Electronics*, vol. 62, no. 2, pp. 7436–7447, 2015.
7. A. Kurs, A. Karalis, R. Moffatt, J.D. Joannopoulos, P.H. Fisher, M. Soljacic, "Wireless power transfer via strongly coupled magnetic resonances," Science, vol. 317, no. 5834, pp. 83–86, 2007.
8. G.A. Covic, J.T. Boys, M.L.G. Kissin and H.G. Lu, "A three-phase inductive power transfer system for roadway-powered vehicles," *IEEE Transactions on Industrial Electronics*, vol. 54, no. 6, pp. 3370–3378, 2007.
9. P. Sample, D.T. Meyer and J.R. Smith, "Analysis, experimental results, and range adaptation of magnetically coupled resonators for wireless power transfer," *IEEE Transactions on Industrial Electronics*, vol. 58, no. 2, pp. 544–554, 2011.
10. J. Dai and D.C. Ludois, "A survey of wireless power transfer and a critical comparison of inductive and capacitive coupling for small gap applications," *IEEE Transactions on Power Electronics*, vol. 30, no. 11, pp. 6017–6029, 2015.
11. J. Kim, D.-H. Kim and Y.-J. Park, "Analysis of capacitive impedance matching networks for simultaneous wireless power transfer to multiple devices," *IEEE Transactions on Industrial Electronics*, vol. 62, no. 5, pp. 2807–2813, 2015.
12. W. Zhang, J.C. White, A.M. Abraham and C.C. Mi, "Loosely coupled transformer structure and interoperability study for EV wireless charging systems," *IEEE Transactions on Power Electronics*, vol. 30, no. 11, pp. 6356–6367, 2015. https://chrismi.sdsu.edu/publications/2015_TPEL_30_11_Wei_Zhang_Loosely.pdf.

13. C.C. Mi, G. Buja, S.Y. Choi and C.T. Rim, "Modern advances in wireless power transfer systems for roadway powered electric vehicles," *IEEE Transactions on Industrial Electronics*, vol. 63, pp. 6533–6545, 2016.
14. J.L. Villa, J. Sallan, J.F. Sanz Osorio and A. Llombart, "High-misalignment tolerant compensation topology for ICPT systems," *IEEE Transactions on Industrial Electronics*, vol. 59, pp. 945–951, 2012.
15. J. Sallan, J.L. Villa, A. Llombart and J.F. Sanz, "Optimal design of ICPT systems applied to electric vehicle battery charge", *IEEE Transactions on Industrial Electronics*, vol. 56, pp. 2140–2149, 2009.
16. M. Pinuela, D.C. Yates, S. Lucyszyn and P.D. Mitcheson, "Maximizing DC-to-load efficiency for inductive power transfer", *IEEE Transactions on Power Electronics*, vol. 28, pp. 2437–2447, 2013.
17. F. Lu, H. Zhang and C.C. Mi, "A review on the recent development of capacitive wireless power transfer technology", *Energies*, vol. 10, p. 1752, 2017.
18. W. Zhang and C.C. Mi, "Compensation topologies of high-power wireless power transfer systems", *IEEE Transactions on Vehicular Technology*, vol. 65, pp. 4768–4778, 2016.
19. M.A. Houran, X. Yang and W. Chen, "Magnetically coupled resonance WPT: Review of compensation topologies, resonator structures with misalignment, and EMI diagnostics", *Electronics*, vol. 7, p. 296, 2018, www.mdpi.com/journal/electronics, 45 p.
20. S.Y.R. Hui, W. Zhong and C.K. Lee, "A critical review of recent progress in mid-range wireless power transfer", *IEEE Transactions on Power Electronics*, vol. 29, no. 9, pp. 4500–4511, 2014.

Market Information for EVs and Chargers

1. S. Chan and G. Reig, "EV market and wireless charging", Texas Instruments, January 2018.
2. H. Engel, R. Hensley, S. Krupfer and S. Sahdev, "Charging ahead: Electric vehicle infrastructure demand", McKinsey & Company, https://www.mckinsey.com/industries/automotive-and-assembly/our-insights/charging-ahead-electric-vehicle-infrastructure-demand.
3. A. Arancibia and R. Collins, "The state of the electric vehicle market and its charging infrastructure", https://www.idtechex.com/en/research-report/electric-vehicle-charging-infrastructure-2019-2029-forecast-technologies-and-players/633.

ChargePoint

1. https://www.chargepoint.com/about/leadership/.
2. https://www.cnbc.com/2014/06/17/disruptors-in-2014-chargepoint.html.
3. https://www.siliconindia.com/shownews/Indian-cofounded-Coulomb-Technologies-raises-14-Million-nid-65146-cid-100.html.
4. http://entrepreneur.wiki/Harjinder_Bhade.
5. https://www.engiestorage.com/green-charge-networks-welcomes-harjinder-bhade-new-chief-technology-officer/.
6. https://connection.mit.edu/praveen-mandal.
7. https://www.linkedin.com/in/tom-tormey-69522433.
8. https://en.wikipedia.org/wiki/ChargePoint.

9. https://www.greentechmedia.com/articles/read/chargepoint-acquires-e-mobility-software-company-kisensum.
10. https://www.businesswire.com/news/home/20180628005411/en/ChargePoint-Ushers-Electric-Fleet-Future-Acquisition-Fleet.
11. https://techcrunch.com/2018/11/28/chargepoint-raises-240-million-to-serve-an-anticipated-flood-of-electric-vehicles/.
12. https://techcrunch.com/2018/09/14/chargepoint-is-adding-2-5m-electric-vehicle-chargers-over-the-next-7-years/.
13. https://www.owler.com/company/chargepoint.
14. https://www.crunchbase.com/organization/chargepoint#section-related-hubs.
15. https://patents.justia.com/assignee/chargepoint-inc.

Wiferion

1. https://www.f6s.com/blueinductivegmbh.
2. https://www.linkedin.com/in/benriah-goeldi-21a41a9/.
3. https://www.linkedin.com/in/florianreiners/.
4. https://www.crunchbase.com/organization/blue-inductive#section-overview.
5. http://www.pr.uni-freiburg.de/pm-en/online-magazine/invent-and-establish/charging-on-the-road.
6. https://datacommons.technation.io/companies/blue_inductive.
7. https://www.ise.fraunhofer.de/en/press-media/news/2017/start-up-blue-inductive-wins-business-plan-competition.html.
8. https://www.wiferion.com/en/.
9. https://patents.justia.com/inventor/johannes-tritschler.

Momentum

1. https://www.momentumdynamics.com/.
2. https://www.linkedin.com/in/andrewdaga?trk=public_profile_browsemap_profile-result-card_result-card_full-click.
3. https://www.linkedin.com/in/jonathan-sawyer-820a535.
4. A. Daga, J.M. Miller, B.R. Long, R. Kacergis, P. Schrafel and J. Wolgemuth, "Electric fuel pumps for wireless power transfer: Enabling rapid growth in the electric vehicle market", *IEEE Power Electronics Magazine*, pp. 24–35, June 2017.
5. https://spectrum.ieee.org/cars-that-think/transportation/self-driving/google-wants-its-driverless-cars-to-be-wireless-too.
6. https://www.nrel.gov/news/program/2018/cutting-the-cord-nrel-demonstrates-wirelessly-charged-electric-vehicle.html.
7. https://insideevs.com/news/337383/us-gets-its-first-wireless-fast-charging-bus/.
8. https://chargedevs.com/newswire/momentum-dynamics-installs-200-kw-wireless-charging-system-for-buses/.
9. https://www.inquirer.com/philly/business/energy/malvern-startup-imagines-a-world-where-electric-vehicles-are-recharged-wirelessly-20180427.html.
10. https://www.cnbc.com/2019/01/16/volvo-group-invests-in-philadelphia-based-wireless-charging-firm-.html.
11. https://patents.justia.com/assignee/momentum-dynamics-corporation.
12. https://www.crunchbase.com/organization/momentum-dynamics-corp#section-overview.
13. https://pitchbook.com/profiles/company/62451-55.

WiTricity Corporation

1. http://www.sciencecoalition.org/downloads/1393519967massachusetts2.0companies.pdf.
2. https://sohman.com/witricity-does-the-magic-with-wireless-charging/.
3. https://www.crunchbase.com/person/andre-kurs#section-overview.
4. http://witricity.com/wp-content/uploads/2016/12/White_Paper_20161218.pdf.
5. http://witricity.com/wp-content/uploads/2018/03/WIT_White_Paper_PE_20180321.pdf.
6. https://witricity.com/innovation/ip-portfolio/.
7. https://www.fda.gov/media/86847/download.
8. https://www.total-croatia-news.com/business/34131-witricity-company-founded-by-marin-soljacic-acquired-biggest-rival.
9. https://www.smithsonianmag.com/innovation/wireless-charging-cars-finally-here-180970494/.
10. https://www.comsol.com/story/download/350121/WiTricity_MS15.pdf.
11. https://pdftojpg-converter.online/converted/a19ff26e/making-wireless-electricity-a-reality-faster-andre-kurs/riv0impbzz0tpqgajzji20wblcuhkxhdlybbzbmmpdf.pdf.
12. https://www.techbriefs.com/component/content/article/tb/features/articles/26214.
13. https://www.builtinboston.com/2019/02/14/boston-tech-roundup-021419.
14. https://epc-co.com/epc/EventsandNews/News/ArtMID/1627/ArticleID/470/Efficient-Power-Conversion-EPC-Announces-a-WiTricity%E2%84%A2-Demonstration-System-Featuring-High-Frequency-Gallium-Nitride-eGaN174-FETs.aspx.
15. https://xconomy.com/boston/2013/10/23/witricitys-wireless-power-tech-attracts-25m-intel-foxconn/.
16. https://www.bizjournals.com/boston/news/2016/12/20/gm-to-test-wireless-electric-car-charger-built-by.html.
17. https://www.businesswire.com/news/home/20131023005315/en/WiTricity-Secures-Additional-25-Million-Funding-Bolster.
18. https://www.fastcompany.com/40533008/the-little-company-thats-bringing-wireless-charging-to-electric-cars.
19. K. Tachikawa, M. Kesler and O. Atasoy, "Feasibility study of bi-directional wireless charging for vehicle-to-grid", SAE International, 2018-01-0669, April 2018, http://witricity.com/wp-content/uploads/2018/05/Study-of-Bi-directional-wireless-charging-for-V2G.pdf.

FINSix

1. https://finsix.com.
2. https://finsix.com/assets/files/FINsix_Tech.pdf.
3. https://craft.co/finsix-corporation.
4. https://patents.justia.com/assignee/finsix-corporation.
5. https://www.crunchbase.com/organization/finsix-corporation#section-lists-featuring-this-company.
6. https://equityzen.com/trending/finsixcorporation/.
7. https://www.businesswire.com/news/home/20180508006846/en/FINsix-Announces-Strategic-Partnership-Toyota-Industries-Corporation.
8. https://www.pntpower.com/1665-2/.

PowerSphyr

1. https://www.computerworld.com/article/3235176/wireless-charging-explained-what-is-it-and-how-does-it-work.html.
2. http://www.powersphyr.com.
3. http://ces.vporoom.com/PowerSphyr/about.
4. https://www.prnewswire.com/news-releases/powersphyr-raises-4m-in-series-a-focusing-on--automotive-cockpit-of-the-future-300746142.html.
5. https://markets.businessinsider.com/news/stocks/powersphyr-and-gan-systems-lead-the-wireless-charging-revolution-1027460178.
6. https://www.crunchbase.com/organization/powersphyr#section-overview.
7. https://www.businesswire.com/news/home/20171004005290/en/PowerSphyr-Acquires-Gill-Electronics-Facilitate-Development-Advanced.
8. https://mibiz.com/item/25160-gill-electronics-acquired-by-california-developer-of-wireless-charging-technology.
9. https://www.crunchbase.com/organization/gill-electronics#section-related-hubs.
10. https://patents.justia.com/assignee/powersphyr-inc.
11. https://www.inc.com/kevin-j-ryan/meredith-perry-ubeam-wireless-charging-tech-works.html.
12. https://www.ansys.com/blog/ces-wireless-charging-magnetic-resonance-induction-rf-harvesting.

6 Introduction to Electric Machines and Drive Systems

6.1 INTRODUCTION

Classification of electromagnetic machines (commonly coined as electrical machines) and a description of well-known machines are presented briefly in this chapter. Readers are encouraged to refer to the texts from which they would have learned their electrical machines at the undergraduate level to remind them of the machine terms and their explanations. Only two types of motors and their drive systems are considered for elaboration where most of the entrepreneurship activity is concentrated at this time and described in Chapter 7. They are permanent magnet synchronous motors and drives and switched reluctance motor and drives. Basic description of these motors, their flux orientation in the form of axial, radial and traverse flow, their stators, rotors, modeling of motors and their basic control are all described to understand the basis for emerging technologies and startups in these fields. Some performance simulation results are included for these motors with their drives to illustrate their high-performance capabilities.

6.1.1 Electrical Machines and Their Classification

Electric machines of interest and mostly in practice are shown in Figure 6.1 and commented upon in the following:

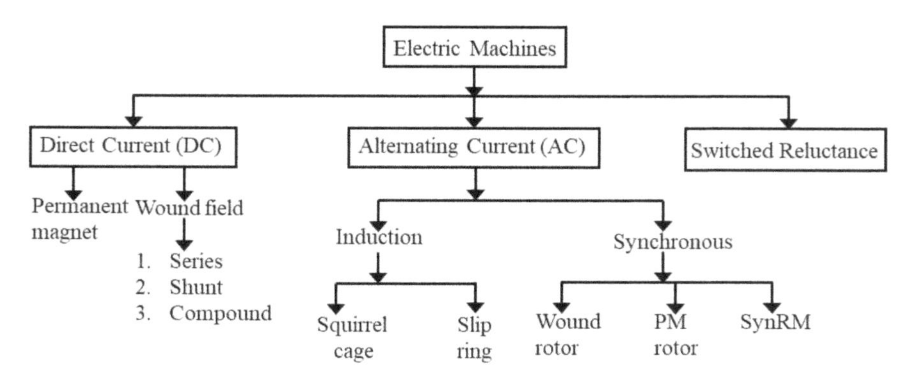

FIGURE 6.1 Classification of electrical machines of interest.

i. Direct Current Machines (DCM): One of the oldest machines with field coils on the stator and the armature in the rotor with commutator and brushes. Because of the armature in the rotor, its power density is limited by the heat transfer capability, and the commutator and brushes limit the voltage and current capability with an additional risk of sparks and need for periodic maintenance. They are classified further based on their field excitation and two types are there: (i) Wound field windings connected in series with the armature, in parallel with the armature called shunt winding, and finally, field windings both in series and in parallel to the armature coming under the name of compound dc machines, and (ii) Permanent magnet excitation from the stator, but the field from it cannot be controlled as in the case of the wound field dc machines and it does not have a wound field winding. But they all have the advantage of simpler armature or field control through a dc converter or ac to dc converter where it is easy to change the magnitude of current and/or voltage and have nothing to do with frequency and phase control as in ac machine drives. They are still in use in variable speed applications up to maximum power levels of 2,500 hp and around and have not been considered in emerging applications and newer installations in advanced countries.
ii. AC Induction Machines (IM): Single-phase and three-phase machines fed with single-phase and three-phase ac supply, respectively, are the most dominant class of machines and constitute about 90% of the motors in use. They do not have commutators and brushes (but a class of induction machines known as slip ring induction machines have slip rings and brushes and only used mostly in higher powers of 3-10 MW). Most in use are squirrel cage induction motors with shorted winding in the rotor and rotor is casted and they have the lowest cost and high reliability among machines. The stator of the three-phase induction machines have three-phase windings displaced in space by 120 electrical degrees (note: electrical degrees = {pairs of pole} × {mechanical degrees}) from each other. When these three-phase windings are fed a three-phase current also displaced from each other by 120 electrical degrees at a frequency of f_s, a rotating magnetic field is created in the stator. The speed of the stator field is equal to the excitation frequency which in this case is equal to the input voltage frequency in electrical rad/sec, w_s, obtained as $2\pi f_s$ and this is known as synchronous speed. That field interacts with the rotor coils inducing a three-phase emf (the basis for the operation of these machines and hence the name induction machines). As the windings in the rotor are shorted, they generate respective three-phase currents which then create a magnetic field in the rotor. The interaction of the rotor currents with the stator magnetic field creates a torque that moves the rotor to be in step with the stator magnetic field. As the rotor speed reaches the synchronous speed, the induced emfs in the rotor coils become zero as the relative speed between the stator and rotor fields becomes zero and hence could not induce emfs in the coils. Hence there

will be no torque, resulting in rotor slowing down which then creates induced emfs and currents in the rotor resulting in torque. A steady-state will be reached with the rotor spinning slightly slower than the synchronous speed, say, a speed of w_r rad/sec. The difference between the stator speed and rotor speed is known as slip speed given as $w_{sl} = w_s - w_r$, rad/sec. But the speed of the rotor magnetic field is given as the sum of the rotor speed and the speed of the field created by slip speed currents from slip speed emfs with the result that its speed will be equal to the synchronous speed, w_s, itself. Note that an induction motor will function always with a slip speed. And the slip speed is positive for motor operation and negative for generator operation. The equivalent circuit, torque and power expressions are not taken up here as the focus of the chapter is on other machines.

iii. Synchronous Machines (SM): These machines (sometimes referred to as wound rotor synchronous machines) are similar to the induction machines in the stator part, but the rotor with a winding provides the excitation with a dc supply input and creating a rotor magnetic field. The rotor is accessed through slip rings and brushes. There is zero slip speed here unlike in the case of the induction machines as the excitation to the rotor is not through the stator supply and its field. This has higher efficiency compared to the induction machines because of this zero slip but has the disadvantage that its rotor excitation has to be provided from an external source. These are machines usually built for high-power operation and generally for generator operation to the tune of 100s of MW capacity. For motoring, they are used up to 60 MW power levels. They are very rarely used in low power mass applications such as in appliances. They will not be considered hereafter in this section.

iv. PM Synchronous Machines (PMSM): They are a class of synchronous machines but with the rotor excitation obtained from permanent magnet poles mounted in the rotor. The advantages of PMSM are:
   i. No need for an external supply for rotor excitation as it is self-contained with permanent magnet excitation.
   ii. No slip rings and brushes to access the rotor as in the SMs.
   iii. No windings on the rotor.
   iv. Compact because of factors (a) to (c), and thus has a high power density.
   v. Losses are mostly confined to the stator and very small losses in the rotor leading to high efficiency.
   vi. Cool rotor requiring minimum cooling overall because of reduced losses in the stator and rotor.
   vii. Low weight and small rotor resulting in low inertia and hence in high torque to unit inertia which amounts to high acceleration capability.
   viii. High power density compared to other machines in most of the applications.

   The disadvantages of the PMSM are:

a. Permanent magnets add on to the manufacturing cost both in materials and in the process itself.
b. Demagnetization of the permanent magnets under stator winding short circuits is an issue for concern in high-reliability applications.
c. Loss of partial magnetism as a function of rotor temperature makes for lower torque, and power outputs and that has to be considered in applications, particularly the effects under allowable maximum temperature for the machine. That will force the reduction of the operating machine torque and output power considerably. Contrast this with induction motor where the temperature has a smaller ill effect on the torque and power output as the torque is directly proportional to rotor resistance and when the rotor temperature increases, rotor resistance increases thus increasing the torque if the rotor current is the same.
d. Maximum rotor speed is affected by the magnets and hence mechanical means to hold them in place in the rotor structure becomes crucial in some PM machine structures; whereas the induction motor is good for 100 K rpm plus because of the robust rotor structure.
e. For high-speed operation, the magnetic field in the air gap is reduced by supplying a stator field that opposes the rotor field. It is usually known as flux weakening. Under this condition, if the control fails, then the phase voltages will become proportional to the rotor speed resulting in a large voltage and large currents leading to machine winding damage and disruption in its operation both of which are undesirable in practice. Such is not the case in induction motors as the magnetizing reactance is a function of the frequency and hence its reactance will contain the excitation current and will not lead to a major disruption in operation.
f. Recycling cost for neodymium boron permanent magnet machines is higher compared to the recycling of induction machines because of the rare earth material components in the machines.

A variation of PMSM is PM brushless dc machine (PMBDCM) which has trapezoidal induced emfs resulting in simpler operating features is fairly common in use at low powers and that will also be considered.

v. Switched Reluctance Machines (SRMs): This is the oldest machine that has not entered into circulation in a big way. The stator and rotor have salient poles, unlike any other electrical machine. The stator has concentric windings around stator poles and the rotor has stacked laminations but with no windings in them at all. The principle of operation is based on the inherent reluctance variation between a set of stator and rotor poles. More of the details will be taken up in the relevant section. Briefly, its advantages and disadvantages are given below. Its advantages are :

a. Simpler stator and rotor structures and easy to build.
b. Minimum coil material as the windings are concentric around the poles in contrast to distributed winding prevalent in ac machines.

c. Lowest rotor inertia contributing to the highest torque to inertia ratio, i.e., acceleration.
d. Mutual coupling between windings is almost nil with the result that fault in one phase does not affect other phases, unlike other ac machines.
e. Capable of working even with only one phase, though with reduced power, when all other phases may have a fault is an attribute that no other machines have. This feature makes it suitable for applications requiring high reliability under worst circumstances, say, such as in emerging electric aircraft.
f. Recycling of the used machines is easy as it is more in line with the conventional machines having no rare earth materials.
g. Cooling is easy as most of the heat is concentrated in the stator.
h. Temperature has no reduced magnetic effect as in the case of PM machines.
i. Low-cost motor compared to all other motors.
j. The number of converter topologies available for this machine is unbelievably high compared to any other machine. This leads to the selection of the right type of minimum device-based (or minimum cost) converter for a given application and that benefit, in general, is not carried over to other motors.

Its disadvantages are:

a. Acoustic noise due to the inherent nature of the machine operation with salient poles. Mitigation steps are essential for high-performance applications and that may slice some of its power density features.
b. High ripple torque if care is not taken in the operation. Measures have been in place to match that of the PMSMs. That requires some extra control without affecting its power density.
c. Number of power devices required for the SRM converter is a function of the number of phases in the machine. The conventional converter for this machine requires two transistors and two diodes per phase for the independent operation of the machine. That means beyond three phases, the number of devices exceeds that of the three-phase inverter count thus making the drive more expensive compared to ac machines. But there are other viable solutions to this issue.

vi. Synchronous Reluctance Machine (SyRM): It is a cross between the SM and SRM with its stator that of the SM and the rotor similar to the SRM and its torque is generated solely due to the reluctance variation in a phase. Its power density is lower than that of the PMSM but uses the three-phase inverter. SyRM is yet to gain traction with the motor manufacturers and has not entered in a big way in the market. It has the positives and negatives of the ac machines, but the rotor structure to gain maximum sinusoidal reluctance variation brings mechanical strength issues. It will not be pursued any further in this chapter.

Only PMSM, PMBDCM and SRM will be further considered given the dominance of the first two at present and the third being so different from them in almost every aspect of its construction, operation, drive system and control. Various types of these motors based on their flux direction, converters in general use for these machines, control for high performance, position sensorless operation, high-efficiency operation and other related features are considered in the following sections.

6.2 PM SYNCHRONOUS AND BRUSHLESS DC MACHINES

These machines have well-known structures of rotary electrical machines: radial, axial and transverse flux paths. Radial structure with the flux path being along the radius of the machine is the most common in machines and is prevalent in practice. Axial structure with the flux path parallel to the shaft of the machine has been scarcely in practice so far but has gained momentum recently. Transverse structure with flux path along the circumference of the machine has a very small presence in the market at this time.

6.2.1 Radial Flux Machines

Some of the rotors with PMs having radial flux paths are shown in Figure 6.2. Axes d and q denote, respectively, the centerline of the PM anchoring the flux flow and the center of the idle space between two PMs and they are referred to as direct and quadrature axes, respectively. The magnet arrangements in the rotor can further be classified as:

i. Surface Mount PM Rotor: The PMs are mounted on the surface of the rotor iron, shown in Figure 6.2a, and holding them in place requires some effort and they are rarely used for high-speed operation in excess

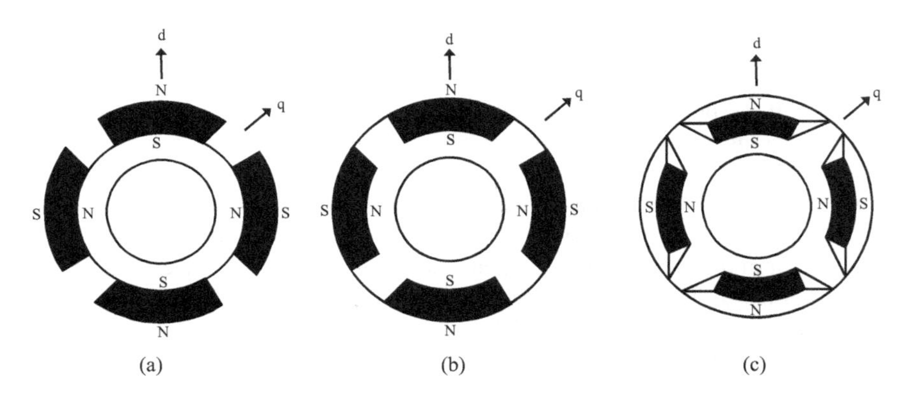

FIGURE 6.2 Permanent magnet arrangement in the PM machines. (a) Surface mount PM rotor, (b) surface inset PM rotor, and (c) interior PM rotor.

of 4,000 rpm in practice. The inductance, when measured on the stator along the direct and quadrature axes, will be almost the same as some of the magnets have a similar relative permeability of the air with the result that the reluctance from direct to quadrature axes will not be very different and hence the inductances along these axes will also be same. It makes the reluctance torque to be negligible as it is directly proportional to the difference between the direct and quadrature axis inductances. Saliency ratio is defined as the ratio between the quadrature and direct axis inductances, which is around one in this type of machine.

ii. Surface Inset PM Rotor: The magnets are buried to their depths in the rotor with the result that the outer side of the magnet is flush with the rotor iron, as shown in Figure 6.2b. This gives a much better mechanical strength to the structure and the rotor can be spun at higher speeds compared to the surface magnet rotor. It has an additional advantage that the q axis inductance will be higher than of its d axis counterpart with the saliency ratio being higher than one, resulting in a nonzero reluctance torque which adds to the synchronous torque thus boosting the net torque of the machine.

iii. Inset PM Rotor: The magnets are placed very much inside from the outer surface of the rotor, as shown in Figure 6.2c, endowing a high mechanical integrity and a larger difference between the q and d axes inductances resulting in higher reluctance torque contributing to a larger net torque of the machine. The saliency ratios for these types of machines can be higher than 2 and even much more than that. There are many variations in placing the magnets for inset configuration influenced by the depth of the steel from the outer surface of the rotor, a number of layers of steel and magnets along the d axis, making the magnet bits v-shaped and placing two of them slanted to each other, thus constituting effectively a single magnet, and many other combinations of placements. Many of the variations in magnet placement have a desirable outcome that could be of crucial importance in specific applications.

The types of windings in these machines are that of traditional electrical machines with no significant differences.

6.2.2 Circumferential Flux Path Machines

A variation of the PM rotor is shown in Figure 6.3 with the poles being placed to have flux along the circumferential orientation of the rotor. One use for such placement is that low flux density PMs can be used here that still can generate a large flux across the air gap and hence high flux density across it to give power levels that are not too far off from rare earth permanent magnet machines. This configuration also provides a high saliency ratio with higher reluctance torque.

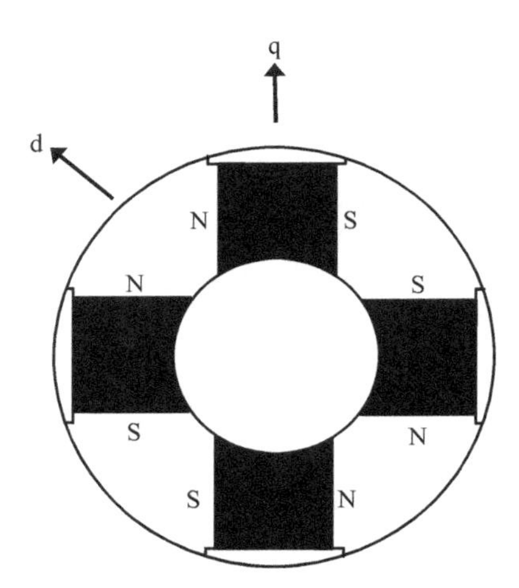

FIGURE 6.3 Circumferential flux path PM rotor.

6.2.3 Axial Flux Machines

The flux path along the axial direction from the magnets, shown in Figure 6.4, leads to a compact machine as the stator and rotor volume shrinks. That contributes to higher power density, a fact well known for a very long time. It has come into the fore recently due to recent industrial interest in high-powered electric vehicles and electric aircrafts in the power range from 100 kW to 1 MW.

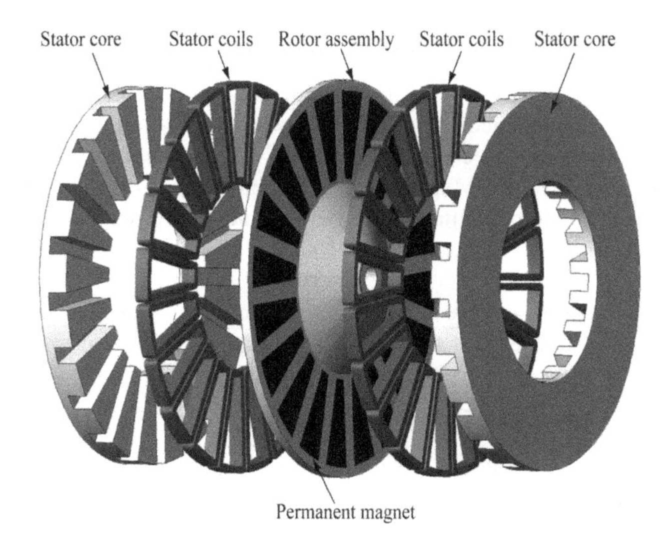

FIGURE 6.4 Axial flux PM machine with two stators and one rotor. (From R. Krishnan, *Permanent Magnet Synchronous and Brushless DC Motor Drives,* Figure 1.73(b), CRC Press, Boca Raton, FL, 2010, With permission.)

Note that the machine in Figure 6.4 has two stators one on each side of the rotor with permanent magnets. There are some other configurations of the axial flux machines with one stator and one rotor, etc., and also with different shapes of the stator and rotor configurations. The most popular one is shown in the figure. Many stator variations are possible, for example, printed circuit board stators type, and toothless stator. If the forces are not balanced along the axial direction it leads to pull and push forces on the rotor resulting in faster bearing wear out and resulting mechanical problems and reduced life of the machine. Twin stators with inside rotor help under this circumstance to an extent.

6.2.4 Model of the PM Synchronous Machine

A simple approach to modeling of the permanent magnet synchronous machine is presented and made use of to understand its operation and control principle. A rigorous derivation and many facets of the modeling are available in reference books cited at the end of this chapter.

Useful preliminary for the modeling is that only a balanced three-phase system is considered hereafter. That means, the line and phase variables such as voltages, currents and mmfs amongst themselves have the relationship that their sum in each category is zero. Consider the three-phase voltages, and for the balanced system they have the relationship,

$$v_{as} + v_{bs} + v_{cs} = 0 \tag{6.1}$$

where the a, b, and c are phases with extra subscript s denoting they are of the stator. It can be seen from this relationship that the third variable can be expressed as the negative sum of the first and second variables implying that there are only two independent variables in a balanced three-phase system. In effect, this gives the basis for a three-phase system to be reduced to a two-phase system forming the basis for the modeling of the system. The approach is to consider a two-phase PMSM, model it in rotor reference frames, find the relationship between rotor and stator reference frames for the variables and transformation to three-phase stator variables to establish the equivalence to three-phase machine from two-phase machine, and derive the electromagnetic torque from the currents.

Model of the two-phase PMSM: The machine in stator reference frames (natural ones) is shown in Figure 6.5. The two phases are q and d that are in quadrature, and each has N turns of winding, and the PM rotor is displaced from stator d axis by an angle θ_r which is varying as a function of rotor electrical speed of ω_r that is related to time t as,

$$\theta_r = \omega_r t \tag{6.2}$$

The rotor magnets produce a flux linkage of λ_{af} between the rotor and armature that is the stator windings, defined very similar to the conventional wound rotor synchronous machine. It is represented to be the phasor on this axis that is

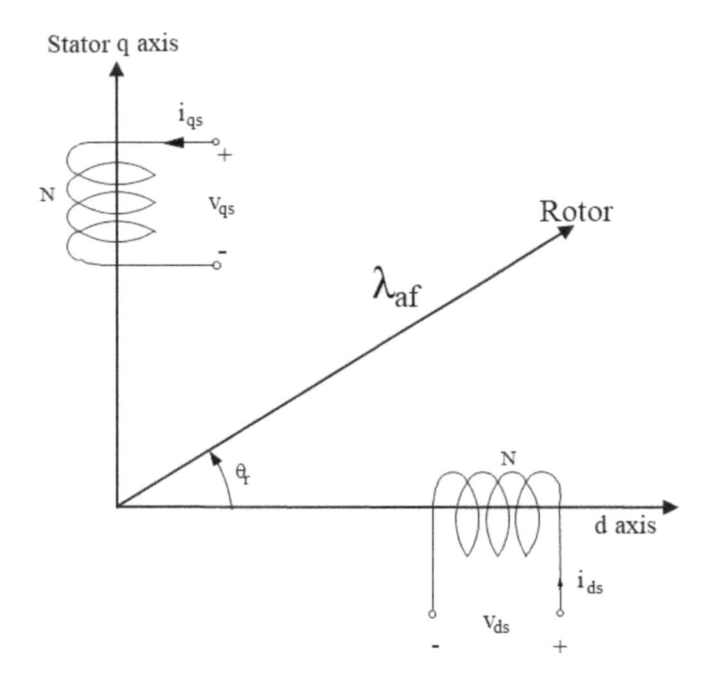

FIGURE 6.5 Two-phase PMSM in stator reference frames.

rotating at the speed of the rotor with its magnitude as λ_{af}. Let this be dr axis. The stator phases can be transformed to be on the rotor related frames but having the quadrature relationship with the new rotational axes of dr and qr as shown in Figure 6.6 with the stator phases and rotor on them. Note that, the qr axis leads the stator q axis by θ_r as does the dr axis from the stator d axis. The phase variables in the rotor reference frames are given by the subscripts qr and dr, while those of the stator reference frames are denoted by qs and ds.

Modeling of the rotor reference frames based PMSM is much simpler from physical understanding and that approach is considered here. Let the self-inductances of the stators on the qr and dr axes be L_q and L_d, respectively, and the resistance in each phase is R_s. The flux linkages along the dr axis is the sum due to the permanent magnets, λ_{af} and that due to the current in the inductance given by $L_d i_{dr}$ (dr axis flux linkages due to the current in the stator dr winding) and the flux linkages along the qr axis is produced by the qr winding current and its self-inductance given by $L_q i_{qr}$. The emf induced in the qr axis winding due to speed and the dr axis flux linkages is given by $\omega_r(\lambda_{af}+L_d i_{dr})$. The terminal voltage across the winding is the sum of the speed voltages and the voltage drop due to its self-impedance of that phase. Similarly, for the dr axis terminal voltage, speed emf given by the qr axis flux linkages is $\omega_r(-L_q i_{qr})$. The minus sign here is given the convention that lagging flux linkages produce a positive speed emf and therefore, leading flux linkages result in a negative speed emf. The stator equations in the rotor reference frames are assembled from this understanding as,

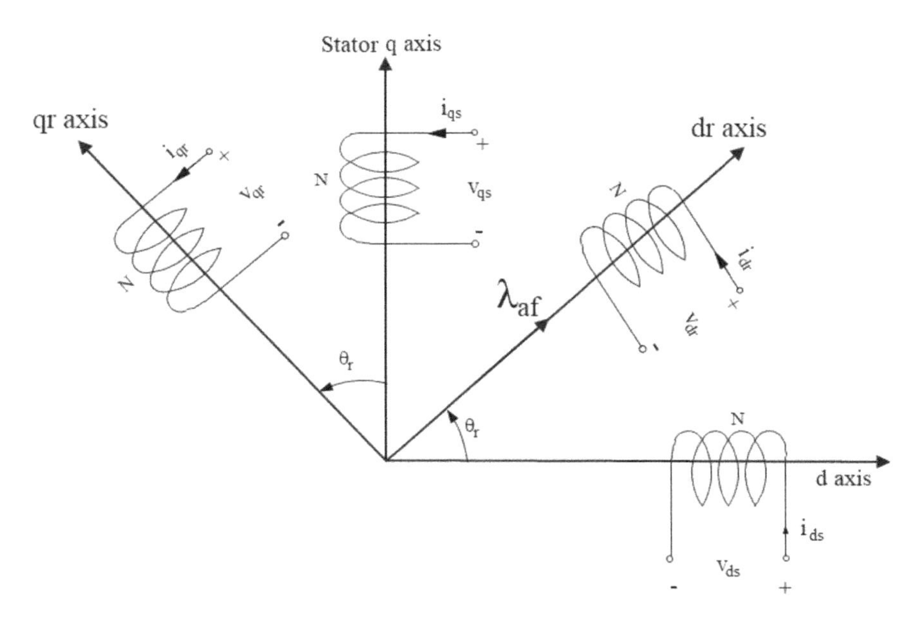

FIGURE 6.6 Two-phase PMSM in stator and rotor reference frames.

$$\begin{bmatrix} v_{qr} \\ v_{dr} \end{bmatrix} = \begin{bmatrix} R_s + L_q p & \omega_r L_d \\ -\omega_r L_q & R_s + L_d p \end{bmatrix} \begin{bmatrix} i_{qr} \\ i_{dr} \end{bmatrix} + \begin{bmatrix} \omega_r \lambda_{af} \\ 0 \end{bmatrix} \quad (6.3)$$

The rotor reference frame and stator reference frame currents are related by projecting the mmfs of q and d axes in stator reference frames on to that of the qr and dr axes, respectively, as in the following:

$$\begin{bmatrix} i_{qr} \\ i_{dr} \end{bmatrix} = \begin{bmatrix} \cos(\theta_r) & -\sin(\theta_r) \\ \sin(\theta_r) & \cos(\theta_r) \end{bmatrix} \begin{bmatrix} i_{qs} \\ i_{ds} \end{bmatrix} = [T_{sr}] \begin{bmatrix} i_{qs} \\ i_{ds} \end{bmatrix} \quad (6.4)$$

where T_{sr} is the transformation matrix to go from stator to rotor reference currents, voltages and mmf in two phases. Note that the number of turns on each side canceled out. Likewise, the rotor referred variables can be transformed to the stator referred variables by using the inverse of T_{sr}.

The next step is to relate the two-phase machine with N turns per phase, shown in Figure 6.5, to three-phase machine with N_1 turns per phase shown in Figure 6.7 by superposing the three phasors of the three-phase machine on to the two phasors of the machines in stator reference frames. For equivalence of mmf produced in the three- and two-phase machine windings, the number of turns in each phase of the two-phase machine has to be 3/2 times that of the phase winding of the three-phase machine which works out to be $N = (3/2)N_1$. After taking that into

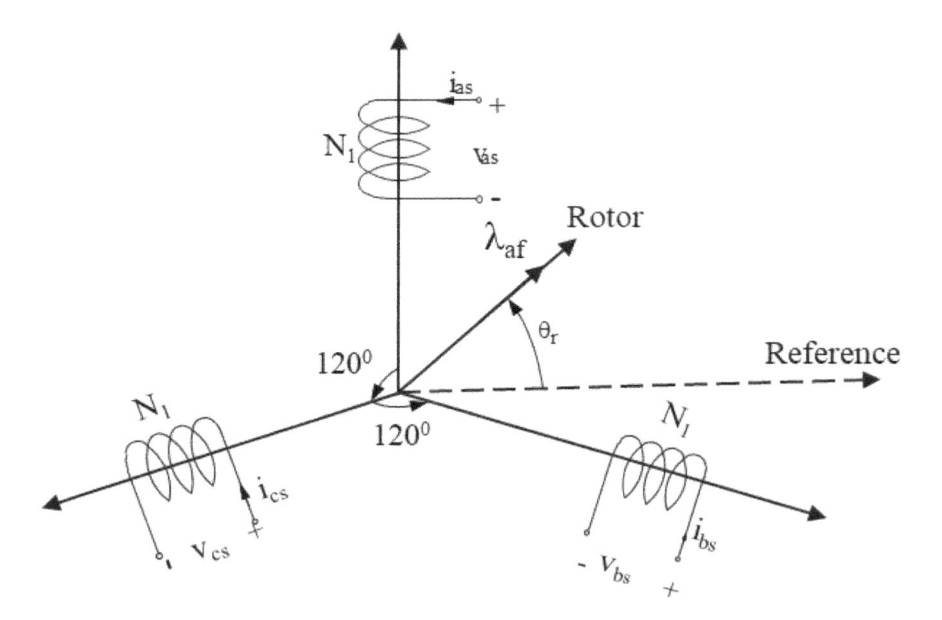

FIGURE 6.7 Three-phase machine in stator reference frames.

account the relationship between the two- and three-phase currents in stator reference frames is,

$$\begin{bmatrix} i_{qs} \\ i_{ds} \end{bmatrix} = \begin{bmatrix} 1 & -1/2 & -1/2 \\ 0 & \sqrt{3}/2 & -\sqrt{3}/2 \end{bmatrix} \begin{bmatrix} i_{as} \\ i_{bs} \\ i_{cs} \end{bmatrix} = [T_{s32}] \begin{bmatrix} i_{as} \\ i_{bs} \\ i_{cs} \end{bmatrix} \tag{6.5}$$

where T_{s32} transforms the three phase stator currents into two-phase currents in stator reference frames. The transformation to the rotor reference frames current in two phases from three-phase stator frame currents, using equations 6.4 and 6.5 are obtained as,

$$\begin{bmatrix} i_{qr} \\ i_{dr} \end{bmatrix} = [T_{sr}][T_{s23}] \begin{bmatrix} i_{as} \\ i_{bs} \\ i_{cs} \end{bmatrix} = [T_r] \begin{bmatrix} i_{as} \\ i_{bs} \\ i_{cs} \end{bmatrix} \tag{6.6}$$

where the final transformation is given by $[T_r] = [T_{sr}][T_{s23}]$ and these matrices are obtained from equations 6.4 and 6.5. This transformation is valid for other variables including voltages and mmf.

Electromagnetic Torque: The sum of the product of q and d axes voltages with their respective currents in the rotor reference frames yields the power input from which the air gap power is given by the product of the speed terms and the currents in equation 6.3. It is

$$P_i = \left[v_{qr} i_{qr} + v_{dr} i_{dr} \right] \tag{6.7}$$

Note that 3/2 is included to account for the three phases in a two-phase machine model as explained earlier. The rate of change current terms along with respective inductances indicates that they correspond to the rate of the stored energy in the self-inductances and it is not associated with air gap power. What is left in the power terms is the air gap power given as,

$$P_a = \omega_r \left\{ \lambda_{af} + \left(L_d - L_q \right) i_{dr} \right\} i_{qr}, \text{ W} \tag{6.8}$$

from which the electromagnetic torque is derived by dividing it with mechanical rotor speed as,

$$T_e = \frac{P_a}{\omega_m} = [P_a] \frac{P/2}{\omega_r} = \frac{3}{2} \frac{P}{2} \left\{ \lambda_{af} + \left(L_d - L_q \right) i_{dr} \right\} i_{qr}, \text{ N.m} = T_{es} + T_{er} \tag{6.9}$$

Note that mechanical and electrical rotor speeds are related by pairs of poles, P/2. The torque expression contains two distinct terms and they are: (i) the synchronous torque, T_{es} and (ii) reluctance torque, T_{er}.

Unbalanced operation in three-phase machines in terms of their currents, voltages have not been dealt with in the modeling and interested students may pursue from the references cited earlier.

Control Basis for PMSM: Electromagnetic torque consists of two components given in equation 6.9: (i) T_{es} is the synchronous electromagnetic torque due to the interaction of the rotor flux linkages and the q axis current stator current in rotor reference frames, and, (ii) T_{er} is the reluctance electromagnetic torque and they are given as,

$$T_{es} = \frac{3}{2} \frac{P}{2} \lambda_{af} i_{qr} \tag{6.10}$$

$$T_{er} = \frac{3}{2} \frac{P}{2} \left(L_d - L_q \right) i_{dr} i_{qr} \tag{6.11}$$

The synchronous torque expression is very similar to the torque in the separately excited dc motor obtained as the product of the field flux linkages and armature current. Such reduction of the PMSM machine into an equivalent armature controlled dc machines is made possible through the transformation of the machine in rotor reference frames and with two phases in quadrature. That forms the basis for control of this machine with $L_q = L_d$.

As for the reluctance torque, it is due to the reluctance variation between the q and d paths and comes into existence only when the q and d phase currents in rotor reference frames exist. Usually, the d axis current is zero in most of the PMSM control, but its controlled quantity comes into play in certain circumstances such as:

i. When the field has to be reduced similar to the dc motors, it is achieved by injecting a d axis current in the rotor reference frames which directly opposes the main field flux linkages of the rotor. It is known as flux weakening and is resorted to operate beyond the base speed to achieve an extended speed of operation in the machines. An application example is the electric car.
ii. Further, this augments the torque in flux weakening mode as i_{dr} is negative and $(L_d - L_q)$ being less than zero gives a positive reluctance torque for positive i_{qr}. This torque augmentation may anywhere be between 3% and 10% in normal machines with inset PM magnets. Higher reluctance torque is possible for particular designs but not encountered much in practice.
iii. In steady-state control such as maximum torque per unit current, both q and d axes currents are controlled to optimize the torque with the result that flux weakening is resorted to intentionally ahead of the base speed of the machine.

The control of the torque components, T_{es} and T_{er} completes the control basis for the PMSM.

Control of PMSM: The basis of the control for these machines has been established from the electromagnetic torque components' expressions. This section is devoted to the control theory and its implementation.

PMSMs cannot directly start from a three-phase ac supply, unlike induction motors. They need a variable frequency source to operate from zero and at any speed. The ac source has to be capable of providing variable voltage and variable frequency and that requires a power electronic inverter. The standard inverter in practice is an H-bridge three-phase inverter and emerging high-power PMSMs may resort to multilevel inverters and both of them are presented in Chapter 3. Therefore, an inverter is assumed for the control theory development.

Consider a current phasor, i_s, having a component on quadrature and direct axes in rotor reference frames, i_{qr} and i_{dr}, respectively is imposed on the machine. The current phasor with its magnitude and phase is given as,

$$|i_s| = \sqrt{i_{qr}^2 + i_{dr}^2} \tag{6.12}$$

$$\delta = \tan^{-1}\left(\frac{i_{qr}}{i_{dr}}\right) \tag{6.13}$$

The angle δ is with reference to d axis and the axis on which the rotor and PM are positioned is revolving at a rotor electrical speed of ω_r with an instantaneous rotor position of θ_r from the reference axis as shown in Figure 6.8. By controlling

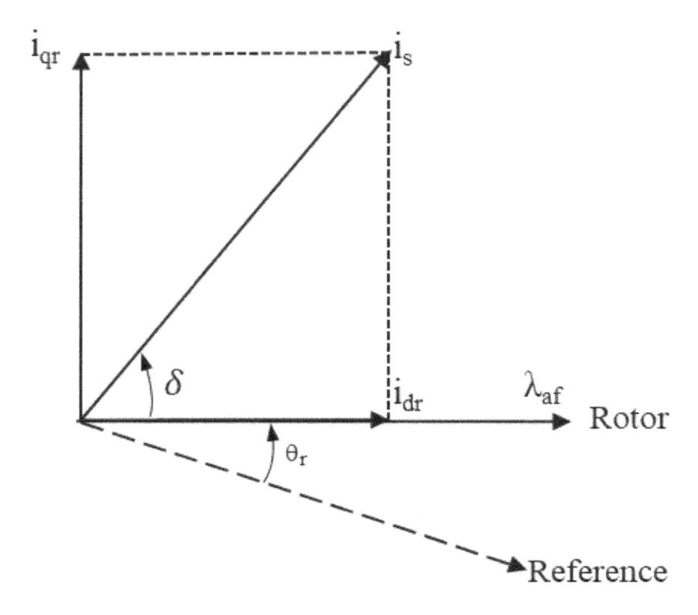

FIGURE 6.8 Vector control of PMSM.

the current vector i_s, the machine torque and its speed are controlled at all times. It generally comes under the broad name of vector control in ac motor control literature. In order to implement it, the current vector and its position with regard to a stationary reference have to be generated in the machine. But the machine has three-phase currents and they are obtained from transforming the current vector into rotor reference frame based q and d axes currents, i_{qr} and i_{dr} which then can be used to obtain the stator phase currents via a transformation. They can be obtained in two steps from earlier derivations 6.4 and 6.5 as,

$$\begin{bmatrix} i_{qs} \\ i_{ds} \end{bmatrix} = \left[T_{sr}^{-1}\right] \begin{bmatrix} i_{qr} \\ i_{dr} \end{bmatrix} \tag{6.14}$$

$$\begin{bmatrix} i_{as} \\ i_{bs} \\ i_{cs} \end{bmatrix} = \begin{bmatrix} 1 & 0 \\ -1/2 & \sqrt{3}/2 \\ -1/2 & -\sqrt{3}/2 \end{bmatrix} \begin{bmatrix} i_{qs} \\ i_{ds} \end{bmatrix} = \left[T_{s32}^{t}\right] \begin{bmatrix} i_{qs} \\ i_{ds} \end{bmatrix} \tag{6.15}$$

Note that T_{sr}^{-1} is the inverse of T_{sr} and T_{s32}^{t} is the transpose of T_{s32}. Therefore, given the rotor position θ_r, i_{qr} and i_{dr}, i_{qs} and i_{ds} are calculated from equation 6.14 and then from the latter to three-phase variables using equation 6.15. Or, in one step they can be computed by combining the two last equations. Then, these computed phase currents can be enforced on the motor via an inverter.

There are many control techniques to obtain various performance objectives such as maximum efficiency, unity power factor operation, maximum torque per unit stator current, etc. Interested readers may consult given references.

Consider a speed-controlled PMSM drive system using the vector control described in the above. Currents are enforced through current feedback control used in all motor drives and speed by the speed control loop which is the outer loop to the inner current control loops. The rotor position information obtained from a sensor or through a sensorless method is fed into the controller to synthesize the current commands to the motor drive. Very many ways of implementing vector controllers exist and interested readers may find them in references. A four-quadrant speed-controlled drive system simulation is shown in Figure 6.9. The y-axis variables are given in per unit (or normalized unit) and are in the following order for each of the subplots: speed command and speed, electromagnetic torque reference, electromagnetic torque produced in the motor, q axis current in rotor reference, d axis current in rotor reference, phase a current, phase b current, and phase c current. Note that the d axis current is zero as there is no flux weakening intended in this example drive system. In that case, the q axis current is directly proportional to the torque reference as seen in the figure. During variations of speed such as rising or falling or reversing speed in command and its response, the command value of torque is set normally at some maximum

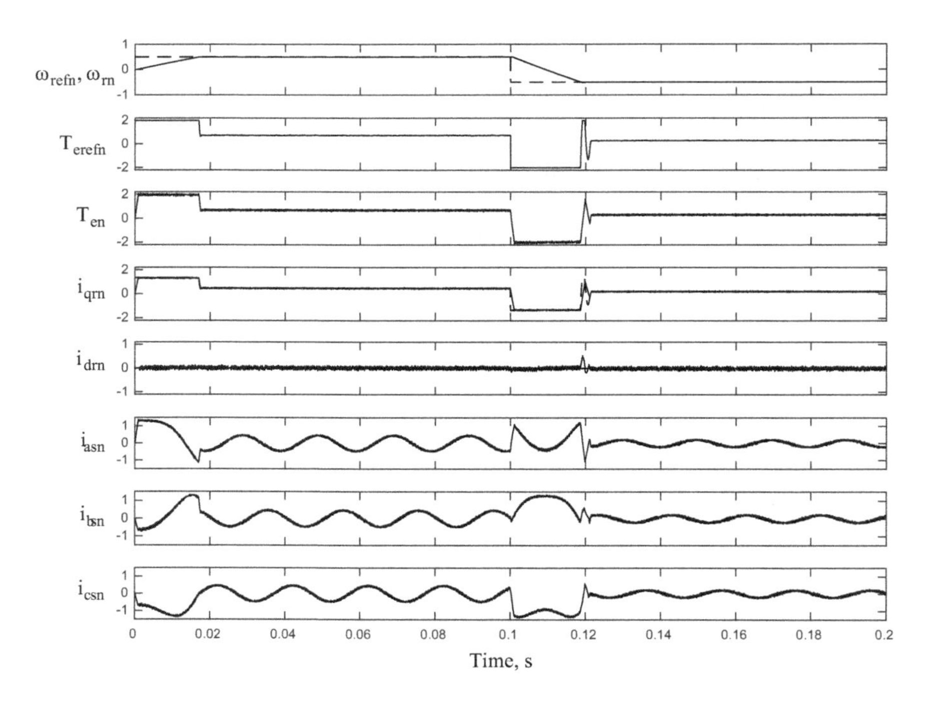

FIGURE 6.9 The dynamic performance of a vector controlled four-quadrant PMSM drive system.

magnitude and in this case at 2 p.u. and note that the q axis current and phase currents reflect that proportionally in their outputs. The torque and various currents follow their respective commands very closely thus guaranteeing a very high dynamic performance of the system, which is a characteristic of the vector controlled motor drives including the PMSM drive system. Note that this allows for safe operational enforcement of the drive all the time as all the key variables such as current and resulting torque are kept under strict control.

6.2.5 PM Brushless DC Machine and Its Control

The PM brushless dc machine induced emfs have a trapezoidal voltage shape, with constant magnitudes over 120° in both positive and negative half-cycles as shown in Figure 6.10 at least for ideal machines. The a, b, c phase induced emfs are e_{as}, e_{bs}, and e_{cs} and likewise phase currents in those phases are i_{as}, i_{bs}, and i_{cs}, respectively. Consider the phase currents are ideally rectangular with constant

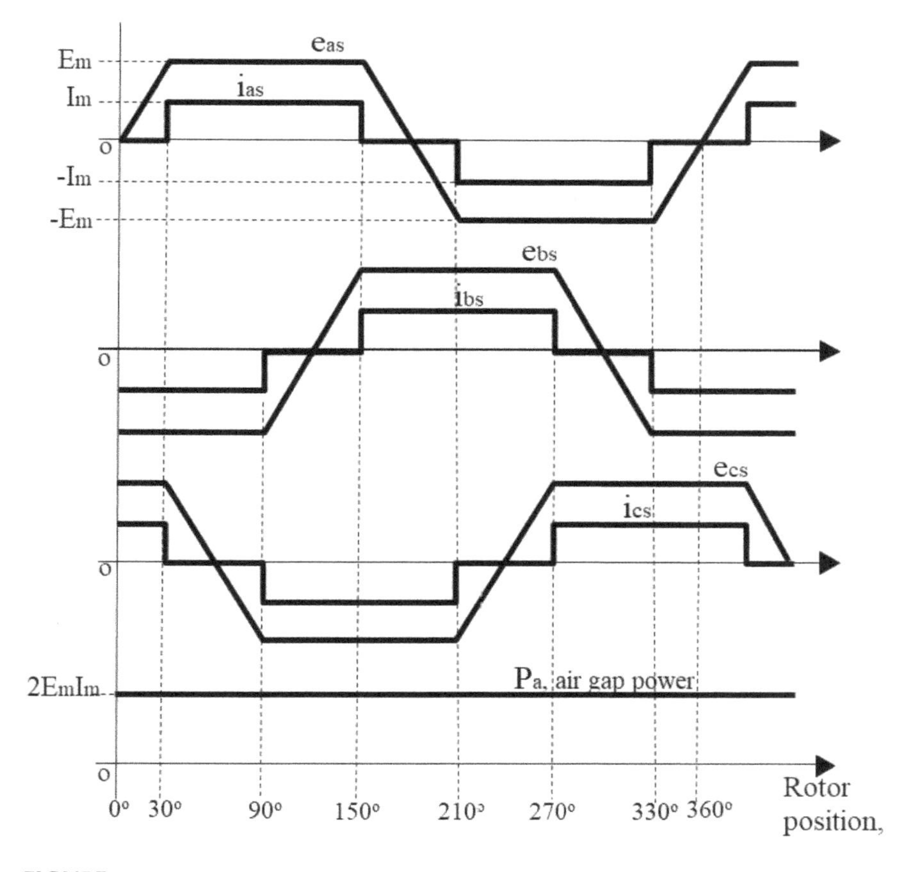

FIGURE 6.10 PM brushless dc machine waveforms.

magnitude (I_m) lasting for the duration where the induced emfs have constant magnitude (E_m) over 120°. The air gap power of each phase in one electrical cycle is equal to $\frac{2}{3}E_mI_m$ and the total for the three phases is then equal to $3(\frac{2}{3}E_mI_m)$ resulting in $2E_mI_m$ as shown in the figure. The air gap power, P_a is constant and accordingly the electromagnetic torque also. This gives a very simple control technique unlike its counterpart, PMSM and is the main reason for its popularity in applications ranging from low power to high power in appliances, and for two- and three-wheel electric vehicles. There are modifications to the control in that there is no way ideal rectangular currents can be generated due to the machine inductances, and therefore, the phases are turned on (say, before 30° in phase A in a positive half cycle) with an advance angle and turned off ahead of or around of the 120° conduction duration (say, ahead of or around 150° in phase a in the positive half cycle). This approach is followed for all the three phases, but it may give way to torque ripples. To completely eliminate the torque ripple, current coordination is resorted to between rising currents in one phase and falling currents in the other phase.

The major advantages of the PMBDC system are:

i. The three phases ideally conduct for 2/3 of the time in a cycle leaving enough time to cool the machine phases.
ii. It is simpler to generate rectangular than sinusoidal currents.
iii. Only two transistors conduct in the inverter at any time thereby reducing the conduction losses and also giving time for the cooling of other devices in the inverter.
iv. The power density of the machine for equivalent resistive losses is higher (1.15 times) than that of PM sinusoidal machines.

Both PM synchronous and brushless dc machines require position information to commutate and control the phase currents. Both position sensor- and position sensorless-based controls have been in practice for some time. The position sensorless control system is very simple for brushless dc machines by extracting the zero crossings of the induced emfs that are made easier by the absence of currents before and after zero crossing for 30° on either side of it. Then that information from all the three phases is used for current initiation and commutation instants.

Generator and Regenerative Operation: The machine is also used as a generator in select applications but not widely used because they do not generate sinusoidal ac voltages. One such select application may be in small wind generator for off-grid dc applications, where the output voltages are rectified and smoothed thus having no sinusoidal requirements on the induced emfs and output currents. When the machine is used in an electric vehicle drive, it can also function as a generator to recover energy during braking by using the regenerative braking operation. During that time, the current in the machine phases is shifted 180° with respect to the motor operation resulting in negative air gap power indicating generating power being fed to the input source of the inverter. The source such as a battery in electric vehicles will absorb and store this energy for later use in motoring region.

Dynamic performance simulation results of a PMBDC motor drive are shown in Figure 6.11 to complete the topic. In this simulation, a four-quadrant operation of the motor drive including load torque disturbances during motoring in positive direction of rotation, and regenerating and motoring in the reverse direction of rotation are all implemented. All the y-axis variables (except rotor position) are in per unit and they are in the following order from the top of the figure: speed command, actual speed, load torque, electromagnetic torque reference command, electromagnetic torque produced in the machine, rotor position in radian, phase a current's reference, and actual phase a current. The maximum torque command is limited to+1.5 p.u. and –2 p.u. which automatically limits the phase current magnitudes and the resulting torque as seen from the figure. Load torque disturbances significantly change the speed profile very much and also the effect of torque ripples that are inherent in the PMBDCM are clear on the torque and their effect in speed responses. Overall, this is capable of very good dynamic performance but not comparable to the PMSM drive is seen from this drive. Much more sophistication of the control can reduce significantly the ripple torque and improve the performance.

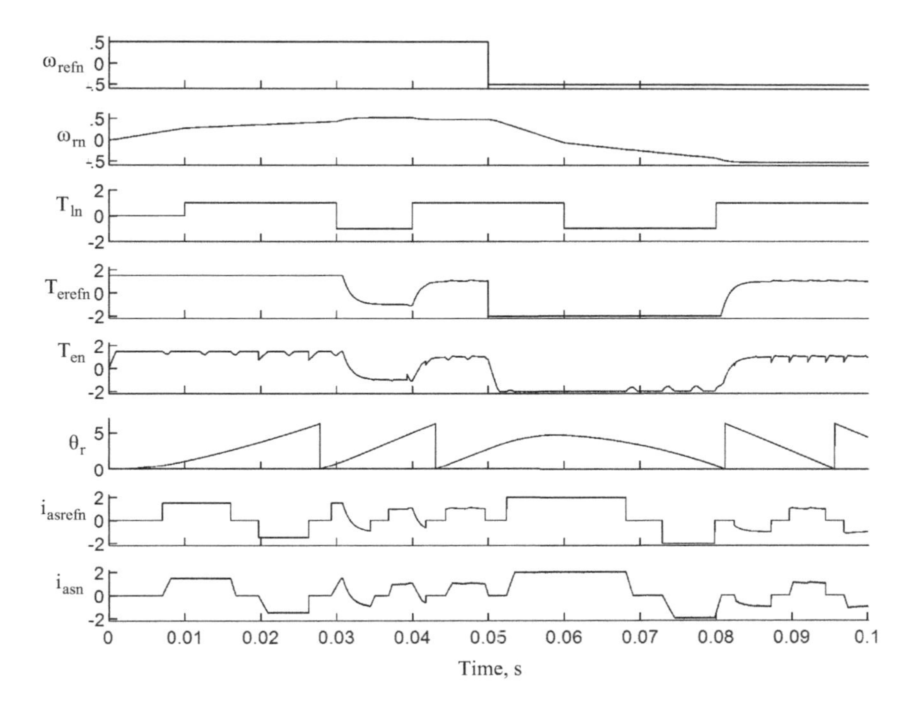

FIGURE 6.11 The dynamic performance of a PMBDC motor drive.

6.3 SWITCHED RELUCTANCE MACHINE (SRM)

A brief introduction to machines and its physical form, principle of operation and requirements for control of this machine, its converters, solutions to its problems, and finally possible applications are presented in this section.

6.3.1 Description of Machine

A rotating SRM is considered and its stator and rotor laminations are shown in Figure 6.12 for illustration. It has four stator and two rotor poles and they are salient (or projecting) poles. Diametrically opposite stator poles A_1 and A_2 with their windings form one phase, say phase A and poles B_1 and B_2 form phase B. The A phase consists of windings on its poles and they are connected in series or in parallel to have minimum input current in the windings or maximum input voltage across each winding, respectively. The windings are concentric as against distributed type such as in PM synchronous machines. Concentric windings are most compact, utilize the least amount of winding material, endow ease of winding and installation, and all of these factors contributing to low cost. The number of stator poles has to be different from that of rotor poles at least by two poles and not to be an integral multiple of stator poles also. For example, if the number of stator poles is 4, the rotor poles can be 2, 6, 10, 14, etc. The same approach applies to SRMs with a higher number of phases.

The rotor has no windings in SRM and in this regard, it is quite distinct from all other machines except PMSMs which have PMs instead of windings in the rotor. This is a significant feature of SRM endowing low cost, simplicity in construction, and suitability for high-speed operation as there is nothing to be secured to the rotor such as PMs or windings as in the case of other machines. Both the rotor and stator cores are of magnetic steel laminations very similar to that of other machines. There is minimum air gap when the stator and rotor poles are aligned and maximum air gap when they are unaligned. This has implications in torque generation and will be explained in the following section.

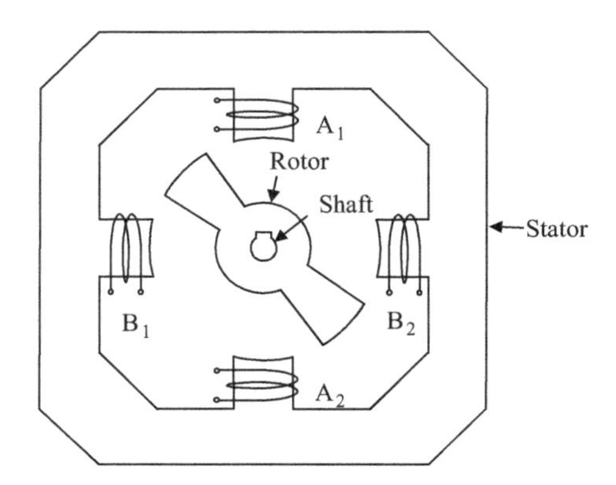

FIGURE 6.12 Switched reluctance machine with four stator and two rotor poles.

6.3.2 Principle of Operation

Let phase A be energized in the machine, as shown in Figure 6.12, by injecting current. A magnetic field is set up with flux that flows from stator pole 1, air gap, nearest top rotor pole, back iron on both sides of the rotor lamination, bottom rotor pole, bottom air gap, bottom stator pole 2, back iron of the stator lamination on both sides to the stator pole 1. This flux creates a force to move the rotor poles to alignment with the stator poles of phase A. Once the poles are close to alignment, the excitation to phase A is removed. As the flux decays to zero, there will not be any force generation and rotor movement slows. Now phase B can be energized which will attract the rotor poles from alignment with phase A to phase B alignment and in the course resulting in rotor movement and torque generation. A continuous motion is produced by alternately going from phase A to phase B energization. Figure 6.13 shows the flux plot of a four-phase SRM having eight-pole stator and six-pole rotor at unaligned, intermediate and fully aligned positions for a phase and the resulting operation in the SRM.

The fully unaligned position for phase A consists of the nearest rotor poles being away from them by 30° from its center and on either side of it. The flux in the stator poles is minimum at the unaligned position having a bigger air gap to cross and maximum at the aligned position where the air gap is minimal. The flux increases more or less linearly increasing in the intermediate intervals from unaligned to aligned positions. Accordingly, the flux linkages (i.e., the flux linking the number of turns in the phase) go from minimum to maximum in 30° in the motoring mode shown in Figure 6.13, and likewise, the opposite happens from maximum to minimum in the succeeding 30° for the machine, when it is in generator mode of operation. The stator flux, ϕ linking the stator winding with N turns has a flux linkage of $N\phi$ and inductance of that phase is obtained as, $L=N\phi/i$ where i is the excitation current of phase A at that instant. Note that the flux and hence the inductance is a function of rotor position and also a function of excitation current as it determines the operating point on the B-H characteristics of the lamination and depending on its magnitude it may push the operation from linear to a nonlinear region (i.e., with saturation). Consider the magnetic circuit is operating in the linear range of the B-H characteristics of the lamination material

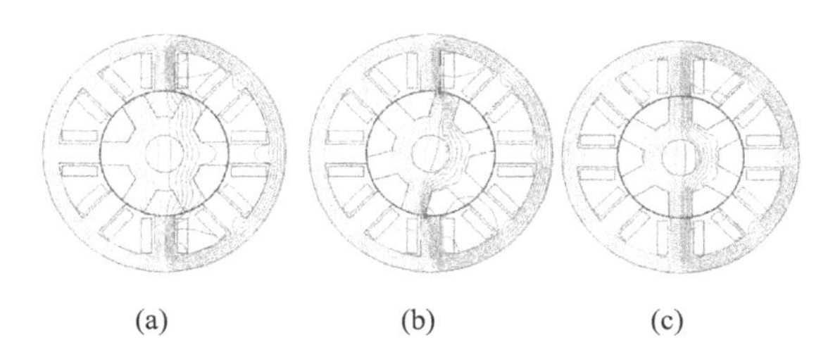

FIGURE 6.13 Flux plot in one phase of an 8/6 SRM at unaligned (a), intermediate (b) and fully aligned (c) positions.

for easier conceptualization and the phase inductance of phase A is as shown in Figure 6.14. The electrical equation for phase A is,

$$v_a = Ri_a + \frac{d}{dt}(Li_a) \tag{6.16}$$

And the input power equation is obtained as,

$$v_a i_a = Ri_a^2 + i_a \frac{d}{dt}(Li_a) = Ri_a^2 + i_a^2 \frac{d}{dt}(L) + Li_a \frac{d}{dt}(i_a) \tag{6.17}$$

It is known that the input power must consist of the resistive losses, rate of change of energy stored in the inductance and output power. The last two terms on the right-hand side are masking the last two variables and can be proven as in the following:

$$\frac{d}{dt}\left(\frac{Li_a^2}{2}\right) = Li_a \frac{d}{dt}(i_a) + \frac{1}{2} i_a^2 \frac{d}{dt}(L) \tag{6.18}$$

From equations (6.17) and (6.18), the input power is obtained as,

$$v_a i_a = Ri_a^2 + \frac{1}{2} i_a^2 \frac{d}{dt}(L) + \frac{d}{dt}\left(\frac{Li_a^2}{2}\right) \tag{6.19}$$

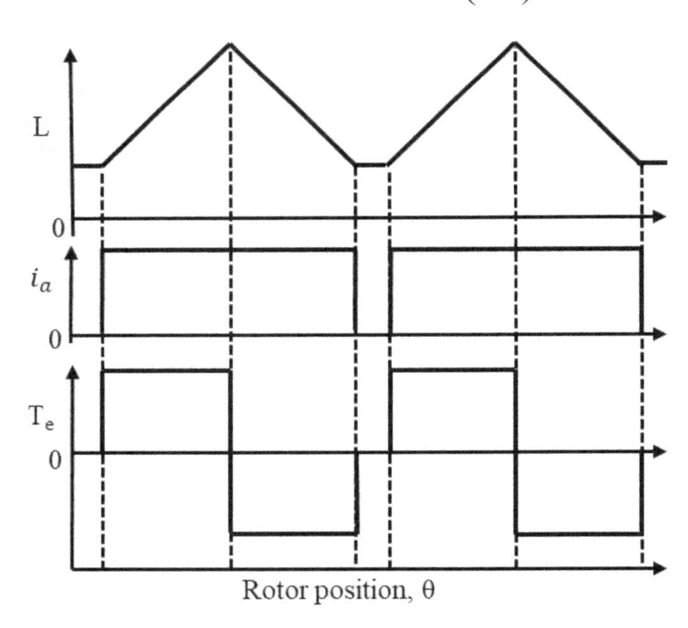

FIGURE 6.14 Principle of torque generation in SRM.

The last term is the power associated with the energy stored in inductance which does not translate into air gap power. Hence, the real air gap power is obtained by substituting for time in terms of rotor speed and hence position as,

$$P_a = \frac{1}{2} i_a^2 \frac{d}{dt}(L) = \frac{1}{2} \omega_m i_a^2 \frac{d}{d\theta}(L) \tag{6.20}$$

from which air gap torque is obtained dividing it by the mechanical rotor speed as,

$$T_e = \frac{1}{2} i_a^2 \frac{d}{d\theta}(L) \tag{6.21}$$

From this equation, it is seen that the air gap torque is positive irrespective of the phase current polarity, but the polarity of the air gap torque is only impacted by the rate of change of inductance with respect to position. The torque is plotted in Figure 6.14 based on this understanding. A current in the same phase is applied both during motoring and generating regions. The inductance is shown to have one peak, but in reality, it will be spread around for a few degrees of rotor position with a maximum value around alignment and note that around the unaligned position, there is a small constant inductance region also as shown in the figure. The figure shows the torque generation only for one phase with a constant current and brings out the motoring and regeneration torque. Note that in general, when the machine has to provide a torque such as in the motoring region, current is injected only during the positive sloped inductance region, and when one phase completes the torque generation, it goes to the next phase to do the same and so on in a multiphase machine. The same logic applies for regenerative torque generation except that the current is injected in the negative sloped inductance regions only. Some observations of interest are:

i. Motoring torque (positive torque) is obtained during the rising inductance region and the generating torque (negative torque) in the decreasing inductance region of the machine inductance vs. position characteristics.
ii. Torque is proportional to the square of current, similar to a dc series motor and that is ideal for large starting torque in applications such as electric vehicles.
iii. Control requires application of voltage/current as a function of rotor position, and therefore, it requires a power converter for its operation. Hence it becomes invariably a variable speed motor drive very similar to PMSMs.
iv. Current control is required for torque and speed control and that is achieved through a power converter that needs to process only dc to dc conversion in contrast to, say PMSMs, requiring inverters processing dc to ac.

v. Only one of the phases conducts most of the time, unlike most ac machines where all the phases conduct all the time. This feature affords a considerable flexibility in selecting the number of phases for the SRMs.

It is noted that linear operation was assumed, but in reality, the machine is highly nonlinear and operates in the nonlinear region for most of its functioning, unlike any other machine. This aspect compounds the complexity of design and control compared to any other machine. Its nonlinearity has strengths too in providing multifarious control techniques and being a dc fed machine opening up the field for a large number of novel converters yet to be matched by any other machine so far.

6.3.3 Converters

A power converter is absolutely essential for the operation of SRM because the current in a phase is not continuous and constant and depends on the rotor position. The requirements imposed by the SRM on the converter are:

i. Ability to supply and/or maintain controlled unidirectional current to the SRM in all its operating regions.
ii. Ability to transfer the energy during regenerative braking from the SRM to the dc source with the unidirectional current.

Given such simple requirements, many converter topologies are available. One well-known circuit shown in Figure 6.15 goes under the name of asymmetric converter. The working of this converter in various operational modes is captured in Table 6.1. Power transfer from dc source to machine occurs when T_1 and T_2 are on. When the current exceeds the set or reference current, then one of the transistors is turned off, say T_1, then the diode D_2 takes over and the phase current is freewheeled through T_2 and D_2 with the voltage across the phase being almost zero. This operation gives the smallest amount of current ripple and hence torque ripple. But there is another option which is to turn off both transistors to limit the current and that enables the diodes D_1 and D_2 to conduct and to carry the phase current to the dc source with the result that the power is transferred from the machine phase to the dc source. Note that the phase voltage is negative of the dc bus voltage for this operation. This is not attractive from the point of view of device turn-off losses in the transistors, current ripple magnitude and hence torque ripple magnitude, and the stress on the dc-link capacitor. It is resorted to generally only when the machine phase has to be turned off to drive the current to zero which it does in the shortest time.

This converter can handle transistor short circuits as the current will be limited by the machine inductance and that provides the control circuit to disable the power supply with the result that the machine is protected. Because of this singular fact, the SRM drive system is considered to be inherently short circuit proof. In contrast, the short circuit of phase transistors in the inverter leads to

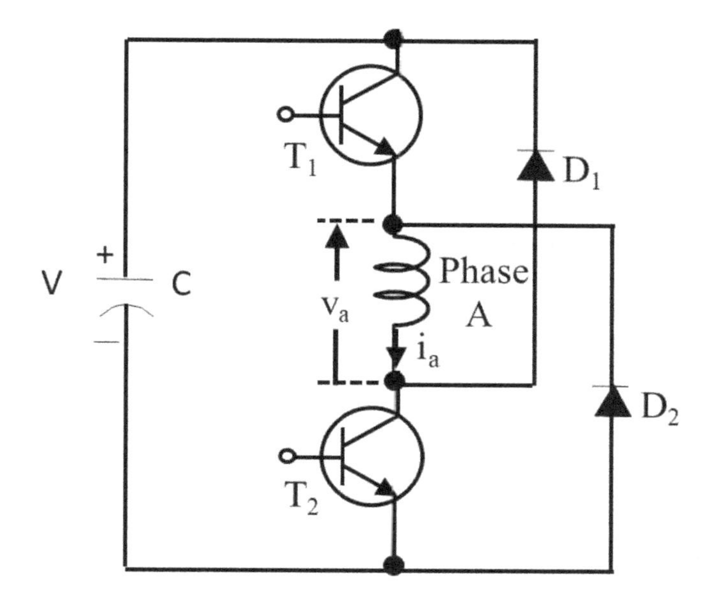

FIGURE 6.15 Asymmetric converter for SRM.

TABLE 6.1
Asymmetric Converter Operational Modes

T_1	T_2	D_1	D_1	i_a	v_a
1	1	0	0	>0	+V
1	0	1	0	>0	0
0	1	0	1	>0	0
0	0	1	1	>0	–V

catastrophe and requires much more protective measures than for SRM. Phase winding faults can be easily isolated using its phase switches. The fault current will not affect the rest of the phases as the mutual inductance between the phases is almost zero unlike ac machines including PMSMs. From these points, the SRM drive system is short circuit proof and provides the highest reliability in operation.

Many converter topologies have emerged and have been patented. For example, a single switch based two-phase SRM with four-quadrant operation has been demonstrated. This is a fundamental breakthrough in motor drives and that is not at all possible with any other motor drive system is well known.

Very many converter topologies for various specific advantages in operation and cost are in existence for SRMs, and they exceed the number of converters for ac machines. This is another distinct feature of the SRM drive systems compared to all other motor drives. Interested readers may find some of them in the references.

6.3.4 Control of SRM

The SRM control is almost as simple as that of the dc motor control except that it has to be coordinated with the position of the rotor poles with respect to the stator poles. The magnitude of current within the conduction period of a phase is regulated by closed-loop current control by comparing the commanded and actual phase currents. The output of the current loop controller error is sampled and compared to a carrier signal whose output is a PWM signal fed to the converter's gate drives. A typical dynamic SRM drive system performance is shown in Figure 6.16. Note that both are enforced by the closed-loop controls very much similar to other PMSM and PMBDC motor drive systems. The variables on the y-axis are all in per unit (except currents and their references) and listed from top to bottom: current reference, phase a current, phase a applied voltage, phase b current, phase b applied voltage, phase a's electromagnetic torque, phase b's electromagnetic torque, rotor speed and rotor position (where 2π radians equal 1 p.u.). All variables of interest are limited to one p.u. and accordingly the speed response is slow. The torque response contains significant ripples which are acceptable for many low-performance applications, but for high performance, they are usually mitigated significantly by control techniques available from patents and technical literature.

Two major issues have been identified with the SRM drive system and they are:

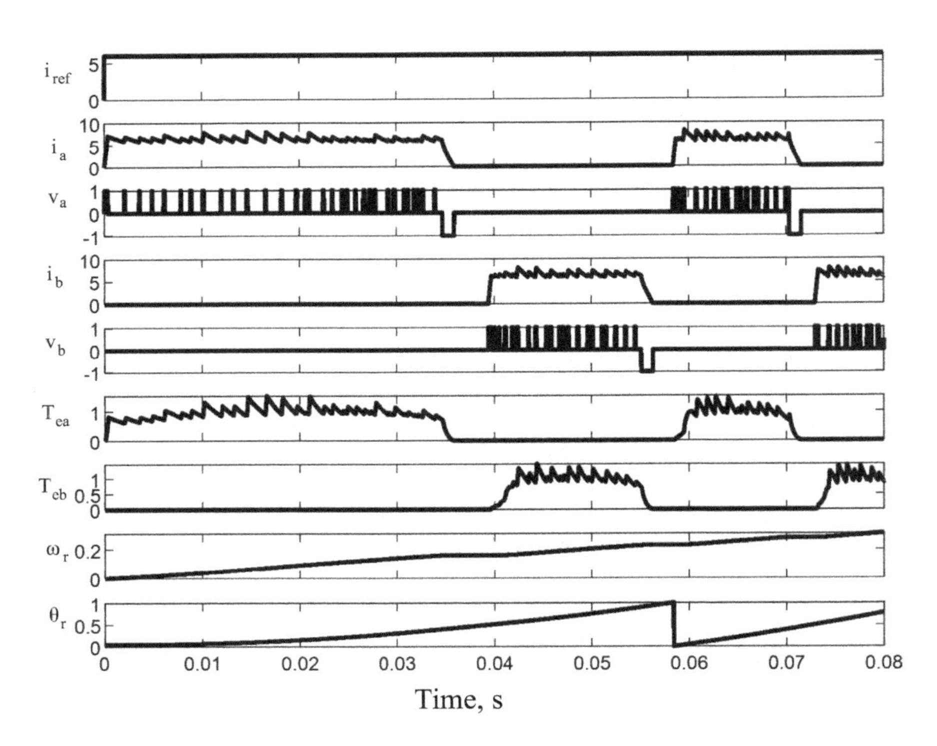

FIGURE 6.16 The simulated dynamic performance of an SRM drive.

i. Torque Ripple: A simple control technique such as mentioned above resulted in significant torque ripple which may not be acceptable in few high-performance applications. They mainly resulted from the inherent nonlinear feature of the machine. Control techniques to overcome this issue have been demonstrated both in theory and practice to the tune of controlling the torque ripple to less than 1% of the rated torque.
ii. Acoustic Noise: Regular control resulted in higher acoustic noise which was unacceptable. This is primarily due to vibration induced by the resonance of switching or operating at speeds near natural frequencies of the machines and many other phenomena such as forces created by the interaction of two salient magnetic parts of a structure such as in SRMs, load-induced noise, cooling related, etc. Ways to control and manage both in the motor design and its realization, and in addition by nonlinear control techniques have been demonstrated by some startups among others.

There is a perception that SRM efficiency is lower than that of PMSM. For the same size and volume of these two machines, equality in efficiency has been experimentally proven for low power applications by large companies with lead in the SRM drives research and development. The solution to this issue lies in the coordinated design of the machine and control strategy.

6.4 DISCUSSION QUESTIONS

1. DC motors are avoided in emerging applications. Why?
2. Line start fixed speed induction motors constitute the major share in the market place. Discuss the reasons.
3. PMSM and PMBDCM have come into play since the 1980s in many applications. List some of them and discuss why they were selected for these applications.
4. PMSM and PMBDCM drives are inherently variable speed motor drives. Explain.
5. PMSM and PMBDCM require converters to operate them. It adds cost to the system. Then why are they preferred in emerging applications?
6. PMBDCM in high volume applications is predominant compared to PMSM. Why?
7. Position information is required for control of PMSM, PMBDCM and SRM drives. That adds additional cost if a position sensor is used or additional control effort is required to acquire the information from other system variables. Will that add cost to the system?
8. Consider question 7 and induction motor drive. Induction motor does not require position information in regular variable speed applications. Will that give induction motor drives an advantage compared to other emerging motor drives?
9. Discuss the fault tolerance of PM motors and drives and compare them to induction motors and their drives.

10. Compare qualitatively SR motors with PM and induction motors in terms of the rotor construction and the resulting cost differentiation among themselves.
11. Discuss the superior reliability of SRM compared to all other machines.
12. Will there be a cost differential between the asymmetric converter for SRM and inverter for PMSM?
13. Discuss why SRM is yet to take off in the market place.
14. Will SRM take off in a big way in the future, say, with the PM becoming expensive and also avoided from a cleaner environment point of view?
15. Is there enough knowledge base in SRM drives and if so, are there students and engineers getting exposed to them?

6.5 EXERCISE PROBLEMS

1. Using reference Krishnan (2009) and using the MATLAB programs given there, assemble and run them to become familiar with dynamic and steady-state operations of PMBDC motor drive both in torque and motoring modes.
2. Vary the machine parameters in problem 1 and run them for various power levels.
3. Vary the speed controller gains and see the effect in problem 1.
4. Run problem 1 in regeneration mode specifically and study the performance including the effect on dc-link capacitor in the inverter.
5. Study the performance of the PMBDC motor drive in problem 1 when one phase winding is open.
6. Solve problems 1-5 but with a PM synchronous motor drive instead of the PMBDC motor drive.
7. Implement a unity power factor operation of the PMSM motor drive and study the performance.
8. Find a legitimate open-access software to simulate SRM and drive. If unable to do so, consider the SRM is linear and develop a software similar to that of PMBDC motor drive. Evaluate its performance using it.
9. Refer to the algorithms given in reference Krishnan (2001) and develop a torque drive through MATLAB simulation and performance evaluation.
10. Close an outer speed loop to the problem 9 and evaluate the SRM drive system performance for a step speed command from zero to rated speed.

REFERENCES

Permanent Magnet Synchronous and Brushless DC Motor Drives

1. R. Krishnan, *Permanent Magnet Synchronous and Brushless dc Motor Drives*, CRC Press, Boca Raton, FL, 2009 (Chinese edition also).
2. D.C. Hanselman, *Brushless Permanent Magnet Motor Design*, McGraw Hill, New York, 1994.

3. J.R. Hendershot and T.J.E. Miller, *Design of Brushless Permanent-Magnet Motors*, Marcel Dekker, New York, 1996.
4. J.F. Gieras, *Permanent Magnet Motor Technology: Design and Applications*, CRC Press, Boca Raton, FL, 2010.
5. C. Xia, *Permanent Magnet Brushless dc Motor Drives and Controls*, Wiley, Hoboken, NJ, 2012.
6. J.F. Gieras, R.-J. Wang and M.J. Kamper, *Axial Flux Permanent Magnet Brushless Machines*, Springer, Berlin, 2008.
7. R. Krishnan, *Electric Motor Drives*, Prentice Hall, Upper Saddle River, NJ, 2001 (Chinese and Greek editions also).

Switched Reluctance Motor Drives

1. R. Krishnan, *Switched Reluctance Motor Drives: Modeling, Simulation, Analysis, Design and Applications*, CRC Press, Boca Raton, FL, 2001.
2. T.J.E. Miller, *Switched Reluctance Motors and Their Control*, Magna Physics Publishing/Clarendon Press, Oxford, 1993.
3. B. Bilgin, J.W. Jiang and A. Emadi (Editors), *Switched Reluctance Motor Drives: Fundamentals to Applications*, CRC Press, Boca Raton, FL, 2018.

7 Entrepreneurship in Electric Machines and Drive Systems

7.1 INTRODUCTION

Startups in permanent magnet synchronous and brushless dc (PMBDC) motors, their drive systems, switched reluctance motors (SRMs) and their drive systems are presented with emphasis on products addressing select markets and technology developments in these fields. Apart from these electromagnetic machines and their drives' startups, a new entrant in the electrostatic machine space is also presented. The technology basis and innovation aspects are presented when individually presenting the startups. The applications cover appliance motor drives, motorbikes, electric vehicle (EV) motors and drives, electric aircraft motors and drive systems, air conditioner drive systems and some general-purpose variable-speed motor drive systems too. The market for these motors and their drive systems are presented briefly to understand the scope for the startups in these topics.

7.2 PRESENT AND FUTURE MARKET SIZE

The market of the motor drives in their three subsystem components is discussed below.

7.2.1 Market for PM Motors

The total global market as predicted by the market research firms for 2019 and for the next 5-year forecasting as given in their news releases correspond from $20B in 2019 to $40B+ in 2024. The markets for various countries may be obtained from their paid reports and also some information from various government reports that usually are free of cost. These figures are likely to be valid when and if the high-volume electric motor bikes and EVs emerge around the globe.

7.2.2 Market for SR Motors

It is yet to evolve given the fact that the variable-speed market is flooded with PM motor drives. A number of factors will influence the market scope for SRMs and some of them are as follows:

i. The growing market for PM motors may force price rises and fluctuations in the permanent magnet materials, and both of these affect the profit margins and uncertainty to the manufacturers. Note that almost all the supply of permanent magnets is mostly coming from China. These factors may encourage the manufacturers and market to be receptive to alternative motor drives.
ii. Temperature sensitivity of PMs and hence their impact on EVs in efficiency and peak torque performance is an additional factor to persuade to look for SRM drives path.
iii. The majority of the emerging EV market is in the two- and three-wheel vehicles around the globe, and they are mostly dominated by low-cost vehicles. That may force the hands of the manufacturers to seek alternative motor drives with SRM taking the top place for that role.

7.2.3 Market for Power Electronic Converters

This market will be as big as the motor market as the PM and SRM drives will require them for their operation. At a very low power level, the converter cost is much lower than the motor cost, whereas that is not the case when the motor rating is in integral hp. This affords a big advantage to SRM drives in comparison to PM drives as there are multiple circuit topologies requiring a lower number of devices than that of the three-phase inverters normally used in PM drives.

7.2.4 Controllers

This discussion equally applies to both the PM and SR motor drives under consideration. The controller for the motor drive system is realized with microprocessors or DSPs and the cost of that is very small compared to that of the motor and power electronic converter. The current, and speed control along with that of their enforcement via respective controllers and then on the power electronic converter through gate control of the power devices is achieved in a small printed circuit board. Sensorless speed control is highly resorted to thus doing away the need for a speed sensor. Position information for the excitation of the machine phases is also obtained with sensorless methods. Overall, the cost of the controller and its packaging in a small board constitutes only a tiny fraction of the total drive system cost. They are hardly a standalone product, so far, in the business as they are usually custom specific and hence comes with the motor and power electronic converter.

7.3 PM MOTOR DRIVE STARTUPS

Most of them are based on innovations in the motors and their control but not necessarily in the power converter. Almost all the startups are specific application-oriented developments except a few. They are described in this section.

7.3.1 Axiflux

Axiflux is a permanent magnet synchronous motor startup in Melbourne, Australia in 2011. Its founders are:

i. David Jahshan, Founder and CTO: Inventor of Axiflux motor drive technology and no additional information is available.
ii. Chris Mosley, Founding investor and CEO: Experienced in software engineering and an experienced entrepreneur in that field.

Technology: The invention is centered around an axial flux permanent magnet synchronous machine. The machine has many independent coils and they may as well be considered as phases as each of them is separately controlled by a single-phase inverter with its control using a microprocessor. The inverters and their controllers for the windings are all next to each other and inside the motor and come as a single package, and some details may be garnered from their patents in its reference section. The claims of this technology are high efficiency, compactness, low weight, higher torque and power density. The comparisons with any existing radial flux PM synchronous machines are not available at this time in public.

Products: Developed motor drives for electric cars through EV Engineering Consortium and was integrated into GM Australia's Holden Commodore vehicles and tested. The claims are 2,808 N.m torque, 30% more efficient, 35.83 Kg lighter, etc. Note that these claims are being made against another product which is not revealed.

Intellectual Property: Three granted US patents with David Jahshan as the sole inventor until this time.

Funding: $850K from the seed round and was part of the electric car consortium that the government funded $3.6M.

Observations:

i. Very few Australian startups in motors and drive systems and this is one of them.
ii. The axial flux version of the PM synchronous motor is well known, but the main innovation (among others) of this company is in the individual power converter for each coil and they may be large in number and, for example, in one of their prototypes it amounted to 23.
iii. While many coils and individual converters have some advantages when it comes to operation under fault conditions, they present significant issues in integration, too many parts and subsystems, reliability in manufacture with them in the motor, and most importantly cost.
iv. The cost is claimed to be lower than the standard EV motor drive, and the comparison has to be studied in detail by researchers and others as it is not available in public.

v. The background of the inventor does not seem to be in motors and drive systems as well as that of the other founder. Whether this factor has helped or hurt them is something that budding entrepreneurs in this area have to look at.
vi. The winding down of the EV car project and not panning into GM's car plans has deprived them of their motor product emerging in a market application. That seems to have left the company quiet for some time.
vii. Interesting to compare this startup with the recent startup in Belgium (Magnax) in axial flux PM synchronous machine mainly targeting EV applications.
viii. An important lesson to take away at this stage from this startup is that product novelty is not only the decisive factor in the company's growth, particularly that of a startup.
ix. Students pursuing thesis research may consider this innovation for analysis, design and testing for an application of their choice and may result in technical contribution and publication as well and will enlighten the general electrical engineering public in a more elaborate way. Note that there is no technical publication in journals on this topic.

7.3.2 Linear Labs, Inc.

Linear Labs, Inc. is founded in 2014 (and in some websites cite that to be in 2010) in Fort Worth, Texas, USA. Its founders are:

i. Fred E. Hunstable, CTO: Focused on technology and development from 2010. He has developed many of the patents and patent applications for the company.
ii. Brad Hunstable, CEO: He has a BS in engineering from US Military Academy and an MBA from Ohio State University. He founded Ustream and successfully exited to IBM. He is the son of Fred Hunstable, the CTO of this startup.

Technology: Consider a PM machine with stator windings that are surrounded by permanent magnets radially from both inside and outside, and to the sides of the stator coils along the axial direction. It is a transverse flux machine configuration with a circular core over which teeth are attached and in between them toroidal coils are placed (Ref: Hunstable's patent). Many variations are given, but the approach is the same. It is claimed that the technology can provide: higher torque due to larger magnet area in the machine, higher use of coil and hence higher efficiency, as they are all under the magnets whereas in other machines end coil portions do not see the magnetic field and hence they only add to the resistive losses lowering efficiency.

There is another part of the technology which is moving the rotors at the ends of the motor, that is the axial kind of version rotors. That allows for the change in the magnetic field strength experienced by the stator with the result that field weakening is introduced. This is achieved mechanically rather than through the control of the stator d axis current excitation in the stator. Note that d axis current will force the reduction of the torque producing component of the stator current and hence torque reduction in the flux weakening mode in PM synchronous machines. Therefore, in this technology, there is no current injection and hence full stator current is used for torque generation with an attendant increase in power output as well. Note that the concept of moving the rotor or stator with respect to the stator, and other methods (Ref: Lipo and others with patents) in the machine design have been in the literature for more than 20 years.

Product: There has been no product release to market at the time of this writing. They claim that their technology can be used in almost all applications including EVs.

Intellectual Property: 21 granted US patents and 29 pending US patent applications.

Funding: A total of $7M from various angel investors mainly and VCs.

Observations:

i. The startup's CEO is a successful founder in a different technology with an exit to one of the behemoths, IBM. The activity since his joining has been huge both in fundraising and publicizing their technology and its potentials.
ii. Funding has been at a level that is in par with similar startups in this domain. Whether that will be sufficient to manufacture the motors for sales is something that will be answered in the future.
iii. They have a significant number of patents and filed applications at this stage. Many of them hover around their basic technology with various interpretations.
iv. The claims of the technology and performance improvements in the motors as against the existing PM synchronous motor drives are significant. Observers will be excited to see those claims to stand vis-à-vis the PM synchronous motor drives. In order for that to happen, the company may at some stage cooperate with independent researchers and universities. Or hopefully, for the benefit of the public and potential users, some university researchers can do an in-depth study of this company's technology and release their results. This is unlikely to happen as western universities may not do studies until they are funded by the sponsors, i.e., government agencies and/or companies.
v. In their enthusiasm like all other entrepreneurs, no negatives and drawbacks are hinted at all and they are not released to the public. It is understandable, in general. But acknowledgment of some of the drawbacks, however minor, and their discussion will only lend credibility to the technology.

vi. In general, everything and anything can be manufactured in high volume. The question is cost. That fact equally applies to all hardware manufacturers and this company is not an exception in this regard, should it choose to go that route.
vii. This is an interesting startup for students of entrepreneurship in this field to follow its developments and learn.

7.3.3 Magnax

Magnax bvba (known as Magnax) is a PM synchronous motor startup in Kortrijk, West-Vlaanderen, Belgium and founded in 2015. Its founders are:

i. Peter Leijnen, Founder and CTO: He came on board first and then the co-founders joined nearly a year later and developed the motor technology. Details of him are scarce even though he has written many of the informative pieces on the technology.
ii. Daan Moreels, Co-founder and CCO: He has an M.E. in aeronautical engineering from KU Leuven, and he developed a business background through his entrepreneurial activities and work experience. He is an early investor in Magnax and works on the product development directed toward their principal market of EV applications.
iii. Kester Goh, Co-founder and CFO: He has a Master in Law from Ghent University and an MBA from Ehsal Management School. He is an experienced entrepreneur as a cofounder in many startups, fundraising, and a legal and accounting expert.

Technology: Axial flux high-power-density PM synchronous machine structures is the technology behind this startup. Novel cooling of the machine is incorporated which is not discussed in detail in their publications. A marketing type exposure of the technology and innovation is made available on their website as well as in promotional articles in IEEE as well. The packaging of the machine with twin permanent magnet rotors with the stator in between them providing flux flow in the axial direction is the base structure. Note that there is no back iron or yoke in this stator as the stator core is along the axial direction and the flux passes through them from the rotors. Whereas in radial flux machines, the flux from the top stator teeth goes through the air gap to the rotor PM, then to the rotor back iron, back to the air gap and then enters the stator back through their teeth in the bottom, then travels via stator back iron (yoke) to the top stator teeth. The twin-rotor single stator axial flux machine does not have back iron in the stator and because of that, it is termed as a yokeless motor. Therefore, there is a saving of stator back iron and hence the resulting weight reduction in these machines. It is important to note that it is a well-known fact for the last 40 years or so and does not constitute an innovation from the machines point of view. This implementation is claimed to have a high peak power density of 12.1 kW/Kg, and it is about four times the standard radial flux PM motor. There is no clear definition for peak

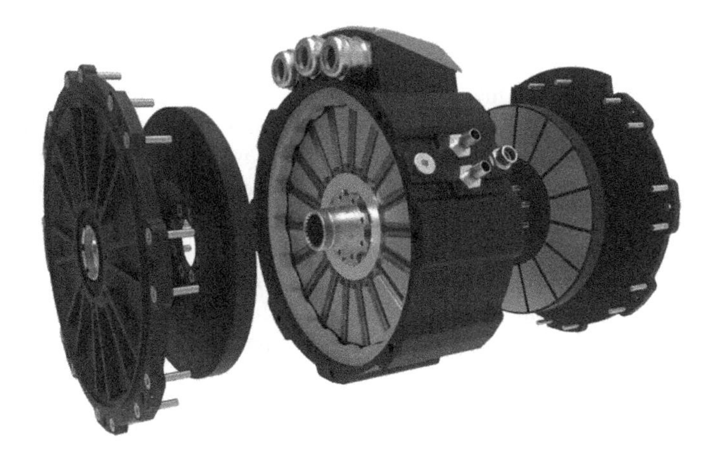

FIGURE 7.1 A sample motor. (Courtesy of Magnax.)

power density and whether it is for sustained or intermittent or once in a certain time period operation is not clear, and therefore, more information may be needed to interpret this data. The research and development of the motor is in collaboration with University Ghent, Belgium.

Products: The axial flux PM motor, drive and controls are the end product, or just motor alone could be a product for integration by drives manufacturers. A sample of its motor is shown in Figure 7.1. These products are being promoted for EV applications, and some comparisons with existing EV motor drives of a limited level have been published by Magnax.

Intellectual Property: Only a few patents have been identified in US from them, and there could be some with EU patent office. The startup has one granted patent and one application pending with US patent office with Peter Liejnen as inventor and co-inventor, respectively. As this is a recent company, it may take some more years before they can build up their IP portfolio.

Funding: $4.1M and has come from seed rounds and unknown investors.

Observations:

i. The company is focused on the yokeless PM synchronous motors for EV applications.
ii. Additional novelty and probably the most significant novelty may be in the packaging and cooling of the motor thus giving their product a very high power density.
iii. Attractive level of peak power density is an asset for marketing and sales for their products.
iv. The startup team has one core technical person, one product developer and a fund raiser with the result that basic requirements for a successful startup are amply and ideally satisfied.
v. Their ability to build and bring one product or a few samples into comparison testing with existing radial flux PM synchronous machines for

EV applications within 3 years is a good achievement. That too with limited funding and personnel.

vi. While the axial flux machines have distinct advantages, they also carry with them some issues such as the uneven pull of the rotors due to manufacturing deviations from specifications and unequal magnetic field due to uneven or manufacturing tolerances in the magnetization of the permanent magnets resulting in shorter bearing life.

vii. There is a cost premium in these machines due to a major change in the stator lamination packaging and resulting manufacturing of the stator drastically different from the well-established and highly reliable low-cost radial flux machine production. They have to be addressed for successful penetration, should they choose, into a high-volume market place where cost and price may dominate over some of the advantages. If the market targets are EV and electric aircraft sector, then they can effortlessly absorb the higher cost of the motor.

7.3.4 YASA

YASA (Yokeless And Segmented Armature) was founded in 2008 and incorporated in 2009 and located in Kidlington, Oxfordshire, UK. It emerged from the University of Oxford's graduate and has close ties to the university and may be considered as a spinoff of the university.

Founder: Dr. Tim Woolmer is also its CTO. He obtained his DPhil degree from Oxford University and also his general engineering degree from the same. His doctoral research was focused on axial flux stator and motor innovations forming the basis for this startup.

CEO: Dr. Chris Harris joined YASA in 2011 and he served as leader of high technology companies in UK, USA and Europe. He has a physics degree from the University of Oxford and a doctoral degree from Imperial College.

Technology: It is based on the axial field permanent magnet synchronous machine without the yoke or back iron for the stator. The removal of the yoke from the stator has resulted in: considerable weight reduction, elimination of core losses which would have generated in the back iron, and higher efficiency because of the lowering of losses and higher power density. The structure of the motor is two rotors with PMs on either side of the yokeless stator with windings. There have been many axial field PM synchronous machines and the uniqueness of the startup is in its innovations in realizing the yokeless stator in detail, about the tooth embedding on the base lamination, embedding the windings, and air cooling directly the windings. The result of the innovations can be seen from their patents. There does not seem to be significant new converter topology or control methods in their technology as the PM synchronous motor drives have been well developed over a period of 40 years. But the packing of the converter to obtain the lowest volume and weight with cooling and use of newer power devices have been resorted to by this startup like all other startups in this domain.

Products: The startup has three products in two categories:

i. Two YASA PM synchronous motors, named as YASA 750 with a continuous torque of 400 N.m. with continuous power of 70 kW but with an axial length of 9.8 cm and diameter of 36.8 cm with a total weight of 37 kg, and YASA P400 about 75% rating of the other motor.
ii. The third is the motor controller, YASA Si400 capable of handling an input voltage of 50–400 V dc, 210 A rms continuous, 96% efficiency, entire package in 5 liters and weighing 5.75 kg.

A custom made YASA's motor product has been employed in Ferrari's first hybrid sports car, SF90 Stradle claiming a power density of 14 kW/kg.

Intellectual Property: YASA has obtained five US patents (granted), ten UK patents and three European patents. Their US patents are duplications of some of their UK patents and likewise the European patents as well. The patents that are central to their IP are relating to stator plate molding, stator pole piece bonding, and cooling of the winding and pole shoes. Pending applications cover about semiconductor's cooling, arrangement, and module in the controller, and also manufacture of the motor.

Funding: $56.2 M over a period of 11 years and it came from Parkwalk Advisors, Inovia Capital, Universal Partners, Oxford Sciences Innovation, and others.

Observations:

i. This technology spinoff from the University of Oxford has steered quickly into competitive industrial/market space thanks to the leadership which it has plenty in its management.
ii. It is one of the highly funded motor drives startups, and that particularly is a striking factor for a company in UK. This has allowed them to develop and release focused products, though limited in number, but significant in performance capabilities.
iii. Its association with famous sports car maker Ferrari in embedding their custom made motor in its upcoming hybrid electric car is a great win for this startup at this early stage.
iv. Their product focus is where the power density and efficiency matters most such as in EVs and aerospace applications where they have already connected themselves recently. Even with limited product releases and their demonstrated willingness to develop a custom product for industries seems to go well for them.
v. This is one of the startups with data sheets of their products to answer most of the observers' queries and that shows their professional capabilities and certainty of their product performance.
vi. The claim of power densities as much as 15 kW/kg is one of the arenas where the basis for that numbers and/or the definition of power density is not clear. This is the way every startup in this space goes by.

vii. The motors and drives for futuristic electric aircraft are all aiming for power densities which may be within the reach of this startup as well as its competitive startups in the axial flux PM synchronous motor and drives category. But these applications may require output power capabilities of hundreds of kilowatts and even 1 MW and above. Whether they will drive their development to reach that level of power with their technology is to be seen.

viii. The intellectual property portfolio matches the size and age of the startup. With newer products and improvements required to make them will certainly result in additional property and the team has that demonstrated capability to achieve it. The progress of their technology path and IP path is worth studying by the entrepreneurship students to appreciate that systematic approach to go from one major development to another with limited startup's intellectual manpower and financial resources.

ix. The prices of their products are not available to the interested readers in open and how much they differ from radial flux PM synchronous motors and drives will determine their success in the conventional market applications which is more cost-sensitive than the emerging applications in the EV and aerospace. It may be that this startup may not go after conventional applications at all.

7.3.5 Guina ePropulsion

Guina Energy Research was founded in 2016 in Surfers Paradise, Queensland, Australia and renamed Guina ePropulsion since 2019.

Founder: Ante Guina, managing director and scientist. He has 25+ years of research, development and entrepreneurial experience in homopolar machines and superconducting machines. He also founded a successful company in 2009 which currently is known as Magnix and that has products for aerospace electric motor drives up to 560 kW rating.

Technology: The only mention about that is axial field permanent magnet (AFPM) motors. No further details are revealed. It is hinted that the motor has a multi-stator structure facilitating electromechanical gearing. Because of multi-stators, it comes with higher reliability.

Products: Yet to be released to the market, but a glimpse of what is to come is available on their website with the ratings of the air-cooled and liquid-cooled AFPM motors. The claim is that air-cooled AFPMs are expected to be 98% efficient for machines with ratings of 48, 138, 269 and 484 kW all at 6,000 rpm, while the same machines with liquid cooling to have more than the twice the power but with efficiencies in the 94%–95% range at max power. It is assumed that the stated efficiencies are only for the motors and not for the motor drive system. Power densities range from 4 to 6.2 kW/kg.

Intellectual Property: Claim to have filed applications from 2018 onward.

Funding: Unknown at this time.

Observations:

i. The power density of AFPMs is in the mid-range of what is being claimed by their competitors.
ii. The most striking about the products is their efficiency claim with air cooling. This when and if it is proved in the product line, will position them well in EV products to start with.
iii. There has been no mention of the power electronic controller for the motor in their product line up. Usually, the motor performance is optimized with the drive control and system efficiency is very much impacted by it.
iv. This is a company to watch because of their earlier startup's success and the current claims of their products.
v. It is a potential takeover candidate for large companies or investment houses should their technology claims can be verified by them.
vi. It is worth watching on many fronts and most importantly what their underlying technology is in this venture and how it differs from what is known and implemented by many researchers and many competitive companies in this domain.
vii. The only reference to this company is *http://www.guina.com.au/*.

7.3.6 QM Power

QM power is a hybrid PM synchronous motor startup in Kansas City, MI, USA and founded in 2006. Its founders are:

i. Charles J. Flynn, CTO: He is the inventor of the technology and named it as parallel path magnetic technology. He has been associated with motor engineering work for a long time much before this startup.
ii. P. J. Piper, former CEO: He remained as CEO for nearly 10 years from inception. He has been a prolific entrepreneur since leaving the company.

The current President and CEO is Hari Harikumar with operational experience from various corporations in India and USA and holds a Ph.D. in aerospace engineering from Indian Institute of Sciences, Bengaluru, India.

Technology: It is based on hybrid permanent magnet motor type. The hybrid motor versions usually take the permanent magnets on the stator itself along with the stator windings and leave the rotor with salient poles made of laminations with no windings on it. Consider a stator with permanent magnets inserted in its back iron as shown conceptually in Figure 7.2 and let the number of the stator PMs equal the stator windings wound on the sections between these PMs. On both sides of each PM on the stator segment (or section), it carries two salient poles on the stator. The rotor has only salient poles and their number is 1.5 times that of the stator segments and coils, in this illustration. The north-south (NS) of PM poles are in the same direction around the stator back iron. Unexcited, the flux goes around the back iron out skirting the rotor poles. Let phase A consist of coils, A_1 and A_2. Consider coil A_1 is excited so that it produces north and

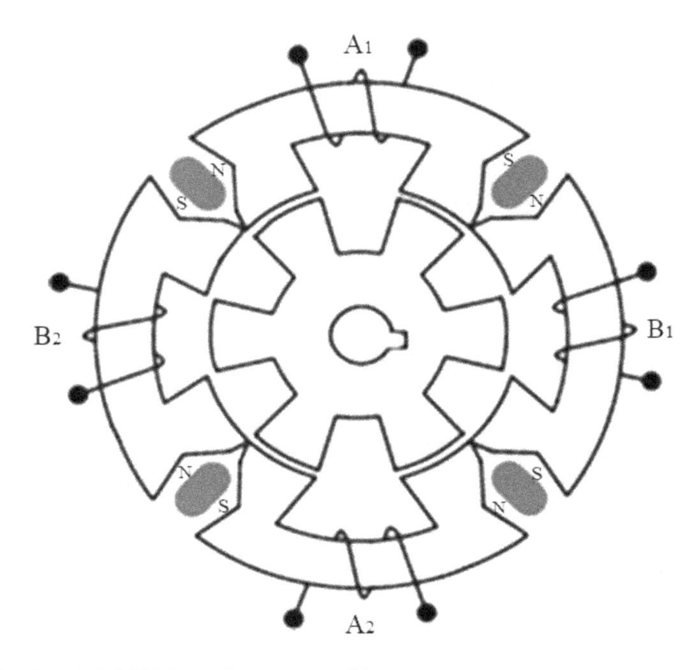

FIGURE 7.2 Hybrid PM synchronous machine.

south poles, north pole facing the north of the PM on its right and south pole facing the south of the PM on its left. That will force the magnet flux to bend toward the stator pole on its left and likewise its own flux due to coil excitation with the result the flux will attract the nearest rotor pole. A similar effect comes in to being with the left side of the coil A_1 and the south pole on its left contributing to attraction to the nearest rotor pole with its stator pole on the left of the stator coil A_1. Simultaneously coil A_2 is also excited with the result that the flux is directed likewise as in the case of A_1 and its corresponding stator and rotor poles. This enables the movement of the respective rotor poles toward the excited stator poles as shown in the figure. A similar operation on coils B_1 and B_2 results in the movement of the rotor poles toward their excited stator poles. Note that flux from excited coils and the adjoining magnets contribute to the generation of force and hence torque. Many different versions of the machine based on the hybrid concept can be realized (and the company may have many more types of realization in its chest), and this is one concept where the company has innovation, patent and its realization in practice. For one-directional running, a simple converter with a rectifier bridge along with a transistor switch can supply each winding.

Product: After about 10 years of founding, the company released a refrigerator fan with a rating in the 38-50 W range in the market place, as shown in Figure 7.3. The claims made for this motor are as follows:

FIGURE 7.3 Refrigerator fan motor. (Courtesy of QM Power.)

i. Efficiency is higher than electronically commutated motors by as much as 20% or so and over the conventional constant-speed motors as much as 80%.
ii. Energy savings will be significant compared to all other motor drives and to the compared systems.

They are based on a consulting firm's study of the motors in situ testing in a plant.

Intellectual Property: 22 granted US patents so far.

Funding: $15M+; both in grants from US agencies and venture capital companies (VCs) and most of the amount from the former than the latter category. This is spread over about 14 years. In some references, the total funding is listed to be $30M.

Observations:

i. The founders are not well known in this field of entrepreneurship prior to starting this venture and, given that, it is commendable that they have achieved a product that is into the market.
ii. Their product is limited to refrigerator fan motor at this time. It is interesting to observe how this technology can be used in other applications and what strategic route this company may take toward that end.

iii. The technology has merits, while at the same time has limitations also. Its main limitation is in the ease of manufacturing, and use of PMs from which many countries want to walk away from in high-volume products, and when scaled up whether the claims of efficiency improvement over PM synchronous motor drives and SRM drives will hold. The rotor construction is very identical to the SR motor's rotor and that part is easy to manufacture is noted. They may have a proprietary manufacturing method to make it low cost.

iv. The IP portfolio is fairly large for a company of this size. Perusing their granted patents, it can be observed that there seem to be repetitions of their basic patents with added claims filed and granted adding to the portfolio numbers. There could be other interpretations to it can be recognized.

v. The funding has been mostly from public/government agencies (whom they list as partners) and with very little from VCs as seen from their website and other sources. This is very characteristic of this field as VCs around the globe stay away from hardware related startups in general and the company's record on that is not different from this industrial segment. It is expected that this trend may change in the near future and some signs are evident with huge investments in EV startups for which the motor drives form a significant cost of the total system.

vi. Their motor drive system has not claimed any four-quadrant variable-speed operation in their portfolio and it will be worth watching how this may be tackled in their future product line when and if they have an opening or want to enter such a market.

vii. The company is ambitious and hinted at an initial public offering (IPO) sideways on the topic. This step, when and if taken, is worth watching and learning more about this industry sector and also this startup in the future.

7.4 THERMAL COOLING TECHNOLOGY BASED ELECTRIC MOTOR STARTUPS

Size reduction of electric machines is required in many existing applications such as aerospace and electric vehicle motor drives. It is achieved by cooling methods with natural, air, and liquid cooling. While all of them are in practice, there is an evolution of newer methods contributing to higher performance with claims that have attracted the motor manufacturing companies to use them in getting to high-power-density electric motors. These technologies are equally applicable to all kinds of electric machines. There are multiple advantages to getting the machine size reduced and they are lower material contents, compactness, efficiency enhancement that may come from lower resistive losses as the windings are maintained at a temperature with default values, and most of all safe operational guarantee even under stressful conditions.

Two startups, Zero E Technologies and LC Drives, Inc., in this category are considered for study here.

7.4.1 Zero E Technologies

This startup founded in 2009 and located in Wheat Ridge, CO, USA, is in PM synchronous electric motor. Its founder is Tom Hopkins. He has a BE from the University of Queensland in Civil Engineering, an M.Eng.Sc from the University of New South Wales in Structural Engineering, and an MBA from the University of Chicago. He has previous startup experiences, served as a board consultant and also taught capstone courses in business at Colorado University, Boulder.

Technology: High-efficiency permanent magnet synchronous machine for industrial applications from 15 hp and higher powers. The machine is a conventional radial flux PMSM, low speed in the range of 1,800 rpm but does not preclude higher speeds. Each stator tooth is a separate unit in the form of a T which can be linked to other teeth by a locking mechanism. The advantage of such teeth is the ease of winding with higher winding fill factor. The cooling mechanism is claimed to be original giving it a higher power density, and this may be the central aspect of their technology base.

Product: Their product which has been tested is a 15 hp PMSM. The claims of its performance are many and they are contrasted to an equivalent induction motor:

i. 26.3% of the volume of the induction motor.
ii. Weighs 54.43 kg as against 158.75 kg of an induction motor.
iii. Motor efficiency is 96.4% and when operated with inverter has an efficiency of 95.5%, much higher than the induction motor as much as by 4%.
iv. The motor is nearing the future standard of the International Electrotechnical Commission's standard IEC 7 (yet to appear and only up to IEC 4 have come to the market place) and hence far ahead of the induction motor which is doubtful to achieve even IEC 5 standard. Hence the future for their PMSM motor is in replacing the induction motor in some applications requiring the higher standard fulfillment.
v. Cost savings due to energy savings with high-efficiency motor.

Intellectual Property: Five granted US patents and five pending applications with the founder as the lead inventor in all of them. Intellectual property is clustered around the cooling method for permanent magnets in the rotor and cooling of the stator.

Funding: From 2009 to 2016, $6.4M has been raised and recent data is not available.

Observations:

i. The motor startup is on the premise that PMSMs can be designed to have higher efficiencies, lower volume and lower weight than what is normally in practice and in the market.

ii. Standard radial flux type PMSM is good enough rather than going after axial field configuration from the point of view of simplicity in manufacturing.
iii. They have hired motor design experts to assist them with design, manufacture and testing. By doing that, they have effectively complemented the in-house expertise. The founder and the lead inventor is not formally educated in motor design has not stopped the team to come up with working prototypes and intellectual properties.
iv. Their entire innovation in the motor seems to hang around the cooling of the rotor PM and stator windings. But there are very many companies having cooling technologies they use in day-to-day production of their motors. It is intriguing to follow and learn how they will not be able to produce motors of similar capacity with similar advantages.
v. There are a reasonable number of granted patents and pending patent applications for a company of 10 years.
vi. Funding has been very average for this company until 2016 and whether that has improved since then will determine the go to market strategy implementation for this company.
vii. It seems that the company's focus is only on motors and not on the inverter and motor control subsystems. Whether they will enter into a relationship with those subsystem manufacturers or develop them independently is unknown at this time.
viii. Interested manufacturers of inverter drive and control systems may collaborate or partner or come to own this company is a distinct possibility and that may be one of the future options for this company.
ix. They tried to raise funds in 2017 to the tune of $10M. There is no information as to whether they have achieved this target. That will determine their future plans to bring various products into the market.
x. They have a well-known member on the board of directors, a three-time governor of Colorado, Richard Lamm. He is also the co-director of public policy studies in the University of Denver, an attorney and a public accountant. His visibility will greatly assist this company in its efforts.
xi. The company is at the time of this writing is quiet in the public domain but is active in filing additional patent applications.

7.4.2 LC Drives, Inc.

LC Drives, Inc. (hereafter referred to as LC Drives) was founded in 2012 in Potsdam, NY, USA and initially located in Clarkson University's Peyton Hall business incubator. It may be considered as a Clarkson University spinoff. Its founders are:

i. Dr. Russel Marvin, CEO: He obtained his B.S. and Ph.D. from Clarkson University in mechanical engineering. He has industrial and entrepreneurial experiences, founded and served as CEO of Optiwind for 5 years and was the CTO of TRC, a fan and motor technology company for 4 years.

ii. Dr. David Leach, Director of Systems Engineering: He obtained his B.S. in electrical engineering from the University of Utah, and Ph.D. in applied physics from Yale University. He has varied industrial experience. He worked for Optiwind startup as director of systems engineering for 5 years and director of controls and electronics for MTI MicroFuel Cells for about 7 years and as a product development engineer for more than 7 years.

Technology: The technology is based on the concept of windings being cooled with liquid glycol circulating in loops in the stator slots, as shown in Figure 7.4 and in comparison with the conventional cooling prevalent in the industry using a cooling jacket over the surface of the stator core. The advantage of LC Drives' method of cooling is that the coolant is exactly placed in and around the stator windings where the most heat is generated. The ways of realizing this concept constitute their basic intellectual property leading to specific products using permanent magnet synchronous machines for the most. The concept of cooling is equally applicable to other kinds of electrical machines too. Additional technology comes in the form of forming teeth and supporting it to the stator back iron for rigidity and strength including forming the coolant manifold in the slot between the winding and stator back iron. The improvements mainly hover around the stator, and their construction in regard to teeth, windings and cooling mechanisms using ethylene glycol and also the construction of a permanent magnet rotor and also in reliable actuator and its construction. Many of their improvements and innovations are applicable to all sizes but more ideally positioned for

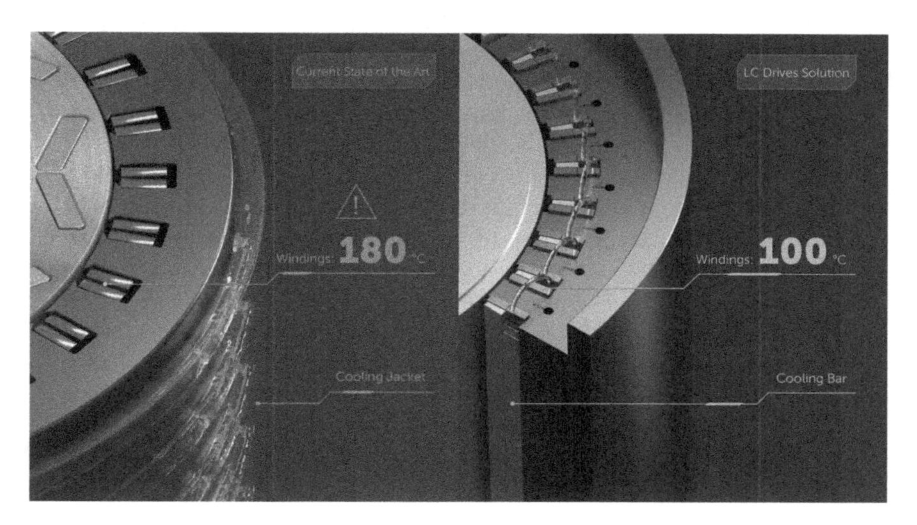

FIGURE 7.4 LC drives motor cooling technology. (Courtesy of LC Drives.)

large electric motors. All of their innovations claimed to lead to lighter machines with high efficiency.

Product: As of date, there has not been a product release, but some product datasheets are made available to the public. Their focus is in high-power large PM synchronous motors for targeted applications in oil and gas drilling at 2,000-3,000 hp level, and marine and vehicle propulsion at 2,280 hp level.

Intellectual Property: Nine granted US patents such as in liquid cooling of PM machines with glycol, stator cooling, high-reliability actuator, winding construction for high efficiency, efficient permanent magnet machine with ceramic magnet sections, and half-bridge switching circuit in power electronics.

Funding: Claimed to be around $20M in some sources. But open sources indicate that major funding came from Koch Engineered Solutions for $15M (in November 2019), and a state grant of $0.55M, a research grant of $75K and others worth about $15.8M.

Observations:

i. Both the founders have industrial backgrounds with doctoral degrees is quite fitting to this startup's technologies and to its benefit.
ii. The chosen cooling with ethylene glycol and its realization with their own method seems to be an improvement even though placing some other means very near to the center of the heat is something that other startups may also be pursuing.
iii. The startup is venturing into domains that very large power companies such as Siemens, ABB, GE and others have a long track record of supplying the targeted industries and customers. Their products have satisfied the customers for a very long time. Note that these companies will not be static in their own inventions and improvements and given that how their customers will be weaned away by the startup will be worth learning.
iv. If these giant companies do not have adequate IP in the cooling aspect (which is doubtful looking from the outside), then that may be an opportunity for this startup to be acquired if and when they are near to a credible competitive product threatening their market space.
v. The real building of products is beginning from this year (2020) with the funding from Koch Engineered Solutions. At least 2 years may be required before the building, testing and certification process is over before they can be with the customers. They have a reasonable amount of funding to sustain themselves for this period.
vi. The company survived with very limited funds for nearly 8 years and its fortune turned with the recent large investment. It is a valuable lesson for startups in that success in funding may take more than normal time for some and sustaining the enthusiasm and work going and not giving up hope. Such strength has to come from the founders and friends of the startup and the surrounding environment that they operate on a day-to-day basis. It seems from that point of view, university incubation centers and/or being near a university is a big help to startups.

vii. They have not made sky-high claims and therefore, they are unlikely to disappoint anyone when products are tested and introduced into the market place.
viii. It is quite likely that they do not have IP in the power electronic converters for the PM synchronous machines at that high power level that they are positioning their electric motors and therefore, they or their customers may depend on another supplier for the electronics and control. That may be a blessing in disguise at this stage because with the limited funding and management bandwidth they can focus only on machines and make that a success.
ix. The intellectual property in the number of patents is very reasonable for a startup of this size and many more are likely to be pursued after this major funding intake.
x. Overall the startup has the right ingredients for success as seen from the list of things viewed. But success will be attained only when the next steps are in the right direction of product development and marketing from this time on.
xi. Students of motor drive entrepreneurship will benefit by following this startup for the next 3 years to see the strategy and marketing techniques rolled down to capture market space.

7.5 PRINTED CIRCUIT STATOR-BASED MOTOR STARTUPS

There has been considerable interest in simplifying stator construction in electric machines to reduce the weight and volume of the machine, save materials, and make them attractive to some applications such as electric bicycles, HVAC compressors, and medical applications. Earlier, the stator iron was removed and substituted with epoxy to hold the windings, but thermal capability became an issue and it could not transcend to higher power greater than a few hundred watts. Further weight reduction and compactness came with a printed circuit board approach and it was also successful for very small power ratings. But to go to kilowatts levels required innovations in this approach and a few startups around the globe have come to address the earlier challenges within the last decade. Two of the startups in this domain are considered in this section.

There is also technology development in 3D printing of entire machines and its feasibility in 2018 by two researchers from Chemnitz University of Technology, Germany. The researchers plan to launch a startup on this technology. This means both the PCB stator and a PCB rotor also possible with this technology thus possibly advancing the PCB technologies further in the future.

7.5.1 ECM Software and Technology

ECM Software & Technology (hereafter referred to as ECM) was founded in 2014 in Concord, MA, USA. (It is also known as ECM's PCB Stator Technology and E-Circuit Motors in some publications.)

Its founder is Dr. Steven Shaw. He obtained his bachelor, master and doctoral degrees from MIT in electrical engineering. He developed printing software and the technology related to the stator. He is a full professor in electrical engineering at Montana State University. He serves as Chief Scientist of ECM. He has developed all the patents for ECM.

Technology: The company has developed comprehensive technology for the composite configuration, optimization, layout of axial flux PM synchronous machine stator and the design of the entire machine including rotors and other structures for final production, on a software which is cloud based with all the features required for professional execution. The name of the software is PrintStator. This software ensures high efficiency and high power density. A sample design of their own 900 W machine stator 1 from Gerber file is shown in Figure 7.5. They also facilitate the integration of the inverter and motor controller in their comprehensive design effort to lead to the optimal design of the motor drive system, just not the motor only. Their technology as seen (from the materials provided by the company to the author and) from patents shows that they have considered the structure of the axial flux PM machine, thermal control of the PCB, loss control in the printed circuit board and magnetic assembly which are all covered by their patents.

Figure 7.6 shows the parts involved in their axial PM synchronous motor with PCB stator in the center, two rotors with magnets on both sides of the stator, and

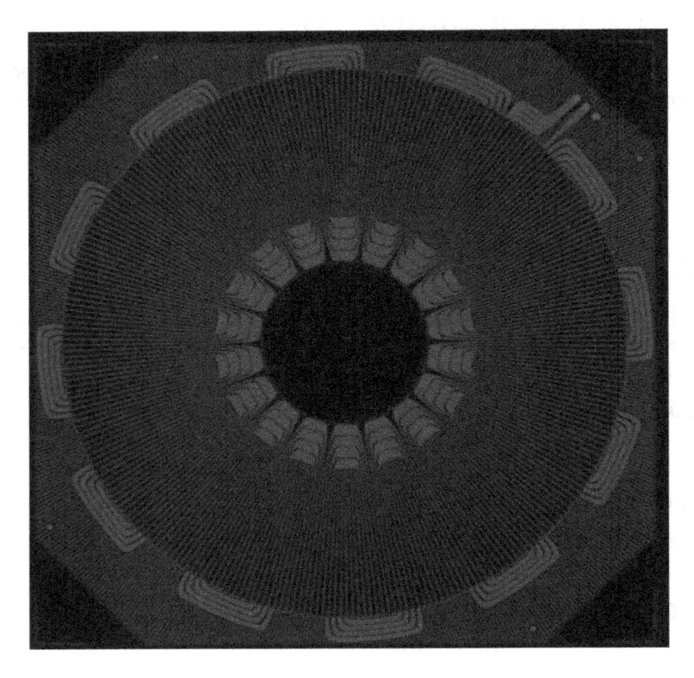

FIGURE 7.5 Stator 1 of ECM's own machine design. (Courtesy of ECM Software & Technology.)

FIGURE 7.6 Axial flux PCB motor. (Courtesy of ECM Software & Technology.)

the end bells are the outermost. The inverter and motor controller could be separate or can be part of the end bells.

Product: The startup has made impressive PM synchronous motor products and comparisons to equivalent induction motors. A 2.2 kW machine is considered and the winding copper weight is 5 kg for the induction motor whereas ECM's motor has 0.6 kg, total induction motor weight is 45 kg whereas ECM motor weight is 15 kg. The power density of a PMSM is much higher than that of the induction motor is well known and that needs to be taken into account and even with that the saving in material and reduction in overall motor weight is considerable. They have product prototypes for autonomous underwater vehicles in 100 and 900 W power levels, and many other applications described in their website.

Intellectual Property: Eight granted US patents and two pending applications so far. The founder is the sole inventor of all the patents.

Funding: Raised in four rounds, about $13.84M with the latest being in February 2019.

Observations:

i. ECM is a startup not from a university campus even though its founder is an academic. The intellectual property has been solely from his own work as seen from the inventor list in the patents.
ii. The technology has been centered around the motor but with a twist of a particular composite of the stator in a printed circuit board and everything else is fitted to suit that. Therefore, it requires the knowledge of motor, and this founder coming from a background of computer science and electrical engineering made use of the knowledge in both fields to realize the machine with optimization and design using modern software tools. That is part of the startup's technology and made available to customers to fit customer applications into the design.
iii. Because of the optimization and layout software, the startup may reach a wider audience of motor drives customers. The attention of the large motor and variable-speed drive companies will come due to end-user demands which can only be satisfied with ECM's motors.

iv. The stator is an axial flux-based system with two rotors on each side of the stator. The savings come from the copper and weight reductions due to the printed circuit board stator. Though the copper is reduced very much, the heat generated in the stator needs to be handled by thermal design. Without forced cooling, higher power densities are not reachable. Natural cooling methods have been introduced for thermal control in this design.
v. Applications that can afford forced cooling may enhance the performance of the startup's motors and drive systems. That may also open up an opportunity for IP generation too. Whether that route will be taken to test the performance limits of the technology is worth watching.
vi. There is a high-volume market in HVAC compressor motors and drive systems. As they have also become highly variable-speed based and cost-sensitive, the PCB motors will receive quite a reception provided there is a cost-saving to the manufacturer and not necessarily to the customers. Such market penetration will determine the market scope of these motors in traditional markets.
vii. There is another market that is emerging in electric two-wheelers such as scooters and motorbikes and the lower power ones can be an opportunity for this technology. But that market is mostly in India and China, and not in the USA. It will be interesting to watch how the company comes up to meet these challenges.
viii. To meet these challenges, good strategy and tactics are required. From that point of view, the most important component in the company to meet these challenges in this regard, next to the technology, is the management team. This startup has plenty of that in its pocket with a management team and board that has proven experience and demonstrated success.
ix. Interested entrepreneurs should follow this company, its strategies, products and whether it will produce and enter the market place for high-volume applications on its own or partner with others. If the latter route happens, then a successful exit is almost guaranteed for this company.

7.5.2 Infinitum Electric

Infinitum Electric is a printed circuit based motor startup founded in 2016 in Round Rock, Texas, USA. Its founder is Bernhard L. Schuler and listed as founder in some and co-founder in some other documents. The company originated from a 2013 startup co-founded by Schuler known as Fanergies, Inc. and names associated with both the entities are Schuler and John Pitts. A few lines about Schuler in the following:

Bernhard L. Schuler is the founder and CEO of Infinitum Electric. The previous company, Fanergies, Inc., that transitioned to Infinitum Electric was involved with recovering energy from the exhaust condenser of the HVAC system. He studied Mathematics at Oregon State University and business in the University of Phoenix. He gained experience in ATS Automation as an electro-mechanical technician and subsequently in engineering and manufacturing with other firms in Japan and Taiwan and also in USA.

Technology: Axial flux PM synchronous and/or brushless dc motor stators are made from the PCBs. It follows the most logical form of twin PM rotors on both sides of the PCB stator and the power electronic inverter and controller is housed near the right-hand-side end bell. The end bell also carries the heat sink necessary for the inverter. Such a configuration is shown in Figure 7.7. The innovation comes in the form of realizing the winding patterns on a PCB and assembly of one motor and then adding many of such unit motors in a single package thus compounding the output power of the system. It is to take them to a higher power such as 30 hp and even higher depending on the possible market opportunities that may open up. This stands out distinctly as their technology.

Product: Products have not been released, but a few prototypes have been discussed in their website and press releases. Their main focus is on the high-volume market in HVAC compressor applications. The product release is expected to be in 2020.

Intellectual Property: Five granted US patents and five pending US patent applications. The patents basically fall into two categories of realizing axial field motor and modular construction of the axial field motor and listed in the references with three patents. Each of the patents is an addition of improvement or additional claims from the original group of the patents. Schuler, the CEO is the first author in all the patents.

Funding: $15.2 M in three rounds, the last of which netted 12.5M from three investors one of which is the well-known Chevron Technology Ventures.

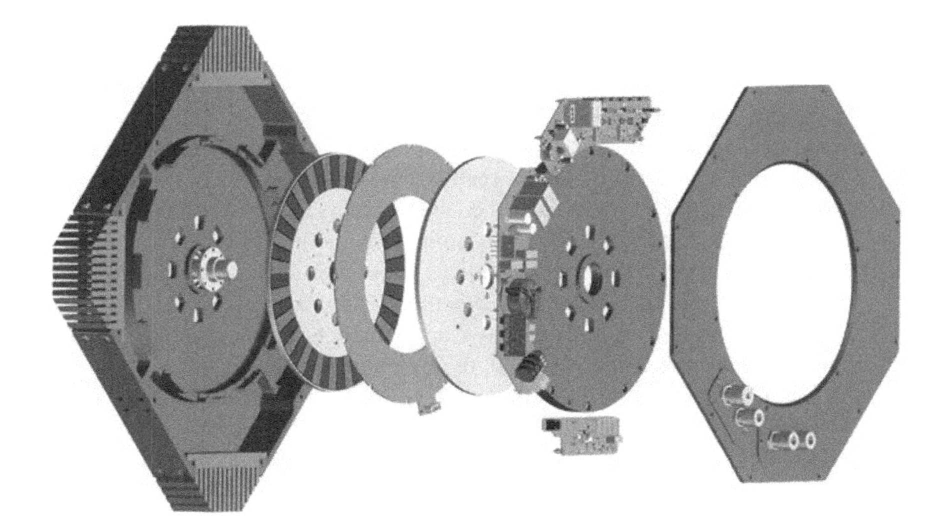

FIGURE 7.7 Configuration of the PCB based motor drive system. (Courtesy of Infinitum Electric.)

Observations:

i. This startup is born around the time when other PCB stator based technology startups started showing their presence. The motor technology has claims to higher efficiency, smaller size and hence higher power density, savings in copper and steel laminations and hence lower cost, and higher power levels up to single-digit hp compared to standard machines.
ii. Much further increase in power rating and its feasibility is being claimed for future applications by this startup. If and when that happens, there may be a stiff competition between this technology's products and conventional motor products depending on the cost that is most important, for example, in HVAC application.
iii. The CEO has no academic background in the motor drives but seems to have acquired the knowledge and innovation skills benefitting from his practical work experience in various employments.
iv. The management has very experienced people. For example, the startup has acquired a very knowledgeable person, Mr. Paulo Guedes-Pinto, as VP Technology, with work experience in motor drive companies, and experienced in IP generation. Further, the team has Bhavnesh Patel as VP of business development, an entrepreneur and an experienced business executive from Rockwell Automation that is a big company.
v. The number of patents and grants is compatible with the size of the company. Further development of their portfolio is very likely from their personnel recruitment in the last 1 year and funding to support such activity.
vi. The fundraising has been good so far for this startup and that too within its birth of 4 years. The funding has also been made possible, among other factors due to its location also, i.e., the state of Texas with its concentration of large industries, VCs and high net worth individuals.
vii. The company has a strong competitor, ECM Software & Technology with its strong engineering basis, product demonstration, management team and founder and also the board of directors in taking the companies to heights. Needless to say, the budding entrepreneurs in the motor drives domain will be watching the developments in this space in regard to both these companies as well as around the globe.
viii. The entrepreneurship scene in China is not clear to outside readers in motors and motor drives. That is something that startups have to come to grips with and find ways to possibly work with them on a non-competitive basis if that is possible to increase their own impact.
ix. Does this require low-cost manufacturing facilities such as in China for them to bring their products into the market place? If it is so, they need to raise higher capital but not necessarily to the same extent as to the manufacture in USA. What forms that will take to manufacture abroad and its requirements are to be strategized and planned ahead. Their strategies may not be available to the public, but by following their news and connecting the dots, one may get an outline of them.

x. Will the startup penetrate the market space by itself? This question has to be raised as it enters a space that is dominated by large companies around the globe and with huge manufacturing facilities and distribution and market clout. They can be a partner or in the process of time if their technology can erode their market even by a little percentage, then it may quite likely be acquired by one of the large players.

7.6 ELECTRIC AIRCRAFT BASED MOTOR DRIVE STARTUP

There has been interest in electric aircraft from small to big passenger planes in recent times. A number of factors such as low weight and volume of the electric motors and drives, i.e., with the highest power density that is yet to be achieved in the range of 5-25 kW/kg, battery capabilities and weight, charging, and redundancy in the subsystems for failure-proof operation in the flight posed the challenges. The first effort is directed toward small passenger planes capable of carrying six to eight personnel. Very few startups in this domain have emerged and one of them in the motor drives segment with an electric motor drive capable of 5 kW/kg power density, known as Magnix is presented in this section. There are efforts by major motor drives manufacturers such as Siemens are in place and newer entrants may be Raytheon and many other manufacturers.

7.6.1 Magnix

Magnix from 2014 was under a different name, Guina Research & Development Pty. Ltd, founded in 2009 in Queensland, Australia. Now it is headquartered in Redmond, Washington State, USA and also located in Gold Coast, Queensland, Australia (where it was born).

Founder: Ante Guina, the scientist founded in the mid-1990s Guina Research & Development and served as its Managing Director. Heron Energy of Singapore became a majority partner in 2009. His invention on electromagnetic turbine motor in 2010 with high power density positioned it for aerospace applications and also the later inventions relating to the same. He left the startup in 2014 which was then named Magnix and started another motor drives company in 2016 which has been described in an earlier section.

Technology: There was R&D for a very long time on superconducting materials and machines, but it did not reach the market place, even though various applications have been targeted from wind generators, ship propulsion to electric vehicle motors in recent time. They have been developed up to 10 MW level. The superconducting materials require to be operated near 4 K or high-temperature superconductors in and around 100-170 K and at that temperatures, the windings made of them do not have any resistance at all leading to high-efficiency operation of the motors with zero heat generated in them. That also increases the power density of the machines leading to lower volume compared to all other machines in use. The biggest concern is the cooling to keep the wires at that temperature and the infrastructure required. Magnix invented ways to make the superconducting synchronous motors (SSMs) with the field coils made of superconducting materials and the armature windings also with the same materials or also with conventional copper with cooling to keep the stator at the proper temperature. The field windings can be stationary and the armature windings can be on the rotor but powered via slip rings or vice versa. They also could be built without the superconducting coils but with regular copper wires and permanent magnets for the field. The founder and his team have made inventions that are available mainly through the patents, but they do not pin which inventions are used in the product positioned for aircraft electric motor. Interested readers may further investigate the references containing their patents and patent applications. Their claim on power density for their motors is about 5 kW/kg and likely to go up to 15 kW/kg in the future. Note that the research community forecasts that from 2030 to 2035, the power densities may be much higher than 15 kW/kg.

Product: Two kinds of products have been made, and they are the motors and their drive system controllers and they are as follows:

i. Motors Magni250 and Magni500 are rated for continuous power output of 280 and 560 kW, respectively, and with power densities of 3.89 and 4.15 kW/kg. They all have base speeds of 1,900 rpm and a max speed of 3,000 rpm, and operating on 450 to 750 V dc link and with efficiencies of >93%.
ii. The drive system named MagniDrive has an output power of 170 kW, efficiency of 98.9%, and operates from a 400 to 800 V dc source.

Though with one drive, the product line is complete for both the motors. The way it is managed is that two and four controllers are used for Magni250 and Magni500 motors, respectively. This approach further endows reliability in case if one of the drive systems fails, the other drive system will function with part power. Both Magni250 motor and magniDrive are shown in Figure 7.8.

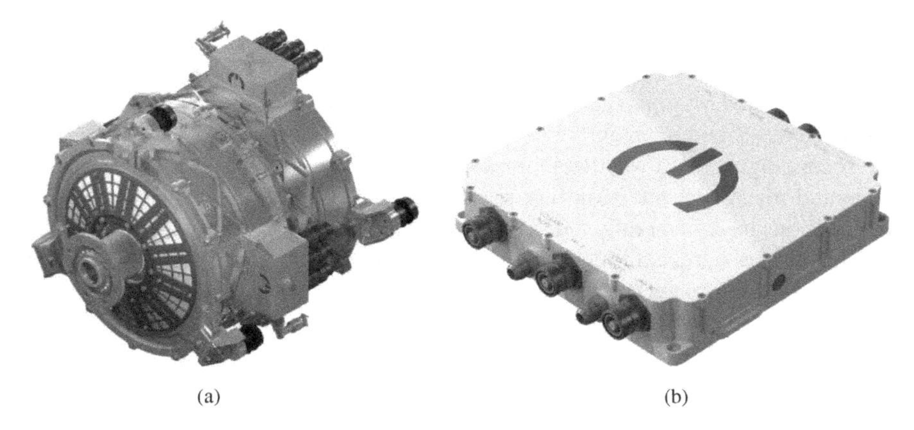

(a) (b)

FIGURE 7.8 Magnix250 motor of 280 kW (a) and Magnidrive system of 170 kW (b) output. (Courtesy of Magnix.)

Their 560 kW motor and drive system products have been used in the successful first all-electric commercial air flight by Harbour Air in December 2019 in a six-passenger DHC-2 de Havilland Beaver aircraft.

Intellectual Property: Three granted US patents and two pending applications in the USA. The Australian and/or other filings and patents are not tracked here.

Funding: Not listed in the public domain. The company is backed by Clermont Group, owned by billionaire entrepreneur Richard F. Chandler. It is stated that they do not have to worry about funding their expansion, with good enough reason.

Observations:

i. The IP came from a strong R&D team of the original founding startup in Australia. The later focus to electric aircraft application has come from a spin-off technical team from the original team.
ii. Their focus and success are much to be owed to the strong funding base provided by the owners of this startup and note that very few startups can claim to be in that position in regard to funding.
iii. The company is very non-revealing about the basic technology behind their products and keep it as a trade secret so far.
iv. Their success in being the first on the test flight with Harbour Air puts them at an advantage that only can be met by very large companies such as Siemens, GE and others of similar caliber and rank. Their competitors may not be far off in their R&D in this direction has to be noted.
v. Their IP portfolio has been confined to three US-granted patents and two pending applications and that may be too small for a company of this magnitude. But the three granted US patents cover a lot of ground is noted.
vi. Their investor has invested into Eviation, an electric aircraft startup company in Israel, in a big way (to the tune of about $80M) thereby assuring that Magnix's product will find, hopefully, a niche market with them. They already are collaborating to make the EV aircraft happen for shorter flights with smaller passenger capability.
vii. A niche in defense applications will be a big opportunity for mega growth that may surpass even their civilian aircraft market space in USA.
viii. Nothing is known about the cost of its products. It is known that the aircraft market can afford much higher prices than commercial applications. That being the case, it is worthwhile to watch and learn if they may have any desire to target the large market in commercial applications ranging from large HVAC to electric truck drive systems, all in the same power range of their current products.
ix. Much more interesting to see how the competition evolves from major players and whether this startup also exits to one of them in the future.

7.7 SWITCHED RELUCTANCE MOTOR DRIVE STARTUPS

There are very few startups in this domain. A number of factors can be attributed and they are: (i) knowledge base for them exist in very few research universities and very few but big motor drives companies; (ii) difficult to generate knowledge base by small startups as it is not part of the wider curricula in academic institutions and therefore very little chance of students in the form of potential employees of the startups coming out with knowledge of SRMs; (iii) issues such as acoustic noise, vibration, and torque ripple have not been helpful and easy to solve; (iv) motor is highly nonlinear and hence its analysis, design and control are not as easy and time-efficient as other motor drive systems; and (v) fluctuations in permanent magnet costs have stabilized in the recent period and there is no threat of their availability at this time, at least to western countries from China, the major producing country and it is bound to change when other markets open such as India that could drive PM prices far higher. Given these factors, it is no wonder that there are few startups in this sector. Four of them are chosen for presentation in this section, all of them with a strong university base.

7.7.1 Panaphase Technologies

Panaphase Technologies LLC (hereafter referred to as PT in this section) is mainly an SRM drives startup founded in Blacksburg, VA, USA in 2002. It is a spinoff from Virginia Tech intellectual properties developed by the founder and his researchers.

Founder: Krishnan Ramu, Chairman of the Board and CTO. The background of the founder may be found from the author's bio in the book.

Technology: Prior to 2002, the SRM technology was focused on competing with induction motors and drive systems. Then the emerging PM motors and drives posed a big challenge in terms of their high efficiency, power density and operational features. Unless the SRMs can match them and not necessarily exceed in some applications, their right to be on the scene, rightfully, became questionable. The technology of PT brought those desirable features, listed in the above, in its innovations and only some are given in the following for brevity:

i. Single Switch Circuit for SRM Drive System: For many applications, the cost is a major factor and one way to reduce the cost is to use a minimum number of power electronic switches such as MOSFETs or transistors. This startup's technology contribution is one such power electronic circuit (among many others) to work with a two-phase machine, shown in Figure 7.9. The SRM has one main phase contributing major power and the other being auxiliary with much reduced power capability. The operation is simple and for energizing the machine, the main phase is energized at the appropriate rotor position and turned off when the auxiliary phase becomes eligible for torque production. During that time, the energy stored in the main phase is transferred to the capacitor C2 which when its voltage exceeds the dc bus voltage, energizes the auxiliary phase. This circuit is a fundamental breakthrough from the point of view of SRM and PM brushless motor drives as it uses only one switch.

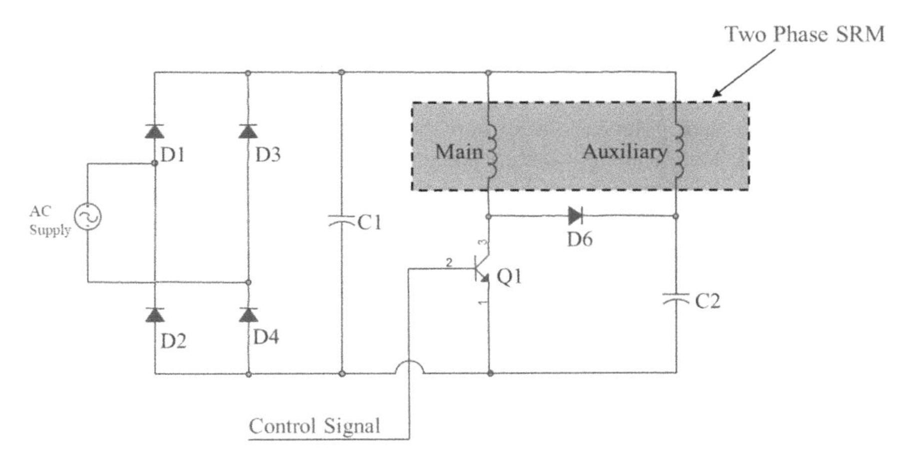

FIGURE 7.9 SRM with a single switch.

This provides the lowest cost and packaging volume for the power electronics and motor.

ii. Four quadrant operation with the single switch SRM drive system: The above motor drive is capable of four-quadrant operation, i.e., both directions of rotation with both positive and negative torque production are demonstrated experimentally in Figure 7.10. This is the fundamental breakthrough which has never been achieved for any of the other motor drive systems such as dc, induction and PM brushless dc. Interested readers may find additional information in references given for this relevant section.

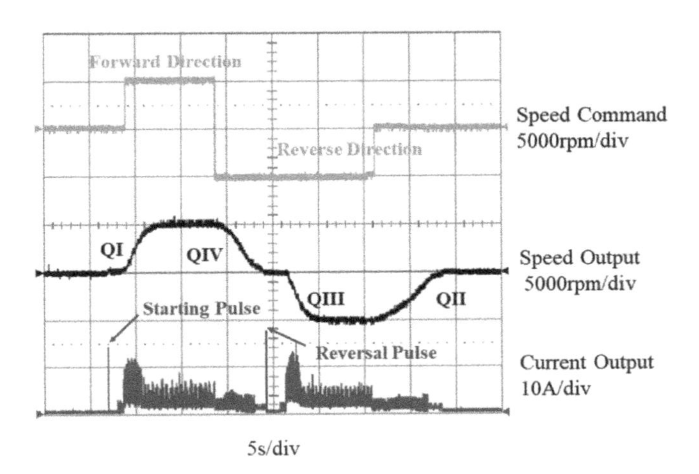

FIGURE 7.10 Four quadrant operation of single switch SRM drive.

iii. SR Motor with No Flux Reversal in the Back Iron (Yoke): SRM back iron experiences usually flux reversal when switching from one phase to another in some sections. In a two-phase machine with four stator poles and two rotor poles, a total of half the back iron experiences such flux reversal. That leads to more core losses and undesirable noise. A motor structure to overcome that has been the contribution of this startup and one such machine is shown in Figure 7.11 for an SRM with six stator and three rotor poles with their flux paths for each phase excitation. Note that there is no flux reversal in the back iron of this machine.

Product: Products have not been released even though one was nearer to customer application in HVAC.

Intellectual Property: A total of eight US patents, on their own one US patent and seven US patents from Virginia Tech had been licensed whose inventors are the founder and his research associates.

Funding: Not listed. Patent and corporate attorneys helped out this firm immensely when it did not have the funds to pay them and without which it would not have survived.

Exit: PT was acquired by Magnetic Torque International (MTI) of Midlothian, Virginia in May 2007 and retained its name. MTI was later named as DeltaGee.

Observations:

i. This seems the first SRM startup in USA where not many research groups existed from the 1980s in the USA. It is a university spin-off in the sense most of its IP was created in a university laboratory and were licensed to this startup. The university centered IPs were created from the outcomes of the research funded by this startup.

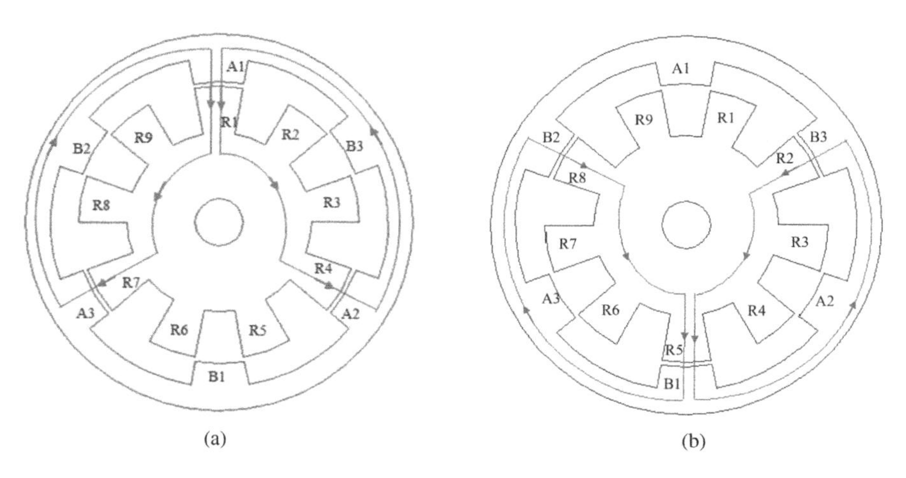

FIGURE 7.11 Flux reversal free operation of a two-phase SRM for phase A (a) and phase B (b) operation.

ii. The dominance of very large motors and drives companies has not made it easy for startups such as this to rise, particularly in USA.
iii. The economic downturn in 2002 did not exactly help this startup as well as others, and many venture capital firms did not have the expertise to even evaluate these kinds of startups in this technology sector.
iv. For a company of its size, it produced significant and fairly a large number of intellectual properties.
v. PT's many of the IPs were groundbreaking and remain so to this day, which garnered some traction with companies not only from USA but from far away India also.
vi. PT may be considered as an R&D company from its outcome point of view, though it had a board and management with industrial experience and backing from a major company.
vii. This startup did not have even a website, was physically located away from all main centers in USA with only access to a small airport about 40 miles away from it and far away from VCs location. In spite of all of that, the startup survived and successfully exited. The primary reasons are in the IPs and its technically outstanding full-time employees.

7.7.2 ePower Motors ApS

ePower Motors ApS (hereafter referred to as ePM in this section for brevity) founded in 2018 is a successor of ePower Technology ApS from 2008 which in turn is a successor to AWS Technology of Aalborg, Denmark from 2003 to 2007. ePM is a spinoff of Aalborg University, Aalborg, Denmark. Its current location is in Copenhagen, Denmark.

Founder: Both ePM and ePower Technology are founded by Flemming Bruhun, and he serves as the CEO. He is a mechanical engineer and has a significant and successful entrepreneurial experience.

The CTO of this startup is Anders Pedersen with specialized experience in SRM power converter and control.

Technology Source: ePM and its previous startups worked closely with Prof. P. O. Rasmussen of Aalborg University. His work culminated in patents owned by Aalborg University and they are licensed to these startups.

Technology: The topology for the switched reluctance motor is chosen to be of transverse type. The transverse machine has the stator cores made of E or U type cores as used in transformers, but the E or U is facing with their winding are 90° from the conventional view of the machine as shown in Figure 7.12.

The advantages of this machine are:

i. End windings are within the stator frame along the axial direction and hence reduced shorter stack length with resulting lower core losses; the other side of the story is that it cuts the circumferential use of the iron, and winding space also.

FIGURE 7.12 E core transverse SRM. (Courtesy of ePower Technology ApS, Denmark.)

ii. Use of discrete E cores and stator and rotor construction that may reduce the cost of the system.
iii. Claimed to have higher power density for this configuration compared to the traditional SRM structure.

Product: The startup's product has been in use in gym80 International GmBH's Synergy Indoor Bike exercise machine. The motor is a five-phase E core Transverse Flux Machine (ETFM). Another product developed is for EV drive train with a continuous output of 17.5 kW at 3,150 rpm speed and maximum power output is twice that at 7,500 rpm. It has a maximum efficiency of 93.5% and most of its operating region has efficiencies greater than 90%. There is no indication that this product is in use at this time.

Intellectual Property: Five EU granted and/or filed patents and one US-granted patent. The patents cover E core transverse SRM, and U core SRM which are of significant value. The rest are in control aspects.

Funding: 4 million Euro was raised for the original startup of ePM, i.e., AWS Technology between 2003 and 2007 to develop their technology and products.

Observations:

i. A good combination of a university faculty/researcher with a business executive has resulted in this startup.
ii. The first 5 years was devoted to a single product development leading to deployment in an indoor exercise bike.
iii. The technology using E core laminates in transverse configuration has been novel with its positives and some issues concerning manufacturing. Note that it deviates from normal machines, their configurations, experienced manufacturing methods and customer acceptance for a century.
iv. Comparison to normal machine configuration in a particular application on a strict academic basis is unavailable to comment upon the chosen structure's distinct advantages. A study of that may be feasible in academia as it may be published for general use and may benefit the industry.
v. The startup had sufficient funding in the first 5 years and that is very fortunate for a company in a nook and corner of Europe with very little interest in SRM and its drive system.
vi. Its intellectual property seems to come primarily from Aalborg University and is a university spin-off.
vii. The intellectual property is not large for a company with nearly 17 years of existence and that too being primarily an R&D motor company. It may be due to limited manpower and significant dependence on the university.
viii. The other product proposal is for the EV drive train and their interest in scaling up in power output for other applications is an excellent approach and it is worth following to see where they go in the future.
ix. It has remained, to the outside observer very much an R&D company. Will that remain so in the future or go aggressively to be a motor drives manufacturing company? If it remains at the same level as it is, then will that tempt many established companies to acquire it? All these are worth following for entrepreneurs in this domain.

7.7.3 Ramu, Inc.

Ramu, Inc. (hereafter referred to as RI in this section) was founded in April 2008 by Krishnan Ramu in Blacksburg, VA, USA. The background of the founder may be found from the author's bio in the book. Ramu served as CEO for the first 2 years and later as CTO and member of the Board of Directors.

Technology: The focus was on the rotating radial flux-based switched reluctance motors, power electronic converters and control systems. The following are some of the technologies that were invented and perfected by RI for potential market applications:

i. Single switch-based SRM drive topologies and their controllers for low-cost applications: Many configurations to suit various applications including applications to PM brushless dc motor drives.

ii. Switched reluctance machine with minimum core of shapes E and/or L based switched reluctance machine structures shown in Figures 7.13 for a two-phase motor. The windings on the diametrically opposite poles constitute a phase and the middle teeth serves to complete the flux path for both phases. Note that nearly 40% and 45% of the stator steel lamination savings in E and L core machines, respectively. The motor packaging becomes compact also with these machines.
iii. A power converter for EV SRMs with an inherent battery charger having no extra switches and no separate inductor.
iv. Item in (iii) is capable of use as uninterruptible motor drives required in essential services such as in-hospital equipment, etc.
v. A high-power-density hybrid machine version of switched reluctance machine plus permanent magnets in the stator teeth. It is shown in Figure 7.14 for a two-phase machine configuration with magnets on the common poles with polarities as indicated. The common poles provide the flux return path for the excited phases. Note that diametrically opposite poles with their windings constitute a phase.
vi. High-density windings for SRMs whereby the packing factor is high compared to conventionally wound and packed machines. Two advantages accrue and they are lower copper losses and high torque at low speeds.
vii. Proprietary motor structures, converters, and position sensorless, noise mitigation and lowest torque ripple controls to provide high efficiency to match that of PM brushless/synchronous motor drives for equivalent motor size are the intellectual properties of this startup.

Product: Developed products for various high-volume applications.

Intellectual Property: Ten granted US patents, and they were filed before and just after the sale of the company.

Funding: Funded by Khosla Ventures and the figures are not released.

Exit: RI was acquired by Regal Beloit Corporation, a multibillion motors and drives company and spread over globally with immense technology in dc, induction, PM synchronous and PM brushless dc categories, in April 2011.

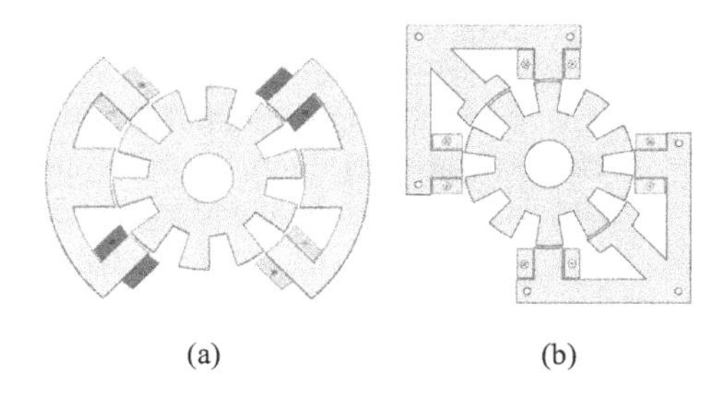

FIGURE 7.13 SRM with stators having (a) E Core and (b) L core laminations.

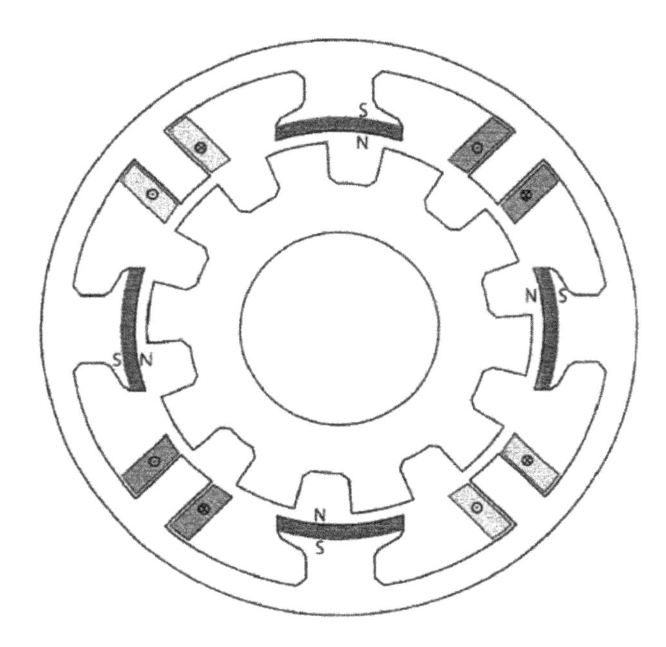

FIGURE 7.14 Hybrid SRM with permanent magnets.

Observations:

i. RI started with funding from Khosla Ventures from day 1 which is very unusual in practice and that allowed to work on the research and development from day 1 of the company startup.
ii. RI created a large number of IP within a short period of time, and all of them geared toward the high-volume applications in motors and drive systems.
iii. Most of its engineers and manager have doctoral degrees in electrical engineering and they are mostly from Virginia Tech. Having the university as a neighbor of the university, RI did have exposure to excellent engineering graduates and benefited from that in its growth.
iv. Due to the high visibility of the funding VC, it gained access to customers much more easily than could have been possible and also its invaluable advice to the management to steer toward profitable outcomes.
v. Lack of its own manufacturing facility inhibited its product release even on a limited scale and it depended on contract manufacturing houses which to an extent skewed the cost scenario and created time delays.
vi. The technology developed is independent of any other technologies and hence self- contained thus bringing value to large industrial motors and drives companies. The technology is all-embracing in SR motors, power converters and controls including noise mitigation and position and speed sensorless systems.

vii. Like Panaphase Technologies (a previous SRM company from the same location), RI also did not have a website throughout the period of its existence. Neither did it seek advertisements or plant articles in trade magazines or work with them to attract the attention of the industry players. These are all unusual in the industry then and now as well. Hopefully, nobody follows this path in their own interest in this era of constant advertisement and promotion!
viii. A successful exit to Regal Beloit Corporation proved the value of RI's intellectual property, their verification and validation, and a high potential for conversion into products.
ix. The hybrid SRM and PM machine invention has high potential for EV application can be seen from Tesla's EV motor drive (wrongly named in the public as PMSRM) which is a PM synchronous motor with high reluctance. It could be replaced with the RI's hybrid machine which might save a lot of PMs (but still using a small amount of PMs) and bringing ease of construction and operation compared to the PM synchronous motor in use.
x. There is a large number of EV applications that could take advantage of RI's patents (including Regal Beloit's patents in later times for which the founder worked for 3 years after the acquisition and produced their patents in this line) that are ideal for this sector.

7.7.4 Software Motor Corporation

Software Motor Corporation (also known as SMC and it is used in this section) is founded in 2013 in San Francisco Bay Area, CA, USA with SRM as its focus at present. Its founders are:

i. Mark Johnston, President and CEO: He has Bachelor's and Master's degrees in physics from the University of Auckland. He is experienced in leadership positions in Asian, European and American markets and is a technology strategist.
ii. Trevor Creary, CTO: He has Master of Science degrees in electrical engineering and computer sciences. He has about 25 years of experience in system and product development.
iii. Piyush Desai, Vice President of Motor Design: He has Master's and Doctoral degrees in electrical engineering from Illinois Institute of Technology (IIT), Chicago. His research work in IIT is related to SMC's focus. He has about 20+ years of experience mostly related to motors and drives.

Technology: It may be divided into three sections:

i. High rotor pole combination with stator poles: Higher number of rotor poles decreases noise which is intrinsically arrived at as it increases the frequency of torque for a given speed of the machine. This is usually not used in practice even though others (for example, PT and RI) developed such machines with different types of stator core shapes also.

ii. The main claim is in their SRM being an axial type, with one stator straddled on each side by a rotor and extending it to two stator and one rotor in the middle, three stators and two rotors, etc. The poles on the stator and rotor are projecting with air gap between the stator and rotor and each having back irons. The stators have windings in between the two projecting poles of the stator and another version with three rotors and two stators (refer to their US Patent 9,960, 662).
iii. For the machine in (ii), rotor position estimation is based on the self-inductance and mutual-inductances of machine phases for use in the control of the SRM drive system.

All the three components listed in the above forms broadly their technology.

Products: SMC has listed 1, 1.5, 2 and 3 hp SRM drive systems with many features for HVAC and fan type of applications claiming to outperform induction motors and drive systems. Their claims of 50% more efficient than the standard motors and 80% saving in energy in fan applications compared to conventional motors have no reference machine for comparison to fully appreciate their products. They have announced that they will be developing EV motor drives also (probably due to BMW's investment in them).

Intellectual Property: 16 granted US patents and nearly 8 of them licensed from IIT, Chicago, IL, most of which are related to hybrid/electric vehicles and not directly related to do with SRM excepting one. This license was owned by HEVT Llc., a startup that was acquired by SMC in 2014.

Funding: $86.4M in five rounds with the latest sizeable funding coming from BMW. Most of the funding (about 65% of the total) came in the year 2019.

Observations:

i. This is one of the few motor drives companies that raised a phenomenal amount of money within 7 years of its existence and that too for an SRM startup that is not in the mainstream in the USA and everywhere else.
ii. The board and its chairman are very good assets to this company particularly in raising capital for it.
iii. Its focus, so far, has been in SRMs for high-volume applications of fans and HVAC applications with a few products on their listing.
iv. The products have not been compared to their main competitor in the market, i.e., permanent magnet brushless and synchronous motors and drive systems that occupy this space in variable-speed motor drives market.
v. SRM advocacy, for fixed speed or variable-speed application, in the name of the efficiency and savings in energy is equally applicable to other motors and drives and their products has to be noted by the observers.

vi. Technology with axial field motor configuration and a high number of rotor poles is very much promoted by the startup. Axial field construction is not new for SRMs and PMSMs as they have been studied, but some details of winding placements and salient poles being fixed to rotor and stator surfaces may come from their contributions. What is interesting will be to see how the construction is achieved with the complexity of the laminations and placement of windings in this structure and how it plays into unique advantages, and also in the domain of manufacturing cost. Note that the PMSM startups have been more or less uniformly taken up with axial flux type structure that could be seen from the startups in the previous section of this chapter.

vii. BMW's investment may open an opportunity for this company to target the automotive market. This market is not as cost-sensitive as other high-volume markets, and this fact will help startups in this sector including this one.

viii. The patent portfolio number wise looks reasonable even though some may not be relevant to their business focus today and if and when it changes over time, that alters the picture in this aspect.

7.8 ELECTROSTATIC MACHINES

Electrostatic machines have been researched from the late eighteenth century starting from Benjamin Franklin. It is based on the principle of electric field which is the counterpart of the electromagnetic field. The latter principle forms the basis for electrical machines of all kinds so far and they occupy the market space, without any technical threat from electrostatic machines so far. Here the electromagnetic machine torque expression is used to arrive at the torque in rotating electrostatic machine. The torque in a rotating electromagnetic machine is given in terms of the self-inductance, L and current, i_a in equation (6.20) from Chapter 6 as,

$$T_e = \frac{1}{2} i_a^2 \frac{d}{d\theta}(L) \tag{7.1}$$

Likewise, the torque in a rotating electrostatic machine, by the counterpart argument, can be derived with current being substituted by voltage, and inductance by capacitance as follows:

$$T_e = \frac{1}{2} v_a^2 \frac{d}{d\theta}(c) \tag{7.2}$$

where v_a is the applied voltage between the stator and rotor plates (aluminum) and c is the position-dependent capacitance between them. If the capacitance between the stator and rotor is made to vary sinusoidally with respect to rotor position and the applied voltage is sinusoidal, then a synchronous electrostatic machine results. The torque is independent of the polarity of the applied voltage and its polarity is solely determined by the slope of the capacitance variation. A few observations are in order:

i. The maximum voltage that is applied between the rotor and stator plates has to be lower than the breakdown voltage of the dielectric medium between the plates. For example, air has a breakdown voltage of about 3 Million volts/meter (3 kv/mm) and pure vacuum will not breakdown at all. The plate metals, such as mica, also have different dielectric strength much greater than air is noted.
ii. The medium in which the electrostatic machine is placed can enhance the dielectric strength, thereby allowing higher voltage operation and hence higher torque and power output.
iii. There are some distinct advantages in the electrostatic machines compared to the electromagnetic machines and they are: (i) with no windings, the stator is much compact and hence the machine could be smaller in size provided the energy stored in these machines approach that of its counterpart, (ii) very little resistive losses, low core losses and hence higher motor efficiency, (iii) with no windings and hence no fear of its short circuit or thermal failure in the long run leading to rewinding maintenance with the beneficial take on simpler maintenance, (iv) simpler construction with stacking rotor and stator plates next to each other, (v) lamination core with its process and manufacturing is replaced by the simpler plates manufacturing, (vi) environmentally friendly compared to PM machines and (vii) simpler machine construction and operation.

Given the information about them, it would be natural to wonder as to what are the major concerns that prohibited the making of electrostatic machines. Then the explanation is as energy stored in an electric medium with then available dielectric materials such as between two plates in a dielectric medium is not comparable to electromagnetic energy stored in its medium and it is widely being held until the opposite is proved in practice. This translates into the amount of voltage that can be applied between the static and rotor plates with a gap between them. That should be less than the voltage that can breakdown that gap and the gap consists of the distance between the stator and outer rotor surfaces. The dielectric materials have limited capability. The use of vacuum inside the machine can change it, but it may not be very practical. Very few electrostatic (hereafter referred to as capacitive) machines have been made in the past and interested readers are directed to reference under this section.

7.8.1 C-Motive Technologies

C-Motive Technologies (hereafter referred to as CMT in this section) was founded in 2012 and located in the state of Wisconsin, USA. Its founders are:

i. Dr. Justin Reed, Co-founder, President & CEO: Graduate of Univ. of Wisconsin, Madison in EE. Developed the technology and the founding patent for this startup with the other co-founder.
ii. Dr. Dan Ludois, Co-founder and Chief Science Officer: Graduate of Univ. of Wisconsin, Madison in EE and currently is an associate professor in the same department. His interests are in electrostatic machines, power converters and control for the same. He is one of the co-inventors and patent authors for the startup. He is building up the research in all aspects of these motor drives with his graduate students.

Technology: Capacitive machines capable of higher power than demonstrated earlier times with multiple aluminum plates in the stator and rotor arranged together with particular designs of the stator surface with insulators on them patched and patterned to yield a flow of dielectric liquid and to create the varying capacitance forms the base technology of this startup. Methods to realize such construction is one of the approaches taken in this startup. Further, the arrangement of the various types of realizing stator and rotor plates and arranging them to form a rotating machine has been developed in this startup which is one of the backbones for building the capacitive synchronous motor. The technology is specifically directed to the practical problems of realization of the capacitive machine with newer machine construction methods and materials to overcome the power density problem. One of their motor prototypes is shown in Figure 7.15 and information about its rating etc. is not available to the general public.

FIGURE 7.15 C-Motive's capacitive motor. (Courtesy of C-Motive Technologies.)

Products: Yet to release product for general public use and it is indicated that a fan motor may be available for customer evaluation in 2020.

Intellectual Property: Three granted US patents so far.

Funding: $2.8M has been raised with the last round contribution of $2M from Prime Impact Fund of Boston, Clean Energy Trust of Chicago and others in 2019.

Observations:

i. This is the first electrostatic motor startup in USA in the past decade. At the same time, there were never too many startups in this arena.
ii. While there are a large number of motor startups with large investment funding in recent times, they are all in the electromagnetic field. To some, it may appear that their technologies are incremental as the electromagnetic motors have been in place for more than 150 years. It is important to realize that the newer startups claim to have significantly increased the power densities for aerospace and some targeted consumer markets such as HVAC motors and drives applications. All of these current developments make the challenge for capacitive motors even much harder particularly in meeting the performance, cost and power density. Most of all the challenge may be customer acceptance and willingness to walk away, from a technology that has delivered reliably for more than a century and that too, without significant disadvantages.
iii. The founders come from a similar educational background without much industrial experience but with a great conviction and continuous pursuit of their objective of realizing a capacitive motor for the market.
iv. The startup has a reasonable number of granted US patents for a company of its size within the period of existence of 8+ years so far. The potential to build intellectual property is fairly high given its close link to academia.
v. Product release with very small funding for the first 7 years is impossible in this line of business of motors and drive systems. This picture may change after the new round in the last year.
vi. Electrical engineers trained in electrostatic machines hardly exist and that may be a big bottleneck to overcome to build a cohesive engineering group to build on to their initial machines. The founders' connection to academia may alleviate some of the problems, but the students may not have industrial exposure in many aspects of the motors and drive systems and that may slow their progress.
vii. If and when the company becomes successful in proving all its claims on capacitive motors with products line up, a market avalanche in this sector can be imagined.
viii. Given some of the concerns and great impact it may have if and when it is successful, it is very critical for all electrical engineers to follow and watch the unfolding of the future of capacitive motors from C-Motive Technologies.

7.9 CONCLUSION

The following observations, in general, are applicable to motor drive startups:

i. The startups in this domain are not, say, as many as in the power electronics area. Reasons that may be attributed are: Motor and drives area is much more complex in that there are three major subsystems involved in its conception and design as against other fields studied in this book. Because of the complexity, the startups with limited resources find very constrained to enter the field. Moreover, the competition from established 200+ motors and drives companies/manufacturers around the world is a big threat to the startups.
ii. The major applications aimed by the motors and drives startups are related to electric vehicles of all kinds starting from two- and three-wheelers, cars, electric trucks to electric aircraft. The second set belongs to appliance applications including HVAC compressor motors and drives for high efficiency. The third set of startups aim at high-power applications in MW.
iii. Technologies in motors range from axial flux motors to thermal cooling enhanced high-power-density machines of small to larger sizes.
iv. Electric aircraft applications of motor drives have negligible startup activity except one or two as the domain is dominated by big and established companies of Siemens size and their equals.
v. Power density is of utmost importance in the above applications but also with very high power as much as in one or two MW level. Startups to work at this level of power require a huge investment and many VCs may not be open to it as they are very shy of hardware-related projects. They prefer software-related startups as they are easy to scale up and their products require very limited tests and certification requirements, almost none compared to hardware-based products even if they are software-driven.
vi. Most of the motors and drive system startups are in the permanent magnet synchronous category and very few are in the switched reluctance motor drives category. That will change considerably: (i) when permanent magnet materials availability becomes a serious issue, (ii) safety under phase short circuits in the motor and/or power electronic drive becomes number one issue such as in aircraft and defense applications, (iii) cost becomes a significant factor, and (iv) complete recyclability for a clean environment is a serious concern. They are not yet affecting the market place significantly.
vii. In the last 5 years (since 2015), the funding to motor drive startups has become significant in USA but still a long way to catch up with other hardware-based startups.
viii. Technology based on transverse flux machines in the PM category has not made any inroads into startups and products even though they have demonstrated advantages.

ix. Compared to extensive academic research, the startups are small in number. But many of the startups have strong connections to or emanate from University laboratories and intellectual properties.
x. Thanks to the emergence of SiC and GaN power semiconductor devices and their introduction into motor drives applications is leading to newer converter design, packaging and control. They provide immense opportunities for newer startups. Interested readers are encouraged to do research more on this opportunity.
xi. Huge developments in linear machines for high rise building elevators, high-speed rails, and other related applications are waiting to be employed and exploited with appropriate motor drives technologies and they are not covered in this section. (Interested readers may follow the author's work in the research domain). Note that these applications require an immense financial investment that is usually not possible from VCs but only from mammoth multi-billion dollar companies and public investment from governments.

7.10 DISCUSSION QUESTIONS (BEWARE: THE DISCUSSION MAY LEAD TO INNOVATIONS AND POSSIBLE STARTUP IDEAS!!)

1. Newer topologies or arrangement is one way to go for innovation in motors. Discuss the flux path based topologies.
2. Are the motor topologies exhausted or still more to come?
3. PM motors are yet to come out in transverse flux arrangement in product lines. Why is that not happening?
4. Is there a way that transverse PM motors can be made very economically on the same footing as radial flux machines?
5. Discuss the merits of having PMs in electrical motors.
6. Discuss the merits of not having PMs in electrical motors.
7. Axial flux configuration is very much in startups now even though they have been around for more than 40 years or so. Discuss the recent developments in the fabrication of these machines.
8. Discuss AFPMs (axial flux PM machines) applications other than discussed in the text.
9. Discuss the construction of AFPMs, and its impact on cost.
10. Will AFPMs completely replace the currently available radial field PM machines?
11. SRMs can have axial flux (AF) configuration as well, and it has been in research about 32 years ago. Why it did not take off both in research and practice?
12. Will AFPM development now resuscitate interest in AFSRMs? Discuss both why and why not based on constructional and cost features.

13. AFSRM has come into a recent startup in this domain which is in the main text of this chapter. Will this be ideal for any particular application and if so why?
14. Transverse SRM is in the product space thanks to a startup. Discuss the merits and demerits of such motor topology and particularly with respect to SRMs. Are there applications that can be satisfied only by this motor topology?
15. Compare qualitatively the differential cost of construction between an SRM and PMSM for a given power.
16. Are there any constraints in building MW machines using PMSM or SRM kind? Discuss the reasons based on constructional and cost features.
17. Induction motor is widely used in industry and it is considered to be its workhorse. Will it ever change: now or in the near future or distant future? What are the reasons that may be attributed to it?
18. Appliances have or are shifting from single-phase induction motors to PM brushless dc motors in spite of the differential cost being against the latter. Discuss the reasons for the trend.
19. List all the possible power electronic converters for PMSM and Brushless dc motor drives.
20. Discuss minimum switch power electronic converter for appliance applications and their configurations for PM brushless dc motor drives.
21. The standard converter is a three-phase inverter for PMSMs in practice. Will that change in the future? If not, why?
22. The standard converter for SRMs is an asymmetric converter. Compare its cost to that of the three-phase inverters commonly employed in ac motor drives.
23. SRMs have options to many converter configurations compared to any other motor drives. Discuss the advantages of this feature. Will this help SRMs in the market place?
24. Control is more demanding in SRMs than in ac and dc motor drives. Is that right?

7.11 PROJECTS

1. Team/Individual Project 1: A startup, KLD Technologies in Texas, USA in 2007 claimed to have nano-crystalline material for lamination core that is capable of providing higher performance. Motor build around it and with motor technology secured from Green Ray Technologies was to be their product. They raised around $15M+ in funding. The company did not have a good financial outcome. The references for this company and its activities can be obtained from the following references. From these references and your own research, prepare an outline of this

startup, founder, technology, products, IP, funding, and your observations on the startup including lessons for budding entrepreneurs. Some references on the startup are:

i. https://www.nextbigfuture.com/2011/08/kld-energy-and-electric-motors-with.html
ii. https://www.texastechpulse.com/interview_with_christian_okonsky_kld_energy/s-0052428.html
iii. https://patents.justia.com/assignee/kld-energy-technologies-inc
iv. https://www.prnewswire.com/news-releases/kld-energy-technologies-inc-to-provide-motor-propulsion-system-for-4-wheel-electric-vehicles-176864261.html
v. https://www.businesswire.com/news/home/20090202006135/en/KLD-Energy-Technologies-Secures-Worldwide-Rights-Breakthrough
vi. https://www.benzinga.com/pressreleases/15/09/p5843364/kld-energy-technologies-inc-acquires-texas-based-mobility-vehicle-desig
vii. https://xconomy.com/texas/2011/01/18/kld-energy-technologies-obtains-10000000-new-round/
viii. https://www.autoblog.com/2009/02/10/sufat-announces-new-electric-scooters/

2. Class Project (Comprising Many Teams): Select and study startups in various branches of motors, power converters for motor drives, applications, their specifics, how they are different from existing technologies, costing, calculation of cost-effectiveness, competitiveness, manufacturability, funding, and future growth, whether they will succeed, and if so, why in your opinion (not parroting what the companies say). You may also examine, in addition to other factors, such as IP portfolio and its significance, and product differentiation from their competitors' in the startup domain and with existing established companies.

Project 1: This class project problem can be split up into many subsections so that teams and many individuals can work together and create an atmosphere of a startup in the class itself. Then their study is packaged into reports.

Project 2: The project reports from project 1 can be exchanged with other project reports (complied by other teams) fact-checked, then discussed and improved and learn many more domains in the process.

Project 3: Let the teams choose the most promising startups (say 2 or 3 reports) to succeed and then teams are formed to develop their strategies and if, possible, IPs also, for their success and communicate them with those companies/managements so that the teams can make contributions and learn from them. If the teams come up with novel business ideas in the marketing, manufacturing and IP domains, and believe that they are competitive in that space or a new space, consider the opportunity to take it forward as a new startup. Right there comes a startup company!

3. Team Projects (with the potential to combine into a class project):

   Project 1: Class can be divided into many teams. The objective of the class project is to study, design and build a prototype motor drive system. This is a very ambitious project! Each team chooses, based on their particular interest, one of the following: motor analysis, motor design, motor construction and assembly, power converter analysis, power converter design, PCB layout, assembling power converter and testing, coding and testing of the embedded controller. The funds for motor, power electronic converter and controller realization can be obtained from within the university most of the time, and if unavailable, contact local industries for underwriting the project and for that a plan is required and the entire class can be involved in that. An organization for the team and inter-team coordination, and final project coordination is set up at this stage.

   Project 2: The teams produce their reports and start the construction of the motor drive. At that time, the analysis teams can merge with teams of their choice for fabrication. The teams can prepare their reports for project 1.

   Project 3: Testing of the final subsystems and system at last. A comparison between the predicted and test results is made and discussed for any discrepancies. The final report is prepared, circulated to the class, corrected, modified and report is presented to the class. A section is devoted to clearly identifying the contributions of each team member in the team, then each team's contributions, and then the role of the management of the teams and finally the entire project itself.

For UG Class: This class project need not be done in all its three phases of 1-3, but only the first one is sufficient for most of the classes in a given semester. If there are two semesters available for the project which is ideal, then all the three phases of the project can be attempted.

For Graduate Class: All the three projects within this entire class project can be attempted in a semester.

REFERENCES

MARKET

1. https://www.mordorintelligence.com/industry-reports/permanent-magnet-motor-market.
2. https://insideevs.com/news/379497/magna-powerful-affordable-electric-motor.

3. https://www.prnewswire.com/news-releases/the-global-permanent-magnet-motor-market-was-valued-at-usd-31-03-billion-in-2018--and-is-expected-to-reach-usd-52-8-billion-by-2024--at-a-cagr-of-9-42-during-the-forecast-period-2019-300867574.html.
4. https://drivesncontrols.com/news/fullstory.php/aid/4693/Permanent_magnet_motor_market_heads_above_$45bn.html.
5. https://www.alliedmarketresearch.com/permanent-magnet-motor-market.
6. Intellectual Property: Only a few patents have been identified with respect to US and there could be some with EU patent office. There is one US patent and one application pending with US patent office and all of them with Peter Liejnen as inventor and co-inventor, respectively. As this is a recent company, it may take some more years before they can build up their IP portfolio.

AXIFLUX

1. http://axiflux.com/.
2. https://www.mathworks.com/content/dam/mathworks/mathworks-dot-com/company/events/conferences/matlab-tour-australia/2014/proceedings/embedded-control-of-an-adaptive-magnetic-flux-array.pdf.
3. https://spectrum.ieee.org/at-work/start-ups/startup-profile-axiflux.
4. https://www.smartcompany.com.au/startupsmart/startupsmart-technology/electric-engine-start-up-rising-as-holden-australias-future-looks-bleak/.
5. https://angel.co/company/axiflux.
6. https://www.crunchbase.com/organization/axiflux#section-lists-featuring-this-company.
7. https://www.goauto.com.au/news/general-news/manufacturing/axiflux/2014-05-13/24151.html.
8. http://www.freepatentsonline.com/EP2719059.html.
9. https://patents.justia.com/assignee/axiflux-holdings-pty-ltd.
10. D.E. Jahshan, Electric motor/generator with integrated differential, US Patent 9,509,198 B2, November 29, 2016.
11. D.E. Jahshan, Electric motor/generator, US Patent 10,008,910 B2, June 26, 2018.

LINEAR LABS

1. https://www.linearlabsinc.com/.
2. https://www.linkedin.com/in/fred-e-hunstable-8ba74b47.
3. https://www.linkedin.com/in/bradhunstable?trk=org-employees_profile-result-card_result-card_full-click.
4. https://gust.com/companies/linearlabsinc.
5. https://drivesncontrols.com/news/fullstory.php/aid/5985/$4.5m_funding_will_help_to_market__91revolutionary_92_motor.html.
6. https://www.prnewswire.com/news-releases/linear-labs-closes-4-5-million-seed-round-to-fund-commercialization-of-new-electric-motor-technology-300818574.html.
7. https://spectrum.ieee.org/cars-that-think/transportation/advanced-cars/new-electric-motor-could-boost-efficiency-of-evs-scooters-and-wind-turbines?utm_source=techalert&utm_medium=email&utm_campaign=techalert-08-22-19&mkt_tok=eyJpIjoiWmpBeE1qmxNelEzWVRFdyIsInQiOiJhR0

1xYTdhR0VEQmViNnVHQU5cL2NGdEFsWG9hb01uUDV6ZjN4RElSW-GtPQWFTalZVY2s3cHlpOG52TWNmMnU3c0RXaUZzOCt0QmpBSn-BEWUdjN0N5MCsySURxY2dmK09NQWU2TlBnTzN6WWtkYXhXYmVs-ZkFEb01jUjhZc3lmbU0ifQ%3D%3D.
8. https://insideevs.com/news/364852/linear-labs-hunstable-electric-turbine/.
9. https://www.cnet.com/roadshow/news/linear-labs-electric-turbine-motor/.
10. https://patents.justia.com/assignee/linear-labs-inc.
11. https://www.linearlabsinc.com/our-mission/.
12. F. Hunstable, Multi-tunnel electric motor/generator segment, US Patent 10,476,362 B2, January 21, 2016.
13. F. Hunstable, Multi-tunnel electric motor/generator, US Patent application, US 2019/0199185 A1, June 27, 2019.
14. T. Lipo and M. Aydin, "Field weakening of permanent magnet machines: Design approaches", *EPE Power Electronics and Motion Control Conference (EPE-PEMC 04)*, Riga, Latvia, 2004.
15. X. Ai, B. Mularcik and R. Knepper, Electric motor with field weakening, US Patent 7960888 B2, 2011.
16. B.A. Maslov and Z. Soghomonian, Rotary permanent magnet electric motor with varying air gap between interfacing stator and rotor elements, US Patent No. 6,727,630, 2004.

Magnax

1. https://www.magnax.com/.
2. https://www.crunchbase.com/organization/magnax#section-overview.
3. https://patents.justia.com/inventor/peter-leijnen.
4. https://spectrum.ieee.org/transportation/alternative-transportation/this-insideout-motor-for-evs-is-power-dense-and-finally-practical?utm_source=techalert&utm_medium=email&utm_campaign=techalert-10-03-19&mkt_tok=eyJpIjoiWWpJeE-16QXlaVFptWkRZNSIsInQiOiJcL0xqWTY3bTVRSnloXC8zVjZNOGRJY29qeVFYR0FwN2xYT0NHOWhGazNqVzRWMzJIczFUeG1UdWhjRWpJK2JnUFZEVEpwa3JRYmp1SzJucTN1UTdPeEQ5c2dDWTgydGMwWnJPSjZiUH-dOR285Rk1zanVJQnB3blVxUHZcL29MVDh6RyJ9.
5. https://www.powerelectronics.com/markets/automotive/article/21864194/axialflux-motors-and-generators-shrink-size-weight.
6. https://www.researchgate.net/figure/MAGNAX-motor-series-specifications_fig17_326294480.

YASA

1. https://www.yasa.com.
2. https://en.wikipedia.org/wiki/YASA_Limited.
3. https://www.greencarcongress.com/2019/09/201090912-yasa.html.
4. https://www.businesswire.com/news/home/20190401005923/en/Electric-Motor-Company-YASA-Launches-New-Series.
5. https://patents.justia.com/assignee/yasa-motors-limited.
6. https://patents.justia.com/assignee/yasa-limited.
7. https://www.ipo.gov.uk/patent/p-journal/p-pj?lastResult=0&perPage=10&startYear=2008&startMonth=January&startDay=30th+-+6193&endYear=2020&endMonth=January&endDay=29th+-+6819&filter=YASA%2520&sort=Publication+Date&status=undefined#starttabs.

8. https://www.crunchbase.com/organization/yasa-motors#section-funding-rounds.
9. https://www.businesswire.com/news/home/20190909005993/en/Electric-motor-controller-manufacturer-YASA-closes-%C2%A318m.

QM Power

1. https://www.qmpower.com.
2. https://www.equitynet.com/c/qm-power-inc.
3. https://www.linkedin.com/in/hari-harikumar-2735064?trk=org-employees_profile-result-card_result-card_full-click.
4. https://www.qmpower.com/newproducts/.
5. https://www.energy.gov/eere/buildings/downloads/qm-power-inc-commercial-refrigeration-fan-applications.
6. https://www.bizjournals.com/kansascity/news/2016/03/21/qm-power-electric-motor-pj-piper.html.
7. https://www.bizjournals.com/kansascity/news/2016/05/10/qm-power-q-sync-motor-energy-savings.html.
8. https://www.qmpower.com/2016/04/sdge-report-features-qm-power-permanent-magnet-synchronous-motor/.
9. https://patents.justia.com/assignee/qm-power-inc.
10. https://pitchbook.com/profiles/company/63705-70.
11. https://www.crunchbase.com/organization/qm-power#section-events.

Zero E Technologies

1. http://www.zeroetech.com/.
2. https://c.ymcdn.com/sites/stockmarketmanager.site-ym.com/resource/resmgr/files/Zero_E/Marble_Arch_Research_Valuati.pdf.
3. https://aceee.org/files/proceedings/2017/data/polopoly_fs/1.3687894.1501159070!/fileserver/file/790273/filename/0036_0053_000048.pdf.
4. https://patents.justia.com/inventor/thomas-h-hopkins.

LC Drives Inc.

1. www.lcdrives.com.
2. https://www.clarkson.edu/news/lc-drives-ceo-alumnus-russel-marvin-speak-clarkson-university-december-recognition-ceremony.
3. https://www.linkedin.com/in/david-leach-868176a.
4. https://www.crunchbase.com/organization/lc-drives#section-overview.
5. https://patents.justia.com/assignee/lcdrives-corp.
6. R.H. Marvin and D.H. Leach, Liquid cooled high efficiency permanent magnet machine with glycol cooling, US Patent number: 10348146 B2, July 9, 2019.
7. R.H. Marvin and T. Johnson, Concentrated winding stator construction for high efficiency machine, US Patent number: 9362785, B2, June 7, 2016.
8. https://www.linkedin.com/company/lcdrives-corp.
9. https://www.clarkson.edu/news/lc-drives-spotlight-startups-clarkson-university-shipley-center-innovation.
10. https://www.wwnytv.com/2019/07/01/clarkson-startup-wants-build-factory/.
11. https://www.nny360.com/communitynews/business/lc-drives-still-seeking-funding-after-potsdam-land-purchase/article_d5475d06-0c24-5da9-a313-ab82aafeb02b.html.

Magnix

1. https://www.magnix.aero/.
2. http://www.guina.com.au/history/.
3. https://en.wikipedia.org/wiki/MagniX.
4. https://inis.iaea.org/search/search.aspx?orig_q=RN:47102963.
5. https://drivesncontrols.com/news/fullstory.php/aid/5365/_91Game-changing_92_electric_motors_could_power_aircraft.html.
6. https://www.prnewswire.com/news-releases/aerotec-to-flight-test-magnix-all-electric-propulsion-system-300871502.html.
7. https://www.prnewswire.com/news-releases/magnix-to-unveil-worlds-most-advanced-electric-propulsion-system-for-aircraft-at-2019-paris-air-show-300865068.html.
8. https://www.ansys.com/about-ansys/advantage-magazine/best-of-aerospace-and-defense/engineering-the-fully-electric-airplane.
9. https://www.prnewswire.co.uk/news-releases/harbour-air-and-magnix-announce-successful-flight-of-world-s-first-commercial-electric-airplane-829211646.html.
10. https://www.geekwire.com/2019/eviation-picks-magnix-provide-propulsion-option-electric-airplane/.
11. https://www.forbes.com/sites/jeremybogaisky/2019/02/20/eviation-gains-backing-from-billionaire-richard-chandler-to-make-first-electric-passenger-plane-take-flight/#173258a76ebb.
12. https://www.greenbiz.com/article/6-electric-aviation-companies-watch.
13. https://www.crunchbase.com/organization/magnix#section-interest-signals-by-bombora.
14. https://pitchbook.com/profiles/company/230405-14.
15. https://patents.justia.com/assignee/heron-energy-pte-ltd.

ECM Software and Technology

1. https://pcbstator.com/.
2. https://www.crunchbase.com/organization/e-circuit-motors#section-funding-rounds.
3. https://pitchbook.com/profiles/company/119536-03.
4. https://www.whoisraisingmoney.com/e-circuit-motors-inc.
5. https://patents.justia.com/assignee/e-circuit-motors-inc.
6. S. Shaw, Structures and methods for controlling losses in printed circuit boards, US Patent, US 9,800,109 B2, October 2017.
7. S. Shaw, Apparatus and method for forming a magnet assembly, US Patent, US 9,673,688 B2, June 2017.
8. S. Shaw, Planar composite structures and assemblies for axial flux motors and generators, US Patent, US 10,170,953 B2, January 2019.
9. S. Shaw, Structures and methods for thermal management in printed circuit board motors, US Patent, US 9,673,684 B2, June 2017.

Infinitum Electric

1. https://www.infinitumelectric.com.
2. https://www.americaninno.com/austin/funding/austin-techs-hot-start-to-2018-see-all-the-fundings-acquisitions-and-more/.

3. https://www.linkedin.com/in/ben-schuler-21567a32/.
4. https://pitchbook.com/profiles/company/101509-12.
5. https://www.crunchbase.com/organization/infinitum-electric.
6. https://insideevs.com/news/386297/infinitum-electric-printed-circuit-board-stator/.
7. https://drivesncontrols.com/news/fullstory.php/aid/5902/IE5_motors_with_PCB_stators_slash_costs_and_weights.html.
8. https://patents.justia.com/assignee/infinitum-electric-inc.
9. https://patents.justia.com/assignee/fanergies-inc.
10. B.L. Schuler, R. Lee and J. Rasmussen, System and apparatus for axial field rotary energy device, US Patent, US 10,186,922 B2, January 2019.
11. B.L. Schuler, R. Lee and J. Rasmussen, System and apparatus for segmented axial field rotary energy device, US Patent, US 10,340,760 B2, July 2019.
12. B.L. Schuler, R. Lee and J. Rasmussen, System and apparatus for modular axial field rotary energy device, US Patent, US 10,135,310 B2, November 2018.

Panaphase Technologies

1. https://patents.justia.com/assignee/panaphase-technologies-llc.
2. https://ece.vt.edu/news/article/panaphase-sold-1.
3. http://www.gentekventures.com/downloads/panaphase.pdf.
4. K. Ramu, Method, apparatus, and system for drive control, power conversion, and start-up control in an SRM or PMBDCM drive system, US Patent 7,271,564 B2, September 2007.
5. K. Ramu, Apparatus for drive control, power conversion, and start-up control in a PMBDCM or two-phase SRM drive system, US Patent 7,372,232 B2, May 2008.
6. K. Ramu, K. Ha and S.-Y. Park, System and method for controlling four-quadrant operation of a switched reluctance motor drive through a single controllable switch, US Patent 8,203,302 B2, June 2012.
7. K. Ramu and S.N. Lobo, Apparatus and method that prevent flux reversal in the stator back material of a two-phase SRM (TPSRM), US Patent 7,579,734 B2, August 2009.
8. S.G. Oh and R. Krishnan, "Two phase SRM with flux reversal free stator: Concept, analysis, design and experimental verification", *Conference Proceedings of IEEE Industry Applications Conference Annual Meeting*, vol. 3, pp. 1155–1162, 2006.

ePower Motors ApS

1. http://epower-technology.dk.
2. https://dk.linkedin.com/in/flemming-bruhn-b1628610.
3. For European patents of this startup, https://register.epo.org/advancedSearch?searchMode=advanced&pn=&ap=&pr=&fd=&pd=&prd=&grd=&ufd=&urd=&pa=&in=Peter+Omand+Rasmussen&re=&op=&ic=&ti=&apl=.
4. P.O. Rasmussen, Transverse flux machine with stator made of E-shaped laminates, US Patent 7,312,549 B2, December 2007.

Ramu Inc.

1. https://vtechworks.lib.vt.edu/handle/10919/72294/browse?type=author&value=Ramu%2C+Krishnan.
2. https://patents.justia.com/assignee/ramu-inc.

3. https://patents.justia.com/inventor/krishnan-ramu.
4. https://chargedevs.com/features/a-closer-look-at-switched-reluctance-motors/.
5. K. Ramu and N.S. Lobo, Apparatus and method that prevent flux reversal in the stator back material of a two-phase SRM (TPSRM), US Patent 7,579,734, 2009.
6. C. Lee, R. Krishnan and N.S. Lobo, "Novel two-phase switched reluctance machine using common-pole E-core structure: Concept, analysis, and experimental verification", *IEEE Transactions on Industry Applications*, vol. 45, no. 2, p. 703, 2009.
7. https://www.prnewswire.com/news-releases/regal-beloit-acquires-ramu-inc-a-technology-company-backed-by-khosla-ventures-119243954.html.
8. https://news.thomasnet.com/companystory/regal-beloit-acquires-ramu-inc-a-technology-company-backed-by-khosla-ventures-845411.

Software Motor Company

1. https://softwaremotor.com/.
2. https://softwaremotor.com/patents/.
3. M. Krishnamurthy, M. Johnston, T. Creary and P. Desai, Mirroring of high rotor pole switched reluctance machines, US Patent 9,960,662 B2, May 2018.
4. M. Krishnamurthy and T. Creary, Method for reliable control of high rotor pole switched reluctance machine, US Patent 10,069,449 B2, September 2018.
5. https://astia.org/2014/03/26/software-motor-corporation-smc-completes-asset-based-purchase-of-hevt-llc-astia-alumni/.
6. https://www.chicagobusiness.com/article/20130209/ISSUE01/302099979/hevt-s-inventive-motor-and-its-pitchwoman-ceo-heidi-lubin.
7. https://www.greencarcongress.com/2008/07/hevt-showcases.html.
8. https://softwaremotor.com/wp-content/uploads/2018/06/EnergyCIOInsights_Top10_SMC.pdf.
9. https://drivesncontrols.com/news/fullstory.php/aid/5661/Switched_reluctance_motor_aims_to_disrupt_the_market.html.
10. https://www.crunchbase.com/organization/software-motor-company#section-funding-rounds.
11. https://www.businessinsider.com/software-motor-company-smc-raises-30-million-2019-7.
12. http://www.finsmes.com/2019/06/software-motor-company-raises-31-4m-in-series-a-2-funding.html.

Electrostatic Machines

1. L.W. Matsch, *Capacitors, Magnetic Circuits and Transformers*, Prentice Hall Inc., Upper Saddle River, NJ, 1964.
2. N.J. Felici, "Ten years of research on electrostatics at the University of Grenoble, 1942-1952", *British Journal of Applied Physics*, https://pdfs.semanticscholar.org/7c7c/0da6f60546123dabbd0ab52ebe06637ea10e.pdf.
3. O.D. Jefimenko, "*Electrostatic Motors: Their History, Types, and Principles of Operation*", Electret Scientific Company, Star City, 1973.
4. W. Aston, "The earth's atmosphere as a source for electric power", https://electret-scientific.com/author/Article.pdf.
5. http://www.douglas-self.com/MUSEUM/POWER/electrostaticmotors/electrostaticmotors.htm.
6. http://www.aml.t.u-tokyo.ac.jp/research/es_motor/es_motor_e.html.

C-Motive Technologies

1. https://www.c-motive.com/about-us.
2. https://www.designworldonline.com/innovative-design-comes-to-electric-motors/.
3. https://drivesncontrols.com/news/fullstory.php/aid/4556/Electrostatic_motor__91out-performs_standard_designs_92.html?current_page=1.
4. B. Ge, A.N. Ghule and D.C. Ludois, "High torque density macro-scale electrostatic rotating machines: Electrical design, generalized d-q framework, and demonstration," *IEEE Transactions on Industry Applications*, vol. 55, no. 2, pp. 1225–1238, 2019.
5. J. Dai and D.C. Ludois, "Single active switch power electronics for kilowatt scale capacitive power transfer," IEEE Journal of Emerging and Selected Topics in Power Electronics, vol. 3, no. 1, pp. 315–323, 2015.
6. https://patents.justia.com/assignee/c-motive-technologies-inc.
7. D.C. Ludois and M.J. Erickson, Self-conforming plates for capacitive machines such as electrostatic motors and generators, US patent, US 9,479,085 B1, October 2016.
8. J.K. Reed, et al., Rotary capacitor for shunting high frequency bearing currents and reducing EMI in electric machinery, US Patent, US 9,653,970 B2, May 2017.
9. M.A. Petrowsky, et al., Electrostatic machine system and method of operation, US Patent, US 9,866,148 B2, January 2018.
10. https://xconomy.com/wisconsin/2019/04/04/c-motive-powers-up-with-2m-to-launch-reimagined-electric-motor/.
11. https://www.prnewswire.com/news-releases/motor-pioneer-c-motive-technologies-closes-financing-round-300823503.html.
12. https://www.crunchbase.com/organization/c-motive-technologies.
13. http://c-motiveweb.squarespace.com/technologyproducts.

8 Conclusions

This chapter summarizes the broad observations from the study of startups and their ecosystem in the fields of power semiconductor devices, power electronics and electric machines and drive systems in electrical engineering. They are made under various subheadings as given in the following.

i. Location: There is no single place or region that is dominant for startups in these fields in USA. There is more around Silicon Valley, CA, and the second would be Texas than in other places; both are influenced by a strong venture capital presence. Obviously, no rural location shows up in this aspect. From countrywide locations, there are many more startups in USA than in other western countries and that may be attributable to economic strength, size of the investment community, presence of a large number of companies in these fields, and access to higher education for a larger segment of the population. It will be important to consider Japan, China and India and then compare them to activities in the western hemisphere. Only general comments are possible at this time: (i) China has very much activity in these fields, but India and Japan do not have many startups in these fields, particularly funded by VCs. (ii) Traditional companies seem to have a large share of technical developments in these fields in these countries.

   Power electronic controlled solar application startups dominate in China, India, some countries in the African continent, Bangladesh and USA with great potential in many other countries to blossom. Appears that this is an area where there is still plenty of room for startups in other countries as well as in them too.

ii. Founders: An overwhelming majority of the founders have electrical engineering education and practical background and only very few of them in business. A small minority of founders do not have industrial experience but have graduate education and research experience from well-known universities. They have gained in-depth knowledge on a particular topic through their graduate education and in which they launch startups. Founders from academia are few but most of them have industrial experience is noted. This necessitates the present textbook because electrical engineers/academics dominate the startup scenario and hence learning of that preferably has to be injected from within the electrical engineering education curriculum itself for students and professionals feel motivated. Likewise, other fields in engineering could take the path to encourage and educate students in their specializations.

iii. Management Team: The management team comprising of CEO, CTO and VP engineering to start with play a larger role and together with VP marketing position the company's technology toward a product that is likely to bring the attention of the players and users in the market to the startup. Usually, the founder usually assumes CTO responsibility, and if the startup has more than founder, then others take the role of VP engineering and CEO. First-time founders need the greatest amount of time to land the client and that is where the others such as CEO and marketing VP play a key role. The fundraising is the crucial part of any startup and that is led and coordinated by the CEO and if the founders are prominent in their fields and recognized, then the CEO's role becomes much simpler. Very few startups land ideal management teams and it takes about 1–2 years to achieve that. But founders who have run big companies prior to the startup have a big advantage in terms of their connections with VCs, high-net-worth individuals (HNIs), contacts, and stature. Not all startups emanate from them and not all of them land one and therefore, startups do struggle in this aspect. Those funded by well-known VCs lend a hand in bringing a top team and CEO when the founders do not want that role or lack operational experience. That is where their role is crucial apart from the finance they bring in. Very few startups have such an opportunity to begin with, and this is a fact of life for them.
iv. Role of Universities: Very few of them have offices to assist the entrepreneurs from their own student body. But their help is tangible in two aspects: (i) Licensing off the intellectual property developed in the university to the startups with a little lighter upfront financial term. This happens only if there is no big appetite or interest from established companies and/or the inventors have a say into whom it is licensed. (ii) University's alumni network can connect entrepreneurs to the HNI within them to acquire funding to launch and expand their startups. A small number of such university-based startups have emerged in USA in the past two decades. A few other leading universities around the globe have this group to cater to their own students. As time goes, universities will make this one of their priorities not only to help their students and faculty but also to gain from the financial fallout it entails with the success of the ventures.
v. Technology: Breakthroughs in technology evolve over a period of time and will keep on improving thereafter. It is true in all the technologies discussed in this book. The differences between various startups' technologies may lie in the subtle variations of the circuit, machine and its topologies, devices and their underlying arrangement, various aspects of control, packaging, and thermal control with all of them contributing to higher efficiency, higher power density, high fault tolerance, and a host of other performance criteria with the most important factor of lower cost at the end of the line. Many research studies lead to some variations from the existing technologies in some aspects and not all of

them qualify to become the underlying technology for a startup. It has to be correlated to product lines. While it is easy to see the unifying factors around various startups' technologies, it is important to understand the differences between them to learn of the various gaps that may be filled by one's own newer technology leading to a startup. It is noted that significant differences between the startups' technologies do not lead to the success of the startup with maximum differences but other factors discussed in this section play their roles in that. Technology development does not stop with the day of launch of the startup but it is a continuous process and by day-to-day learning with the interaction of the customers and emerging market trends, it gets modified, strengthened and in some cases deviating away from the original technologies, if necessary. The founders and engineering team have to be alert to these changes and for maintaining and developing the winning technology for the startup.

vi. Products: The ultimate test for a startup, regardless of its technology base, comes in the domain of its products and the target customers. Technology gives an edge but does not lead to the greatest product liked by customers. That is a combination of many aspects: marketing, understanding the clients' needs, emerging markets, sizing of the competition and strategies and tactics to overwhelm both the competition and customers, and engineering team's ability to meet the specifications, and most of all at a price that is acceptable to the market. This is the final battleground for the startups and as seen from the startups' study that it takes a long time to get to a good position in products. All of them require a good funding base to get to this level and not many startups can be fortunate. Even a large funding and staying alive for a decade do not guarantee that success is a fact of life for startups!

vii. Intellectual Property: Intellectual property in the form of patents, trademarks and trade secrets arm companies against their competition infringing their technologies embedded in their products. It is expensive to file patents, get patents granted and maintain them over a period of 20 years and many startups at the beginning can't afford to do them until their funding from investors is secured to a satisfactory level. It is not common to see startups of similar technologies engaging in legal battles but it happens as seen in one of our studied startups. The legal battle is also very costly and may cost the survival of the startup itself. Note that the startups have neither time nor resources to pursue court battles. Startups' proceeding against big competitors is not common and very rarely done for obvious reasons. The big corporations have no need to infringe in general against startups and if they are interested it is much cheaper to buy them than to infringe them! Further, when an exit comes, one of the most important aspects that large corporations are looking in to is the intellectual property of the startup, and from that point of view, it is one of the important assets for the startup. The patents clearly define the technology capture in the implementation aspect and clearly draw what is owned by the startup and therefore establishes the boundary lines on their property. Of course, the final test on its originality and ownership can come from the court of law should a contest happens between two contending parties.

viii. Funding: Except very few of the studied startups in this book, most of them have funding from VCs and/or HNIs. That fact itself puts them at a likely success rate in the range of 50%–70% which is far greater than the ones that are not in that category of having such funding sources. Therefore, to get the startups to the stage where it can be funded or noticed by VCs requires immense efforts by founders and their management team, which may not be in place until the funding takes place in most cases. Funding for scaling up the startups is fairly high (in the tune of 10s of $M in USA) and many are on that track in our study. Very few make revenue at these stages and that puts a large stress on them. An emerging trend is that sizeable VC activities in India with American VCs and possibly in other countries including China. Two things they are aiming for in that they are funding startups to address the local but very large markets there but at nearly one-tenth of the investment that they would otherwise put in American and European startups. Further, the skilled highly educated engineering workforce, many of them trained and educated as well in USA, is plenty in those countries and available at relatively low cost. Even though the startups have not emerged in large numbers in India in the three fields of interest in this book, it is not far off before they take off for the reasons discussed.

ix. Collaboration: Collaboration with established companies in their same line to develop and/or supply products and tie-ups to cater to big customers' demands who are not in their line of business of designing, developing and manufacturing the products are a great hallmark of the success of the startups. Such relationships do not develop overnight and require patient and savvy work on the part of the management, engineering and marketing teams and very few startups are fortunate in that regard. Startups strive to land such relationships and their funding VCs play a key role in initiating such deals. As seen from the startups study, very few companies have such relationships within two to three years of their launch. These deals guarantee financial success for the startups resulting in attracting higher rounds of investment for further growth. But it also enables the collaborating companies to evaluate the suitability and value of the startup to be part of their own companies which may result in acquisition of the startups, thus providing a successful exit to them.

x. Exit: Some of the many ways exit can be qualified are: (i) financial success to the founders and investors which is the primary measure in most cases by outright sale for cash and some incentives, (ii) acquisition by another firm paid with the stocks of the acquiring company, and (iii) assigning the revenue and benefits to a nonprofit at some stage of the startup. The exit happens, or sought or forced upon, by various factors (such as a desire to cash, or lack of capital to move to a higher level of growth and/or development, lack of market opportunities to the level desired by the investors, or failure of the company itself), and they are dependent on the founders and board which may be dominated by the investors in the startups. The readers can benefit immensely by tracking the various stages of development and their exits.

General Comments: This is the first attempt to study startups and their ecosystem as applied to power devices, power electronics and electric motor drives. No two startups are identical just like their founders, but they have many common features as seen from the DNAs relevant to the three fields studied in this text. The journey of startups and hence the stress and excitement on the founders, the management team and all involved cannot be adequately expressed in words and hence left to readers to imagine. Not that everyone is ideal for such a journey or longing for it either. Entrepreneurship studies as applied to startups in the electrical engineering domain are illustrated with a large number of startups, and it is hoped that such an approach lends itself for study and practice in any area and field. It is our hope and belief that the book arms the potential entrepreneurs in their current and future efforts and paves their path with less rockiness and much success.

Complimentary Learning: Through discussions, project exercises, and interaction with their faculty and mentors, students and professionals could learn more than contained in this book.

Global Picture: There is a lack of study at this time on startups in three fields of electrical engineering from other countries and the access to their data is not freely available over the media. It is important for the budding entrepreneurs to study them because of the potential for entrepreneurship is equal to and/or even higher than in western countries from the point of view of their market sizes, growth potential, lower cost structure, the advantage of a large labor pool of comparable skilled and its availability, and independence from the export market until the startup is scaled way high. Given such advantages, it is important for the startups here also to explore and study them. Migrating or opening offices there requires an understanding of the legal system, market, competition and top management team recruitment in product development, marketing, and manufacturing. To a large extent, they can be achieved by recruiting professionals there who have been exposed to western practice either through education, practice and/or working for companies that have their headquarters here. It is hoped that knowledge of the ecosystem, that are outside the reach of even their local VCs which are likely to be substantial in numbers, can be made available by readers well versed in them through conference publications and/or communicating with the author who will make attempts to incorporate such information in the following editions of the book.

I wish the readers a successful launch and growth of their own startups, now or any other time in future!

Index

active power filter 113
  hybrid active filter 116
  series active filter 115
  shunt active filter 113
air gap power 349, 354, 359
angel investors 33
attorney services 35
axial flux permanent magnet synchronous machine 344, 369, 372, 374, 386, 389
axial flux switched reluctance machine 403

battery energy storage systems 246
  startups 250, 254, 260
bipolar junction transistor 47
business
  idea 8
  organization 34, 37
  plan 37

capital (funding) sources 25
charging 280, 281, 284
  automated conductive power transfer 283
  charging pad 304, 310
  conductive power transfer 249, 281, 303
    bidirectional 282
  dual mode charging pad 308
  dual more charging pad 308
  electric vehicle 292, 295, 297, 301
  inductive charger 284, 304
    power transfer 284, 286
  wireless power transfer 284
    capacitive 288
    inductive 284, 286
    magnetic resonant 304
    radio frequency (radiative charger) 304
charging of EV 280
  chargers for electric buses 297
  charging stations 161, 248, 291
  DC fast charger 290
  home flex charger 292
charity funds 27, 30
chartered public accountant (services) 36
circumferential flux machine 4, 343
company incorporation 34
company location 34
control
  current 77
  power factor 72, 120, 140, 147, 261
  reactive power 260
  voltage 68, 85, 262, 265, 268
control circuits 210
control of PM Brushless DC Machine 353
control basis of PM Synchronous Machine 350
controlled rectification 106
converter circuits 53
  AC to AC 108
    cycloconverter 108
    matrix converter 110
  AC to DC
    controlled 106
    uncontrolled 102
  boost, step up 60
  buck boost, step down/up 60
  buck, step down 54
    forward converter 55
    push pull converter 56
    synchronous buck converter 58
    synchronous forward converter 59
  flyback 61
  inverter control 68
    current control 77
    modeling of inverter 76
    over modulation 74
    programmed pulses 68
    pulse width modulation 72
    space vector modulation 84
  multilevel inverters 94
    cascaded H bridge inverter 100
    diode clamped three level inverter 95
    flying capacitor three level inverter 97
  resonant converter 102
  single phase half wave inverter 62
  three phase inverter 84
converters for SRM 360
  asymmetric converter 360
  single switch converter 394
crowd source 33

DC to DC power conversion 53
differential mode choke 272
diode 46
direct axis 343
direct current machine 338
distributed grid controller 265

E core SRM 400
edge of network grid optimization 268

electric aircraft motor drives 391
electric vehicle (EV) charger 292, 295, 297, 301
  market potential 289
electromagnetic torque 348
  reluctance torque 349, 359
  synchronous torque 349
electrostatic machines 404
entrepreneurship 7

finance companies 28
four quadrant operation 355
  SRM with single switch - four quadrant operation 395
funding sources for startups 25

gallium nitride (GaN) FET 132
  cascode 134
  depletion mode (D-mode) 132
  enhancement mode 132
  targeted applications 139
gallium nitride transistor 52, 133
GaN devices 133, 151, 160
government grants 31
grid edge management system 268
grid energy router 262
grid interface 116

high net worth individual 34
home energy management systems 215
  AC coupled system 239
  DC coupled system 239
  market 218
  solar energy systems 220, 221, 239, 241, 243
  startups 219, 221, 225, 228, 230, 234, 236, 238, 241
hybrid PM machine 377
hybrid SRM 400

inductance
  direct axis inductance 343
  quadrature axis inductance 343
induction machine (motor) 338
inset PM rotor 343
insulated gate bipolar transistor 50, 141, 148, 190
intelligent grid platform 276

L core SRM 400
leadership team 35
litigation 8, 266, 267

manufacturing of products 17
market
  GaN and SiC devices 148
  home energy management systems 218
  microgrids 218
market for motor drives 367, 368
metal oxide varistor (MOV) 272
mobile power platforms 244
modeling of PMBDCM 353
modeling of PMSM 345
MOSFET 49
motor 337
  DC 337
  induction 338
  PM synchronous 342
  PM brushless dc 353
  switched reluctance 356
  yokeless 372, 374
motor drives 337, 368, 385, 391, 393

non institutional funds 32
  crowd source 33
  family 33
  friends 33
  self (own) 32

permanent magnet (PM)
  brushless DC machine 353
  synchronous machine 342
philanthropic charity foundation entrepreneurship funds 30
PM brushless dc machine 353
PM synchronous machine 342
  axial flux 344
  cooling technology 380
  printed circuit board stator 385
  radial flux 342
  startups 369, 370, 372, 374, 376, 377, 381, 382, 385, 388
PMSM cooling technology 381, 382
  startups 381, 382
power grid edge control 259
  market 261
  startups 261, 264, 267, 271, 275, 277
  technology 260
Power Guardian 273
power regulator 265
  in-line power regulator 265
power semiconductor devices 45
Power Line Guardian 273
printed circuit board (PCB) stators 385, 388
printed circuit motors 385
  PM synchronous 385, 388
products 16, 17, 20

quadrature axis 342

radial flux machine 342
rotor reference frames 346

sensorless control 354
silicon carbide (SiC) devices 51, 140
  lateral SiC MOSFET 144
  normally on JFET (cascade) 141
  normally off JFET 142
  SiC MOSFET 144
  startups 180, 181, 183, 184, 186, 188
  targeted applications 147
  vertical SiC MOSFET 145
smart valve 271
solid state circuit breaker 212
SRM control 362
  four quadrant operation 395
static VAR compensator 265
stator reference frames 345
surface inset PM rotor 343
surface mount PM rotor 342
switched reluctance motor, SRM 356
  aligned position 356
  asymmetric converter 360
  axial flux 403
  control 362
  converters for SRM 360, 394, 399, 400
  four quadrant operation with single switch converter 395
  hybrid SRM 400
  principle of operation 357
  radial flux 399
  rotor 356
  single switch converter 394
  startups 394, 397, 399, 402
  stator 356
  torque 359
  transverse flux 397
  unaligned position 356
synchronous machine 339

testing products 23, 24
thyristor 48
transverse flux SRM 397

uncontrolled rectification 102
uninterruptible power supply (UPS) 116
university venture funds 29

vacuum series link 272
venture capital 27, 29

wide bandgap materials and devices 131, 140
  GaN startups 149, 150, 152, 155, 157, 159, 163, 164, 166, 169, 171, 173, 177
  SiC startups 180, 181, 183, 184, 186, 188
wireless charging 284
  capacitive 288
  inductive 284
  magnetic resonant 285